Multi-Sensor Filtering Fusion with Censored Data Under a Constrained Network Environment

This book presents the up-to-date research developments and novel methodologies on multi-sensor filtering fusion (MSFF) for a class of complex systems subject to censored data under a constrained network environment. The contents of this book are divided into two parts covering centralized and distributed MSFF design methodologies. The work provides a framework of optimal centralized/distributed filter design and stability and performance analysis for the considered systems along with designed filters. Simulations presented in this book are implemented using MATLAB.

FEATURES:

- Includes concepts, backgrounds and models on censored data, filtering fusion and communication constraints.
- Reviews case studies to provide clear engineering insights into the developed fusion theories and techniques.
- Provides theoretic values and engineering insights of the censored data and constrained network.
- Discusses performance evaluation of the presented multi-sensor fusion algorithms.
- Explores promising research directions on future multi-sensor fusion.

This book is aimed at graduate students and researchers in networked control, sensor networks and data fusion.

Multi-Sensor Filtering Fusion with Censored Data Under a Constrained Network Environment

Hang Geng, Zidong Wang and Yuhua Cheng

CRC Press
Taylor & Francis Group
Boca Raton London New York

CRC Press is an imprint of the
Taylor & Francis Group, an **informa** business

Designed cover image: Shutterstock

First edition published 2025
by CRC Press
2385 NW Executive Center Drive, Suite 320, Boca Raton FL 33431

and by CRC Press
4 Park Square, Milton Park, Abingdon, Oxon, OX14 4RN

CRC Press is an imprint of Taylor & Francis Group, LLC

ISBN: 978-1-032-55550-8 (hbk)
ISBN: 978-1-032-61025-2 (pbk)
ISBN: 978-1-003-46162-3 (ebk)

DOI: 10.1201/9781003461623

Typeset in Times LT Std
by Apex CoVantage, LLC

To my supervisor and my family.

Contents

Contents   ix

Figures

Tables

Symbols

$\odot$	The Hadamard product
$\otimes$	The Kronecker product
$\mathbb{R}^n$	The n-dimensional Euclidean space
$\mathbb{R}^{n\times m}$	The set of all $n \times m$ real matrices
$\mathbb{R}^+$	The set of all positive real numbers
$\mathbb{N}$	The set of natural numbers
$\mathbb{S}^n_+$	The set of $n \times m$ positive definite matrices
A^T or A'	The transpose of matrix A
$A^\dagger$	The Moore-Penrose pseudo inverse of A
$A > 0$	The matrix A is positive definite
$A \geq 0$	The matrix A is positive semidefinite
$A < 0$	The matrix A is negative definite
$A \leq 0$	The matrix A is negative semidefinite
$\|\cdot\|$	The Euclidian norm of real vectors or the spectral norm of real matrices
$\|\cdot\|_{\max}$	The maximum singular value of a matrix
$\delta(\cdot)$	The Kronecker delta function
$\phi(x)$	The cumulative distribution function of normally distributed random variable x
$\Phi(x)$	The cumulative distribution function of normally distributed random variable x
$\mathrm{tr}(A)$	The trace of matrix A
$\|x\|_P^2$	Equals to $x^T P x$ when x is a vector
$\mathbb{P}\{\cdot\}$	The occurrence probability of the event "$\cdot$"
$\mathbb{E}\{x\}$	The expectation of stochastic variable x
$\mathrm{Var}\{x\}$	The variance of stochastic variable x
$\mathbb{E}\{x|y\}$	The conditional expectation of x given y
I	The identity matrix of compatible dimension
0	The zero matrix of compatible dimension
$\mathbf{0}_n$	The $n \times n$ zero matrix
$\mathbf{1}_n$	The $n \times 1$ column vector with all elements equal to 1
$\mathrm{vec}\{x_1, x_2\}$	The column vector $\begin{bmatrix} x_1^T & x_2^T \end{bmatrix}^T$
$\mathrm{vec}_n\{x_i\}$	The column vector $\mathrm{vec}\{x_1^T, x_2^T, \cdots x_n^T\}$
$\mathrm{diag}\{x_i\}$	The block diagonal matrix with ith block being x_i and all other entries being zero
$\mathrm{diag}_n\{A_i\}$	The block diagonal matrix $\mathrm{diag}\{A_1, A_2, \cdots, A_n\}$
$\{M_{ij}\}_{n\times n}$	The partitioned matrix with M_{ij} being (i, j)-th block submatrix

Preface

With the ever-increasing scale and complexity of modern networked systems, how to ensure their reliable implementation and eliminate potential threats has been extensively studied. As a result, it is of great significance to monitor/track the networked systems effectively. Multi-sensor filtering fusion (MSFF) techniques enable us to better understand the working condition of addressed networked systems and compensate for unexpected performance deterioration without changing network structures. In practical engineering, owing to the massive usage of low-cost commercial and off-the-shelf sensors, network sensors are easily prone to a very special kind of measurement nonlinearity named censored data or measurement censoring. Meanwhile, taking into account the limited network resources, data transmission in a networked environment is unavoidably subject to communication constraints. Both the censoring phenomenon and the communication constraint have a great impact on the multi-sensor fusion performance, and the traditional MSFF schemes may be no longer applicable in these cases. Consequently, it is of utmost importance to re-design the MSFF unit or even develop new methodologies for the relatively complex networked systems.

The objective of this book is to present the up-to-date research developments and novel methodologies on MSFF for a class of complex systems subject to censored data under a constrained network environment. The content of this book can be divided into two parts, where the first part (Chapters 2–5) and the second part (Chapters 6–10) present centralized and distributed MSFF design methodologies, respectively. The work provides a framework of (1) optimal centralized/distributed filter design and (2) stability and performance analysis for the considered systems along with designed filters subject to various kinds of complex network constraints, including fading measurements, packet dropouts, redundant channels, communication protocols, etc. Several techniques including the recursive Riccati equation, matrix decomposition, optimal estimation theory and mathematical optimization methods are employed to develop the desired MSFF unit. Additionally, this book provides many valuable reference materials for researchers who wish to explore the MSFF in these complicated cases.

The concise frame and description of the book are given as follows. Chapter 1 introduces the recent advances on MSFF problems for complex systems under network constraints and the outline of the book. Chapter 2 is concerned with the optimal filtering for networked systems with censored data and fading measurements. Chapter 3 studies the minimum-variance recursive filtering with censored data and time-correlated multiplicative sensor noises under redundant channel transmission, where the filtering performance is compared for different channel numbers. Chapter 4 deals with the optimal state estimation problem with censored data under non-Gaussian Lévy and time-correlated additive sensor noises. Chapter 5 considers the Tobit Kalman filtering for networked systems with integral measurements and probabilistic sensor failures, where self-propagated upper and lower bounds are determined on the filtering error covariance. Chapter 6 addresses the distributed filtering fusion problem with both packet dropouts and transmission delays via a probabilistic view. Chapter 7 copes with the Federated Tobit Kalman filtering fusion issue with dead-zone-like censoring and dynamical bias under the Round-Robin protocol. In Chapter 8, a distributed fusion estimator is designed for a class of time-varying systems under parametric uncertainties and measurement censoring. Chapter 9 discusses a class of protocol-based MSFF schemes under state saturation and cyber-attacks. Chapter 10 is concerned with the variance-constrained MSFF for nonlinear cyber-physical systems under the stochastic communication protocol. Chapter 11 gives the conclusion and some possible future research directions. Simulations presented in this book are implemented using the MathWorks MATLAB software package.

This book is a research monograph whose intended audience is graduate and postgraduate students as well as researchers.

Hang Geng
Chengdu, China

Zidong Wang
London, U.K.

Yuhua Cheng
Chengdu, China

Acknowledgement

The authors would like to express their deep appreciation to those who have been directly involved in various aspects of the research leading to this book. Special thanks go to Professor Bo Shen from Donghua University, Shanghai, China, Professor Fuad E. Alsaadi from King Abdulaziz University, Jeddah, Saudi Arabia, Professor Xiaohui Liu from Brunel University London, London, U.K., Professor Abdullah M. Dobaie from King Abdulaziz University, Jeddah, Saudi Arabia, Professor Tingwen Huang from Texas A&M University at Qatar, Doha, Qatar, Professor Hongli Dong from Northeast Petroleum University, Daqing, China, Professor Weiyin Fei from Anhui Polytechnic University, Wuhu, China, and Professor Yurong Liu from Yangzhou University, Yangzhou, China.

The writing of this book was supported in part by the National Natural Science Foundation of China under Grant U2330206, Grant U2230206, Grant 62173068, Grant U21A2019, Grant U2030205, Grant 61973102, Grant 61933007, Grant 61903065 and Grant 61803074; in part by the New Cornerstone Science Foundation through the XPLORER PRIZE; in part by the Natural Science Foundation of Sichuan Province of China under Grant 23NSFSC3603; in part by the Royal Society of the U.K.; and in part by the Alexander von Humboldt Foundation of Germany.

Foreword

I am delighted to introduce this book on *Multi-Sensor Filtering Fusion with Censored Data Under a Constrained Network Environment.* When I came to know about this book project undertaken by three of the most active researchers in the field, I was pleased that this book is coming in an early stage of a field that will need it more than most fields do. In most emerging research fields, a book can play a significant role in bringing some maturity to the field. Research fields advance through research papers. In research papers, however, only a limited perspective could be provided about the field, its application potential, and the techniques required and already developed in the field. A book gives such a chance. I liked the idea that there will be a book that will try to unify the field by bringing in disparate topics already available in several papers that are not easy to find and understand. I was supportive of this book project even before I had seen any material on it. The project was a brilliant and a bold idea by three active researchers. Now that I have it on my screen, it appears to be even a better idea.

Multi-sensor filtering fusion started gaining recognition in the last century as a field. Perception, monitoring, tracking, communication, estimation and fusion technologies had advanced enough that researchers and technologists started building approaches to combine sensor data or data derived from disparate sources such that the resulting state estimation has less uncertainty than would be possible when these sources were used individually. By properly selecting fusion criteria to provide complementary information, the multi-sensor filtering fusion approaches aspire, much like human perception/fusion mechanisms, to create a holistic picture of an interested system status using only partial information from separate sources.

Owing to the massive usage of low-cost commercial and off-the-shelf sensors, multi-sensor filtering fusion could be easily prone to a very special kind of measurement nonlinearity named censored measurements. Meanwhile, taking into account the limited network resources, data transmission in a networked environment is unavoidably subject to communication constraints. As a result, the necessity (of researching multi-sensor filtering fusion theories/methodologies with censored data under a constrained network environment) arises, paving a way for people to hypothesize relationships among data sources and explore support for system monitoring/ tracking. Once successfully developed, such theories/methodologies will find vast applications in many diverse domains and keep getting more applications. In fact, many new filtering/control approaches are a direct outgrowth of network constrained filtering fusion, and it is likely to become a powerful system identification/estimation tool.

Contributors

Khalid H. Alharbi
King Abdulaziz University
Jeddah, Saudi Arabia

Fuad E. Alsaadi
King Abdulaziz University
Jeddah, Saudi Arabia

Yun Chen
Hangzhou Dianzi University
Hangzhou, China

Abdullah M. Dobaie
King Abdulaziz University
Jeddah, Saudi Arabia

Yan Liang
Northwestern Polytechnical
University
Xi'an, China

Hongjian
Liu Anhui Polytechnic University
Wuhu, China

Lifeng Ma
Nanjing University of Science
and Technology
Nanjing, China

Alireza Mousavi
Brunel University London
London, U.K.

Xiaojian Yi
Beijing Institute of Technology
Beijing, China

Lei Zou
Donghua University
Shanghai, China

1 Introduction

Following the tremendous advancement in information sensing, acquisition and processing technology, the last few decades have seen an unprecedented research interest towards multi-sensor filtering fusion (MSFF). The main objective of MSFF is to properly integrate data from all available sensors in order to achieve the optimal state estimation. Compared to the single sensor filtering, MSFF is capable of producing better estimation performance due to its abundant information sources, holistic target perception and strong fault tolerance. In this respect, MSFE has captured vast research attention with broad applications in a wide variety of fields including cyber-attack detection [1–3], fault detection [4, 5], signal processing [6–9], system monitoring [10, 11], target localization [12–14], image processing [15, 16], etc. To date, a wealthy body of work has been devoted to MSFF for multifarious models/ systems where two categories of fusion technique (i.e. centralized and distributed filtering fusion) have been well investigated, see e.g. [17–20].

In the case of centralized filtering fusion, a fusion center gathers measurements from all the available sensors and produces an optimal state estimate in the sense of least mean-squared variances. Although centralized filtering fusion has the smallest data loss, it displays drawbacks like huge communication burdens, high computational overheads and fragile fault tolerance [21–23]. To counter such shortcomings, distributed filtering fusion is proposed, where state estimates given by local estimators are propagated to a fusion center so as to output a global optimal/suboptimal state estimate on the basis of various fusion criteria.

Note that one of the most crucial factors when implementing distributed filtering fusion lies in the selection of a suitable fusion criterion. Some prevalently adopted distributed filtering fusion criteria include but are not limited to consensus fusion [24–26], robust fusion [10, 27, 28], information fusion [29–31], covariance intersection fusion [32], federated fusion [33] and sequential fusion [34].

It is well acknowledged that both centralized and distributed filtering fusion massively utilize cheap commercial and off-the-shelf sensors, and this inescapably gives rise to the so-called phenomenon of *censored measurements*, where sensor measurements become continuous functions with respect to system states within a preset dynamic interval but are constant outside that interval [35]. As a particular type of measurement nonlinearity, censored measurements ubiquitously appear in engineering applications ranging from biological modeling [36] and economics [37] to distributed detection [38] and vision tracking [39]. The occurrence of such censored/ saturated measurements mainly arises from physical limitations of system dynamics and/or external interferences. For instance, the well-known Tobit-type-1 censoring model (described as piecewise-linear transforms in regard to output variables with zero slopes in censored regions) can be taken to characterize sensor saturations and limits on detection regions.

DOI: 10.1201/9781003461623-1

When it comes to censored measurements, conventional fusion approaches, e.g. the Kalman fusion approach, are proven to be futile for the reason that measurement noises (near censoring intervals) turn out to be non-Gaussian of unknown statistics [35, 40, 41]. In this regard, an alternative tool for coping with censored measurements is the Tobit Kalman filter (TKF) featuring a fully recursive structure and nearly equivalent performance to standard Kalman filters [35, 42–51]. As a result, the MSFF problem based on the TKF has stirred a great surge of research enthusiasm, with some initial results presented in [52–54]. To date, the majority of the existing results have been concerned with the *one-side* censoring, and the corresponding *two-side/ dead-zone-like* censoring problem has received much less attention, despite its great applications in engineering systems undergoing detection constraints, image frame effects as well as sensor saturations.

With the fast-growing scale of multi-sensor systems, the limited resources assigned for data communication between sensors/estimators inevitably give rise to certain communication constraints [55–58]. These communication constraints may result in undesirable *network-induced phenomena* (e.g. packet dropout, packet disorder, time delay and measurement degradation), thereby severely deteriorating the fusion performance [59–66]. Meanwhile, such network-induced phenomena could appear in a probabilistic way, which is customarily named randomly occurring incomplete information.

Generally speaking, the aforementioned incomplete information leads to an urgent necessity of establishing new MSFF techniques with a viewpoint to meeting the need of engineering practice. Consequently, it is not surprising that MSFF with incomplete information has appeared, in the past few years, as an imperative and challenging topic that has drawn a large deal of research attention [5, 67–69]. To summarize, different from standard MSFF techniques, the potential difficulty lies in developing MSFF algorithms (with censored measurements under communication constraints) stems from not only the network-induced phenomena but also the complicated coupling between censored measurements and network-induced phenomena.

1.1 CANONICAL MSFF SCHEMES

Two typical MSFF strategies can be found in the literature, namely, centralized and distributed fusion. Generally speaking, the centralized fusion has the smallest data loss and the optimal fusion performance at the cost of huge communication burdens, high computational overheads and fragile fault tolerance. In contrast, the distributed fusion displays friendly communication burdens, low computational overheads and reliable fault-tolerance at the cost of suboptimal fusion accuracy under certain performance criteria.

Before proceeding to the detailed MSFF strategies, let us first introduce the following discrete-time system model with p sensor described by

$$\begin{cases} x_{k+1} = A_k x_k + \omega_k, \\ z_{m,k} = C_{m,k} x_k + \upsilon_{m,k}, m = 1,2,\ldots,p, \end{cases} \tag{1.1}$$

where $x_k \in \mathbb{R}^n$ and $z_{m,k} \in \mathbb{R}$ are, respectively, the state vector and the measurement vector; A_k and $C_{m,k}$ are known matrices; and ω_k and $\upsilon_{m,k}$ are zero-mean white Gaussian noises whose covariances are, respectively, Q_k and $R_{m,k}$. It is assumed that x_0, ω_k and $v_{m,k}$ are mutually independent.

Denote $\hat{x}_k$ as the state estimate, P_k as the filtering error covariance, P_k^- as the prediction error covariance and K_k as the gain matrix. In the sequel, various fusion methods are described one by one and, subsequently, applied to system (1.1) in order to establish different MSFF frameworks.

1.1.1 CENTRALIZED FILTERING FUSION

Figure 1.1 depicts the structure of centralized MSFF. Each sensor m directly sends its outputs to a fusion center (through networks), where the collected measurements are further fused and processed for filtering purpose. An easy way to carry out the fusion process is to augment all received measurements to an equivalent one and later run the traditional Kalman filter based on the augmented model.

Following (1.1), the measurements from all p sensors can be augmented as follows:

$$z_k = C_k x_k + v_k, \tag{1.2}$$

where

$$z_k = \begin{bmatrix} z_{1,k}^T & z_{2,k}^T & \cdots & z_{p,k}^T \end{bmatrix}^T,$$
$$v_k = \begin{bmatrix} v_{1,k}^T & v_{2,k}^T & \cdots & v_{p,k}^T \end{bmatrix}^T,$$
$$C_k = \begin{bmatrix} C_{1,k}^T & C_{2,k}^T & \cdots & C_{p,k}^T \end{bmatrix}^T.$$

According to the centralized fusion rule illustrated in Figure 1.1, the following centralized fusion filter for system (1.1) is established.

$$\begin{cases} P_{k+1}^- = A_k P_k A_k^T + Q_k, \\ P_{k+1} = (I - K_{k+1} C_{k+1}) P_{k+1}^-, \\ K_{k+1} = P_{k+1}^- C_{k+1}^T (C_{k+1} P_{k+1}^- C_{k+1}^T + R_{k+1}), \\ \hat{x}_{k+1} = (I - K_{k+1} C_{k+1}) A_k \hat{x}_k + K_{k+1} z_{k+1}. \end{cases} \tag{1.3}$$

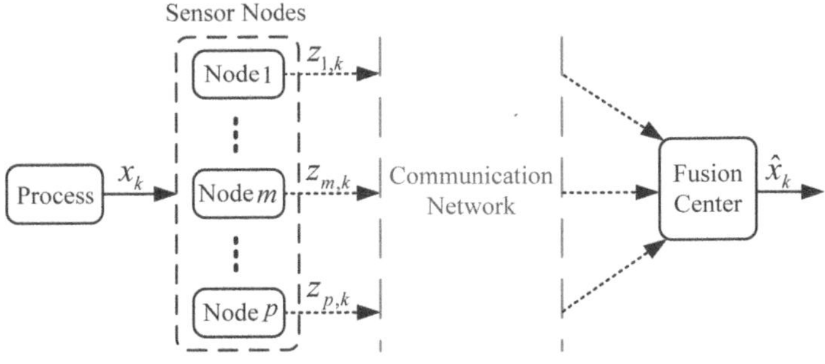

FIGURE 1.1 Schematic diagram of centralized fusion.

1.1.2 INFORMATION FILTERING FUSION

Figure 1.2 depicts the structure of the information based MSFF. This is an inverse implementation of the centralized MSFF, which propagates the inverse of the error covariances defined by $F_k \triangleq P_k^{-1}$ and $F_k^- \triangleq (P_k^-)^{-1}$ rather than propagating P_k and P_k^-. In other words, the information-based MSFF propagates the information matrix F_k of the target system.

According to the information filtering fusion rule illustrated in Figure 1.1, the following information fusion filter for system (1.1) is established.

$$\begin{cases}
\hat{x}_{k+1}^- = A_k \hat{x}_k, \\
\hat{x}_{k+1} = (F_{k+1})^{-1} F_{k+1}^- \hat{x}_{k+1}^- + (F_{k+1})^{-1} \\
\qquad \times \sum_{m=1}^{p} \left[F_{m,k+1} \hat{x}_{m,k+1} - F_{m,k+1}^- \hat{x}_{m,k+1}^- \right], \\
F_{k+1} = F_{k+1}^- + \sum_{m=1}^{p} \left[F_{m,k+1} - F_{m,k+1}^- \right], \\
F_{k+1}^- = Q_k^{-1} - Q_k^{-1} A_k \left[F_k + A_k^T Q_k^{-1} A_k \right]^{-1} A_k^T Q_k^{-1},
\end{cases} \qquad (1.4)$$

where $\hat{x}_{m,k+1}^-$ and $\hat{x}_{m,k+1}$ are the local prediction and estimate; and $F_{m,k} \triangleq P_{m,k}^{-1}$ and $F_{m,k}^- \triangleq (P_{m,k}^-)^{-1}$ are the inverse of local error covariances. Here, $\hat{x}_{m,k+1}^-$, $\hat{x}_{m,k+1}$, $P_{m,k}^-$ and $P_{m,k}$ are all obtained via running a series of local Kalman filters.

1.1.3 SEQUENTIAL FILTERING FUSION

Figure 1.3 sketches the structure of sequential MSFF where at time k, the remote estimator sequentially collects and integrates sensor outputs according to their arriving order. To be more specific, at the current time k, the remote estimator runs a standard Kalman filter to obtain an initial estimate $\hat{x}_{m,k}$ using the arrived measurement $z_{m,k}$ ($m = 1, 2, \ldots, p$). Subsequently, $\hat{x}_{m,k}$ is further fused with the newly arrived sensor output $zn_{,k}$ ($n = 1, 2, \ldots, p$) to update a new estimate $\hat{x}_{n,k}$. The update process continues until all sensors' outputs at time k are adopted for the measurement update in

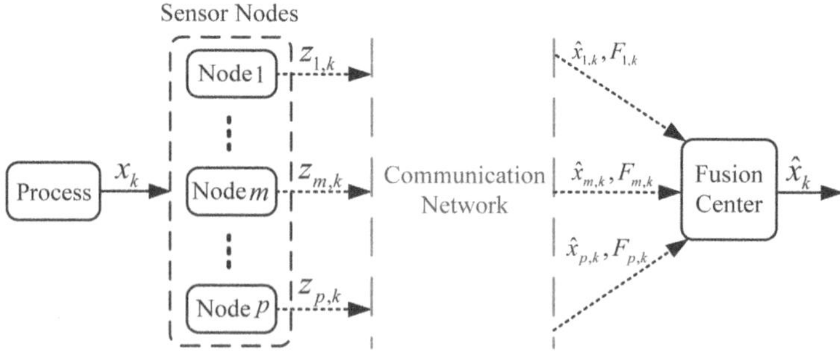

FIGURE 1.2   Schematic diagram of information fusion.

order to achieve the final estimate $\hat{x}_k$ at time k. When the time instant moves from k to $k + 1$, simply repeat the whole fusion process to acquire the state estimate $\hat{x}_{k+1}$ at time $k + 1$.

According to the sequential fusion rule illustrated in Figure 1.3, the following sequential fusion filter for system (1.1) is established.

$$\begin{cases} P_{m,k} = (I - K_{m,k}C_{m,k})P_{m-1,k}, \\ \hat{x}_{m,k} = \hat{x}_{m-1,k} + K_{m,k}(z_{m,k} - C_{m,k}\hat{x}_{m-1,k}), \\ K_{m,k} = P_{m-1,k}C_{m,k}^T(C_{m,k}P_{m-1,k}C_{m,k}^T + R_{m,k})^{-1}, \end{cases} \qquad (1.5)$$

where $x_{0,k} = A_{k-1}x_{k-1}$ and $P_{0,k} = A_{k-1}P_{k-1}A_{k-1}^T + Q_k$; and $x_k = x_{p,k}$ and $P_k = P_{p,k}$ are the optimal state estimate and associated error covariance at time k under the sequential fusion rule.

1.1.4 WEIGHTED FILTERING FUSION

The sequential filtering fusion scheme utilizes measurements directly for the purpose of state estimation. We next introduce the weighted filtering fusion under which the measurements (integrated at the fusion center) are replaced by state estimates that are locally computed.

Figure 1.4 describes the structure of weighted MSFF, which typically consists of two stages. At the first stage, sensor outputs are adopted to form initial state estimates at each local filter. At the second stage, all state estimates from local filters are integrated at the fusion center to form an optimal state estimate in the linear unbiased minimum variance (LUMV) sense. Let $\hat{x}_{m,k}$ and $P_{m,k}$ be the estimate and error covariance given by the mth local filter, $R_{mn,k}$ be the mn-th $(m,n = 1,2,\ldots,p)$ entry of R_k, and $\Pi_k \in \mathbb{R}^{np \times np}$ be a matrix with its mn-th entry being $P_{mn,k}$. Here, $P_{mn,k}$ is defined as the cross-covariance $P_{mn,k}$ between the m-th and n-th local filters. Denote $E \triangleq \begin{bmatrix} I & I & \cdots & I \end{bmatrix}^T$ and $W_k \triangleq \begin{bmatrix} W_{1,k} & W_{2,k} & \cdots & W_{p,k} \end{bmatrix}^T$ where $W_{m,k}$ is the weighted matrix in regard to $\hat{x}_{m,k}$.

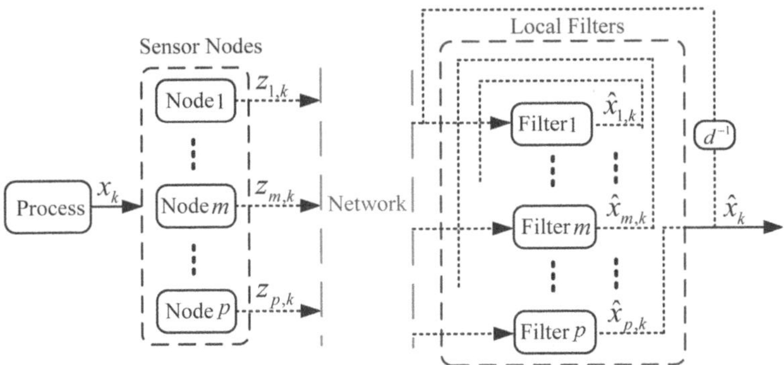

FIGURE 1.3 Schematic diagram of sequential fusion.

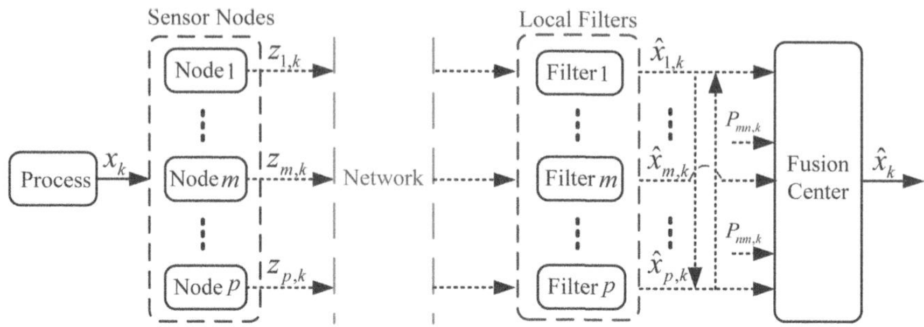

FIGURE 1.4   Schematic diagram of weighted fusion.

According to the weighted fusion structure demonstrated in Figure 1.4, the following fusion filter for system (1.1) is established.

$$\begin{cases} \hat{x}_k = \displaystyle\sum_{m=1}^{p} W_{m,k}\hat{x}_{m,k}, \\[2mm] P_k = \left(E^T \Pi_k^{-1} E\right)^{-1}, \\[2mm] W_k = \Pi_k^{-1} E \left(E^T \Pi_k^{-1} E\right)^{-1}, \end{cases} \tag{1.6}$$

where

$$\begin{aligned} P_{mn,k} = {}&(I - K_{m,k} C_{m,k})(A_{k-1} P_{mn,k-1} A_{k-1}^T + Q_{k-1}) \\ &\times (I - K_{m,k} C_{m,k})^T + K_{m,k} R_{mn,k} K_{n,k}^T, \end{aligned} \tag{1.7}$$

$$\begin{aligned} K_{m,k} = {}&(A_{k-1} P_{m,k-1} A_{k-1}^T + Q_{k-1}) C_{m,k}^T (C_{m,k} \\ &\times (A_{k-1} P_{m,k-1} A_{k-1}^T + Q_{k-1}) C_{m,k}^T + R_{m,k}). \end{aligned} \tag{1.8}$$

Here, $\hat{x}_{m,k}$ and $P_{m,k}$ are the local estimate and error covariance obtained under the centralized fusion rule.

Remark 1.1 It is well acknowledged that the weighted fusion filter given by (1.6)–(1.8) is optimal in the LUMV sense and has advantages of high accuracy, strong robustness and large flexibility. Nonetheless, in most application cases, calculating cross-error covariances is rather challenging and complex, which can hardly be carried out. An alternative approach to solving this problem is the covariance intersection fusion method.

1.1.5 COVARIANCE INTERSECTION FUSION

Figure 1.5 plots the structure of covariance intersection MSFF, which typically consists of two stages. At the first stage, sensor outputs are adopted to form initial state estimates at each local filter. At the second stage, all state estimates from local filters

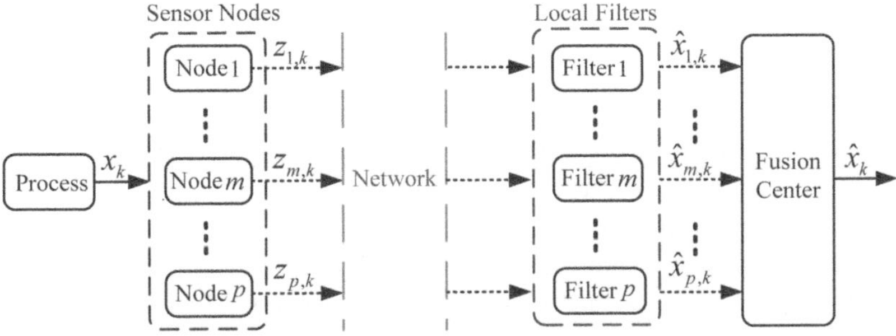

FIGURE 1.5   Schematic diagram of covariance intersection fusion.

are integrated at the fusion center to form a suboptimal state estimate without taking into account the cross-correlation between local estimates.

According to the covariance intersection fusion structure demonstrated in Figure 1.5, the following fusion filter for system (1.1) is established.

$$
\begin{cases}
\hat{x}_k = \sum_{m=1}^{p} w_{m,k} \overline{P}_k P_{m,k}^{-1} \hat{x}_{m,k}, \\
\overline{P}_k = \left(\sum_{m=1}^{p} w_{m,k} P_{m,k}^{-1} \right)^{-1},
\end{cases}
\tag{1.9}
$$

where $\hat{x}_k$ is the fused state estimate, $x_{m,k}$ is the local estimate obtained by a standard Kalman filter based on measurement from sensor m, and $\overline{P}_k$ is the upper bound on the fused error covariance P_k. $w_{m,k} \in [0,1]$ is the scalar weight, which satisfies $\sum_{m=1}^{p} w_{m,k} = 1$ and is acquired by solving the following optimization problem:

$$
\min_{\substack{0 \leq w_{m,k} \leq 1 \\ \sum_{m=1}^{p} w_{m,k} = 1}} \operatorname{tr}(\overline{P}_k)
\tag{1.10}
$$

where $\operatorname{tr}(\overline{P}_k)$ is the trace of $\overline{P}_k$. Moreover, the exact value of the fused error covariance is given by

$$
P_k = \overline{P}_k \left(\sum_{m,n=1}^{p} w_{m,k} w_{n,k} P_{m,k}^{-1} P_{mn,k} P_{n,k}^{-1} \right) \overline{P}_k,
\tag{1.11}
$$

where $P_{m,k}$ and $P_{n,k}$ are the local error covariances obtained by the standard Kalman filtering algorithm; and $P_{mn,k}$ is the associated cross-covariance.

Remark 1.2 Comparing the presented covariance intersection fusion filter (1.10) with the weighted fusion filter (1.10), it can be obviously observed that filter (1.10) has a smaller computational burden than filter (1.9) through avoiding computing cross-covariance $P_{mn,k}$ (at the cost of lower filtering accuracy) and solving the non-linear optimization problem (1.11) (based on function "fmincon" embedded in the optimization toolbox of MATLAB).

As a matter of fact, filter (1.10) is known as the batch covariance intersection filter in the literature, accounting for its direct integration of all local estimates at one single step. Apart from such a batch fusion filter structure, two other commonly used fusion structures are the sequential covariance intersection and the parallel covariance intersection filters. To be more specific, in the sequential covariance intersection filter strategy, estimates from two randomly selected local estimators are firstly selected to form a fused estimate following the batch covariance intersection fusion rule. Then, following the same fusion rule, the fused estimate is further adopted to integrate with the local estimate from any one of the rest $p - 2$ local estimators. Such a fusion process is repeated by sequentially integrating state estimates from all the rest $p - 3$ local filters in order to achieve the final estimate that combines the information from all p local filters.

As to the parallel covariance intersection fusion mechanism, it randomly divides the p local filters into $(p + 1)/2$ or $(p - 1)/2$ groups, with each group containing two local filters. In each group, estimates are combined to generate a fused estimate following the batch covariance intersection fusion rule. The obtained $(p + 1)/2$ $((p - 1)/2)$ groups of fused estimates are again randomly classified into multiple groups, with each group containing two local fused estimates, which are further fused according to 10. Repeat such a fusion process until there is only one fused estimate in hand, and this results in the final fused state estimate comprising the information from all p local filters.

1.1.6 Federated Filtering Fusion

Although the covariance intersection fusion filter works in a distributed manner and avoids calculating cross-covariance, its filtering fusion performance is suboptimal in the LUMV sense. Regarding this weakness, a natural idea arises that: how can we establish a distributed filtering fusion framework with the optimal performance and moderate computation burden? The answer to this question is the famous federated filtering approach initially proposed in [33].

Figure 1.6 plots the structure of the federated MSFF, which typically consists of three stages. At the first stage, initialize the local estimates, noise covariance and error covariance in line with the information-sharing principle [33]. At the second stage, measurements from different sensors are adopted to form an initial state estimate at each local filter. At the third stage, all local estimates are combined at the fusion center to form the fused estimate.

According to the federated fusion structure demonstrated in Figure 1.6, the following fusion filter for system (1.1) is established.

$$
\begin{cases}
\hat{x}_k = P_k \displaystyle\sum_{m=1}^{p} P_{m,k}^{-1} \hat{x}_{m,k}, \\[2ex]
P_k = \left(\displaystyle\sum_{m=1}^{p} P_{m,k}^{-1} \right)^{-1},
\end{cases}
\tag{1.12}
$$

where $x_{m,k}$ and $P_{m,k}$ are the local state estimate and error covariances obtained by the standard Kalman filtering algorithm. Moreover, it is worth pointing out that, abiding

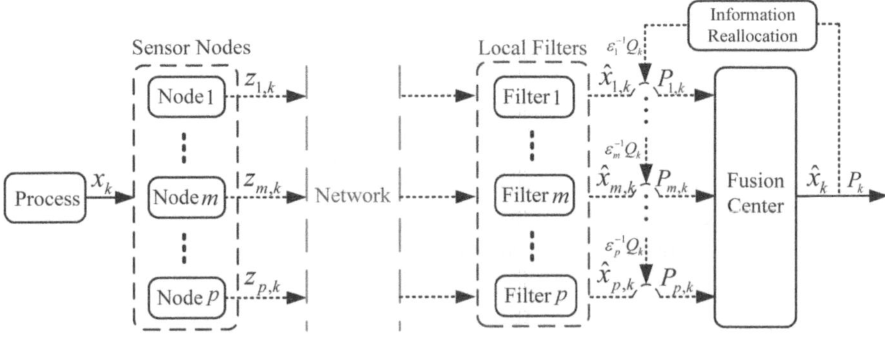

FIGURE 1.6 Schematic diagram of federated fusion.

by the information sharing-principle with respect to the federated fusion rule, at time $k-1$, the local estimates, upper bounds and process noise covariances should be initialized as follows:

$$\begin{cases} \hat{x}_{m,k-1} \triangleq \hat{x}_{k-1}, \\ P_{m,k-1} \triangleq \epsilon_m^{-1} P_{k-1}, \\ Q_{m,k-1} \triangleq \epsilon_m^{-1} Q_k, \end{cases} \tag{1.13}$$

where ϵ_m is the information redistribution weight of the mth local estimator and $\sum_{m=1}^{p} \epsilon_m \triangleq 1$. Note that the federated filtering fusion given by (1.12) with information-sharing principle (1.13) has the same filtering fusion accuracy as the centralized fusion, information fusion and the sequential fusion in (1.3)–(1.5).

Table 1.1 summarizes the inherent properties of the investigated six fusion schemes in terms of fusion accuracy and computation burden.

Remark 1.3 Recalling the aforementioned filtering fusion algorithms, we have made the following observations. (1) In terms of fusion structure, only the centralized fusion falls into the centralized case, while the other five fusion approaches belong to the distributed case. (2) In terms of filtering fusion accuracy, the centralized fusion, information fusion, sequential fusion and federated fusion have the highest (also the same) accuracy, the weighted fusion has moderate accuracy and the covariance intersection has the lowest accuracy. (3) In terms of computation burden, the centralized fusion has the highest burden, the sequential fusion and the weighted fusion have a moderate (also comparable) burden and the information fusion, covariance intersection fusion and federated fusion have the smallest (also comparable) burden.

1.2 CENSORED MEASUREMENTS

Owing to the massive usage of cheap commercial and off-the-shelf sensors, censored measurements arise pervasively in MSFF applications, taking the form of continuous

TABLE 1.1

Comparison in Fusion Accuracy and Computation Burden

Strategy	Structure	Accuracy	Computation cost	References
Centralized	Centralized	High	High	[21–23]
Information	Distributed	High	Low	[29–31]
Sequential	Centralized	High	Moderate	[34]
Weighted	Distributed	Moderate	Moderate	[70, 71]
Covariance	Distributed	Low	Low	[32]
Federated	Distributed	High	Low	[33, 52, 53, 72]

functions with respect to system states within a preset dynamic interval but are constant outside that interval [35]. In this section, two typical kinds of censored measurements are introduced and discussed based on the so-called Tobit measurement model.

1.2.1 One-Side Censored Measurements

According to the Tobit measurement model, the one-side censored measurement is given by the following equation:

$$
y_{m,k} = \begin{cases} z_{m,k}, & z_{m,k} > \mathcal{I}_m, \\ \mathcal{I}_m, & z_{m,k} \le \mathcal{I}_m, \end{cases}
\tag{1.14}
$$

where $y_{m,k} \in \mathbb{R}$ is the one-side censored measurement and $\mathcal{I}_m$ is the one-side censoring threshold.

Based on Tobit observation model (1.14), we define a series of Bernoulli random variables $\gamma_{m,k}$ $(m = 1, 2, \ldots, p)$ to regulate the censoring phenomenon of $y_{m,k}$ as follows:

$$
\gamma_{m,k} = \begin{cases} 1, & z_{m,k} > \mathcal{I}_m, \\ 0, & z_{m,k} \le \mathcal{I}_m, \end{cases}
\tag{1.15}
$$

with the following probability distributions:

$$
\begin{cases} \text{Prob}\{\gamma_{m,k} = 1\} = \bar{\gamma}_{m,k}, \\ \text{Prob}\{\gamma_{m,k} = 0\} = 1 - \bar{\gamma}_{m,k}, \end{cases}
\tag{1.16}
$$

where $\bar{\gamma}_{m,k}$ are known non-negative constants. Additionally, it is supposed that $\gamma_{m,k}$ is uncorrelated with ω_k and the initial system state. Taking advantage of $\gamma_{m,k}$, we rewrite $y_{m,k}$ in (1.14) as follows:

$$
y_{m,k} = \gamma_{m,k} z_{m,k} + (1 - \gamma_{m,k})\mathcal{I}_m.
\tag{1.17}
$$

1.2.2 TWO-SIDE CENSORED MEASUREMENTS

In contrast to the one-side censored measurements, the two-side (also called dead-zone-like) censored measurements are largely overlooked in spite of their pervasive presence in a great variety of physical scenarios suffering from detection limitations, sensor saturations and image frame effects.

According to the Tobit measurement model, the two-side censored measurement is given by the following equation:

$$y_{m,k} = \begin{cases} \mathcal{I}_m^l, & z_{mk} \leq \mathcal{I}_m^l, \\ z_{m,k}, & \mathcal{I}_m^l < z_{m,k} < \mathcal{I}_m^r, \\ \mathcal{I}_m^r, & z_{m,k} \geq \mathcal{I}_m^r, \end{cases} \tag{1.18}$$

where $y_{m,k}$ is the censored measurement, and $\mathcal{I}_m^l$ and $\mathcal{I}_m^r$ are, respectively, the left- and right-censoring thresholds.

Based on the Tobit observation model (1.18), we define a series of Bernoulli random variables $\gamma_{m,k}^l$ and $\gamma_{m,k}^r$ to regulate the two-side censoring phenomenon of $y_{m,k}$ as follows:

$$\gamma_{m,k}^l = \begin{cases} 1, & z_{m,k} \leq \mathcal{I}_m^l, \\ 0, & z_{m,k} > \mathcal{I}_m^l, \end{cases} \tag{1.19}$$

$$\gamma_{m,k}^r = \begin{cases} 1, & z_{m,k} \geq \mathcal{I}_m^r, \\ 0, & z_{m,k} < \mathcal{I}_m^r, \end{cases} \tag{1.20}$$

with the following probability distributions:

$$\begin{cases} Prob\{\gamma_{m,k}^l = 1\} = \bar{\gamma}_{m,k}^l, \\ Prob\{\gamma_{m,k}^r = 1\} = \bar{\gamma}_{m,k}^r, \\ Prob\{\gamma_{m,k}^l = 0\} = 1 - \bar{\gamma}_{m,k}^l, \\ Prob\{\gamma_{m,k}^r = 0\} = 1 - \bar{\gamma}_{m,k}^r. \end{cases} \tag{1.21}$$

Here, $\bar{\gamma}_{m,k}^l$ and $\bar{\gamma}_{m,k}^r$ are known non-negative constants that are uncorrelated with ω_k and the initial system state. Taking advantage of (1.19)–(1.20), $y_{m,k}$ in (1.18) can be rewritten as follows:

$$y_{m,k} = (1 - \gamma_{m,k}^l - \gamma_{m,k}^r)z_{m,k} + \gamma_{m,k}^l \mathcal{I}_m^l + \gamma_{m,k}^r \mathcal{I}_m^r. \tag{1.22}$$

1.2.3 KALMAN FILTERING WITH CENSORED MEASUREMENTS

The Tobit measurement models in (1.14) and (1.18) have been intensively studied in economics. Nonetheless, when it comes to the filtering field, such models have not been given full consideration until recently. Generally speaking, due to the nonlinear nature of the censoring phenomenon, filtering problems with censored measurements can be roughly tackled by some well-known nonlinear filtering techniques such as the particle filter, which certainly gives rise to heavy computation burden, especially under the context of MSFF.

To save computation burden and energy consumption, initial attempts have been made on the Kalman filtering with censored measurements, where censored measurements have been assumed to be uncorrelated with true system states, and hence, measurements within censored regions have been treated as missing. Nonetheless, it is often the case that censored measurements governed by Tobit measurement models *do* depend significantly on true system states [73], and the simple treatment of censored measurements as missing measurements could result in certain biased estimates. The censored and uncensored measurements have been processed separately through an iterative maximum likelihood method in [74], and all previous measurements have been used for implementing the algorithm.

It has been pointed out in [75] that a main challenge when designing a Kalman-like filter with censored measurements is that measurement noise nearing censoring regions becomes non-Gaussian, and this renders the straightforward application of Kalman filtering futile. To solve this problem, a fully recursive filter, named the TKF, has been developed in [35] based on two novel definitions and one local approximation. To be more specific, (1) new Bernoulli random variables are first defined and adopted to characterize the censoring phenomenon of the measurement (see (14) and (18)); (2) new definitions with respect to the measurement expectation and variance are formulated to benefit the calculation of filter gains and covariances, which give rise to the so-called Tobit regression model; and (3) a local approximation (that the state prediction well approximates the real system state in case of small estimation errors) is introduced to determine censoring probabilities.

The following two theorems present the Tobit regression models for one- and two-side censored measurements, respectively.

Theorem 1.1 [35] The expectation and variance of the one-side censored measurement $y_{m,k}$ conditional on the system state x_k are

$$\mathbb{E}\{y_{m,k} \mid x_k\} = \Phi\left(\frac{C_{m,k}x_k - \mathcal{I}_m}{\sqrt{R_{m,k}}}\right)$$
$$\times \left[C_{m,k}x_k + \sqrt{R_{m,k}}\,\lambda\left(\frac{\mathcal{I}_m - C_{m,k}x_k}{\sqrt{R_{m,k}}}\right)\right] \qquad (1.23)$$
$$+ \Phi\left(\frac{\mathcal{I}_m - C_{m,k}x_k}{\sqrt{R_{m,k}}}\right)\mathcal{I}_m,$$

$$Var\{y_{m,k} \mid x_k\} = R_{m,k}\left[1 - \varphi\left(\frac{\mathcal{I}_m - C_{m,k}x_k}{\sqrt{R_{m,k}}}\right)\right], \qquad (1.24)$$

$$\Phi\left(\frac{z_{m,k}-C_{m,k}x_k}{\sqrt{R_{m,k}}}\right)=\int_{-\infty}^{\alpha_{m,k}}\frac{1}{\sqrt{2\pi R_{m,k}}}e^{-\frac{(z_{m,k}-C_{m,k}x_k)^2}{2R_{m,k}}}d_{\alpha_{m,k}},\tag{1.27}$$

$$\begin{aligned}\varphi\left(\frac{\mathcal{I}_m-C_{m,k}x_k}{\sqrt{R_{m,k}}}\right)&=\lambda\left(\frac{\mathcal{I}^m-C_{m,k}x_k}{\sqrt{R_{m,k}}}\right)\\&\times\left[\lambda\left(\frac{\mathcal{I}^m-C_{m,k}x_k}{\sqrt{R_{m,k}}}\right)-\frac{\mathcal{I}_m-C_{m,k}x_k}{\sqrt{R_{m,k}}}\right].\end{aligned}\tag{1.28}$$

where

$$\phi\left(\frac{z_{m,k}-C_{m,k}x_k}{\sqrt{R_{m,k}}}\right)=\frac{1}{\sqrt{2\pi}}e^{-\frac{(z_{m,k}-C_{m,k}x_k)^2}{2R_{m,k}}},\tag{1.25}$$

$$\lambda\left(\frac{\mathcal{I}_m-C_{m,k}x_k}{\sqrt{R_{m,k}}}\right)=\frac{\phi\left(\frac{\mathcal{I}_m-C_{m,k}x_k}{\sqrt{R_{m,k}}}\right)}{1-\Phi\left(\frac{\mathcal{I}_m-C_{m,k}x_k}{\sqrt{R_{m,k}}}\right)},\tag{1.26}$$

and $\Phi(\cdot)$ and $\varphi(\cdot)$ are given by (1.27)–(1.28).

Theorem 1.2 [75] The expectation and variance of the two-side censored measurement $y_{m,k}$ conditional on the system state x_k are

$$\begin{aligned}\mathbb{E}\{y_{m,k}\mid x_k\}=&\Phi\left(\frac{\mathcal{I}_m^l-C_{m,k}x_k}{\sqrt{R_{m,k}}}\right)\mathcal{I}_m^l+\left(1-\Phi\left(\frac{\mathcal{I}_m^r-C_{m,k}x_k}{\sqrt{R_{m,k}}}\right)\right)\mathcal{I}_m^r\\&+\left[\Phi\left(\frac{\mathcal{I}_m^r-C_{m,k}x_k}{\sqrt{R_{m,k}}}\right)-\Phi\left(\frac{\mathcal{I}_m^l-C_{m,k}x_k}{\sqrt{R_{m,k}}}\right)\right]\\&\times\left[C_{m,k}x_k-\sqrt{R_{m,k}}\lambda_{m,k}\right],\end{aligned}\tag{1.29}$$

and

$$var\{y_{m,k}\mid x_k\}=\mathcal{R}_{m,k}\left[1+\varphi_{m,k}\right],\tag{1.30}$$

where

$$\lambda_{m,k}=\frac{\phi\left(\frac{\mathcal{I}_m^r-C_{m,k}x_k}{\sqrt{R_{m,k}}}\right)-\phi\left(\frac{\mathcal{I}_m^l-C_{m,k}x_k}{\sqrt{R_{m,k}}}\right)}{\Phi\left(\frac{\mathcal{I}_m^r-C_{m,k}x_k}{\sqrt{R_{m,k}}}\right)-\Phi\left(\frac{\mathcal{I}_m^l-C_{m,k}x_k}{\sqrt{R_{m,k}}}\right)},\tag{1.31}$$

$$\varphi_{m,k} = \frac{\dfrac{\mathcal{I}_m^l - C_{m,k} x_k}{R_{m,k}} \phi\left(\dfrac{\mathcal{I}_m^l - C_{m,k} x_k}{\sqrt{R_{m,k}}}\right)}{\Phi\left(\dfrac{\mathcal{I}_m^r - C_{m,k} x_k}{\sqrt{R_{m,k}}}\right) - \Phi\left(\dfrac{\mathcal{I}_m^l - C_{m,k} x_k}{\sqrt{R_{m,k}}}\right)}$$

$$- \frac{\dfrac{\mathcal{I}_m^r - C_{m,k} x_k}{R_{m,k}} \phi\left(\dfrac{\mathcal{I}_m^r - C_{m,k} x_k}{\sqrt{R_{m,k}}}\right)}{\Phi\left(\dfrac{\mathcal{I}_m^r - C_{m,k} x_k}{\sqrt{R_{m,k}}}\right) - \Phi\left(\dfrac{\mathcal{I}_m^l - C_{m,k} x_k}{\sqrt{R_{m,k}}}\right)} - \lambda_{m,k}^2. \tag{1.32}$$

Remark 1.4 The enhanced regression model in Theorem 1.2 explicitly supplies us with the specific expressions of the measurement mean and variance in the case of two-side censoring phenomenon. In contrast with its one-side counterpart in Theorem 1.1, it is evidently seen that a suite of terms with respect to $\mathcal{I}_m^l \left(\mathcal{I}_m^r\right)$ appears here that carries the two-side censoring information. In addition, due to the presence of two-side censoring, the structures of many traditional terms (of one-side censoring) (e.g. $\lambda_{m,k}$ and $\varphi_{m,k}$) vary to a great extent. Note that the presence of these new terms and the variation of these traditional terms will inevitably result in a different TKF design and analysis framework, as shown in the following two theorems.

In response to the Tobit regression models presented by Lemmas 1.1–1.2, the following two theorems provide the Tobit Kalman filters for the one- and two-side censoring cases.

Theorem 1.3 [35] The TKF for system (1.1) and censored measurement model 1.17 is given by

$$\hat{x}_{m,k} = A_{k-1} x_{m,k-1} + K_{m,k}(y_{m,k} - \hat{y}_{m,k}^-), \tag{1.33}$$

$$P_{m,k}^- = A_{k-1} P_{m,k-1} A_{k-1}^T + B_{k-1} Q_{k-1} B_{k-1}^T, \tag{1.34}$$

$$P_{m,k} = (I - \bar{\gamma}_{m,k} K_{m,k} C_{m,k}) P_{m,k}^-. \tag{1.35}$$

Here, $K_{m,k}$ and $\hat{y}_{m,k}^-$ are, respectively, the gain matrix and the one-step measurement prediction given by

$$K_{m,k} = P_{xy_{m,k}} P_{y_{m,k}}^{-1}, \tag{1.36}$$

$$\hat{y}_{m,k}^- = (1 - \bar{\gamma}_{m,k}) \mathcal{I}_m + \bar{\gamma}_{m,k}$$

$$\times \left[C_{m,k} \hat{x}_{m,k}^- + \sqrt{R_{m,k}} \lambda\left(\frac{\mathcal{I}_m - C_{m,k} \hat{x}_{m,k}^-}{\sqrt{R_{m,k}}}\right) \right], \tag{1.37}$$

where

$$P_{xy_{m,k}} = P_{m,k}^{-} C_{m,k}^{T} \bar{\gamma}_{m,k}, \tag{1.38}$$

$$\bar{\gamma}_{m,k} = \Phi\left(\frac{C_{m,k}\hat{x}_{m,k}^{-} - \mathcal{I}_{m}}{\sqrt{R_{m,k}}} \right), \tag{1.39}$$

$$P_{y_{m,k}} = \bar{\gamma}_{m,k} C_{m,k} P_{m,k}^{-} C_{m,k}^{T} \bar{\gamma}_{m,k}$$
$$+ R_{m,k}\left[1 - \varphi\left(\frac{\mathcal{I}_{m} - C_{m,k}\hat{x}_{m,k}^{-}}{\sqrt{R_{m,k}}} \right) \right]. \tag{1.40}$$

Theorem 1.4 [53] The TKF for system (1.1) and censored measurement model (1.22) is given by the following structure:

$$\hat{x}_{m,k} = A_{k-1}x_{m,k-1} + K_{m,k}(y_{m,k} - \hat{y}_{m,k}^{-}), \tag{1.41}$$

$$P_{m,k}^{-} = A_{k-1}P_{m,k-1|}A_{k-1}^{T} + B_{k-1}Q_{k-1}B_{k-1}^{T}, \tag{1.42}$$

$$P_{m,k} = \left(I - \left(1 - \bar{\gamma}_{m,k}^{l} - \bar{\gamma}_{m,k}^{r}\right) K_{m,k}C_{m,k} \right) P_{m,k}^{-}. \tag{1.43}$$

Here, $K_{m,k}$ and $\hat{y}_{m,k}^{-}$ are, respectively, the gain matrix and the one-step measurement prediction given by

$$K_{m,k} = P_{xy_{m,k}} P_{y_{m,k}}^{-1}, \tag{1.44}$$

$$\hat{y}_{m,k}^{-} = \bar{\gamma}_{m,k}^{l}\mathcal{I}_{m}^{l} + \bar{\gamma}_{m,k}^{r}\mathcal{I}_{m}^{r} + \left(1 - \bar{\gamma}_{m,k}^{l} - \bar{\gamma}_{m,k}^{r}\right)\left[C_{m,k}\hat{x}_{m,k}^{-} + \sqrt{R_{m,k}}\lambda_{m,k} \right], \tag{1.45}$$

where

$$P_{xy_{m,k}} = P_{m,k}^{-} C_{m,k}^{T}(1 - \bar{\gamma}_{m,k}^{l} - \bar{\gamma}_{m,k}^{r}), \tag{1.46}$$

$$\bar{\gamma}_{m,k}^{l} = \Phi\left(\frac{\mathcal{I}_{m}^{l} - C_{m,k}\hat{x}_{m,k}^{-}}{\sqrt{R_{m,k}}} \right), \tag{1.47}$$

$$\bar{\gamma}_{m,k}^{r} = \Phi\left(\frac{C_{m,k}\hat{x}_{m,k}^{-} - \mathcal{I}_{m}^{r}}{\sqrt{R_{m,k}}} \right), \tag{1.48}$$

$$P_{y_{m,k}} = R_{m,k}\left[1 + \bar{\varphi}_{m,k} \right] + \left(1 - \bar{\gamma}_{m,k}^{l} - \bar{\gamma}_{m,k}^{r}\right)^{2} C_{m,k}P_{m,k}^{-}C_{m,k}^{T}, \tag{1.49}$$

and

$$\overline{\lambda}_{m,k} = \frac{\phi\left(\dfrac{\mathcal{I}_m^r - C_{m,k}\overline{x}_{m,k}^-}{\sqrt{R_{m,k}}}\right) - \phi\left(\dfrac{\mathcal{I}_m^l - C_{m,k}\overline{x}_{m,k}^-}{\sqrt{R_{m,k}}}\right)}{\Phi\left(\dfrac{\mathcal{I}_m^r - C_{m,k}\overline{x}_{m,k}^-}{\sqrt{R_{m,k}}}\right) - \Phi\left(\dfrac{\mathcal{I}_m^l - C_{m,k}\overline{x}_{m,k}^-}{\sqrt{R_{m,k}}}\right)}, \tag{1.50}$$

$$\overline{\varphi}_{m,k} = \frac{\dfrac{\mathcal{I}_m^l - C_{m,k}\overline{x}_{m,k}^-}{R_{m,k}}\phi\left(\dfrac{\mathcal{I}_m^l - C_{m,k}\overline{x}_{m,k}^-}{\sqrt{R_{m,k}}}\right)}{\Phi\left(\dfrac{\mathcal{I}_m^r - C_{m,k}\overline{x}_{m,k}^-}{\sqrt{R_{m,k}}}\right) - \Phi\left(\dfrac{\mathcal{I}_m^l - C_{m,k}\overline{x}_{m,k}^-}{\sqrt{R_{m,k}}}\right)}$$
$$- \frac{\dfrac{\mathcal{I}_m^r - C_{m,k}\overline{x}_{m,k}^-}{R_{m,k}}\phi\left(\dfrac{\mathcal{I}_m^r - C_{m,k}\overline{x}_{m,k}^-}{\sqrt{R_{m,k}}}\right)}{\Phi\left(\dfrac{\mathcal{I}_m^r - C_{m,k}\overline{x}_{m,k}^-}{\sqrt{R_{m,k}}}\right) - \Phi\left(\dfrac{\mathcal{I}_m^l - C_{m,k}\overline{x}_{m,k}^-}{\sqrt{R_{m,k}}}\right)} - \overline{\lambda}_{m,k}^2. \tag{1.51}$$

Remark 1.5 Making a brief comparison between Theorem 1.3 and Theorem 1.4, two prime differences can be observed. One is the replacement of the one-side censoring probability $\gamma_{m,k}$ by the two-side censoring probability $\left(1 - \overline{\gamma}_{m,k}^l - \overline{\gamma}_{m,k}^r\right)$. The other is the replacement of censoring-induced terms $\lambda(\cdot)$ and $\phi(\cdot)$ by the newly structured terms $\overline{\lambda}_{m,k}$ and $\overline{\phi}_{m,k}$. The two replacement terms sufficiently reflect how the one- and two-side measurement censoring influence filter design.

1.3 COMMUNICATION CONSTRAINTS

As the scale of a networked system grows, the limited resources assigned to data communication inevitably render communication constraints [66, 76–80], and this inescapably deteriorates the MSFF performance to some extent. Such communication constraint–induced phenomena cover but are not limited to communication delays, fading measurements, nonlinear disturbances, quantized measurements, disordered measurements, etc. In this section, we are going to provide a full and concrete description of such phenomena and their effects on the MSFF performance.

1.3.1 COMMUNICATION DELAYS

Due mainly to finite switching speed in information processing, communication delays have been observed in a great variety of industrial systems ranging from automobile manufacturing to communication networks [81–83]. Hereunto, in order to reduce the conservatism originated from communication delays, many efficient methods have been proposed in the literature, such as the bounding method [84], delay fractioning method [85], descriptor system method [86] and slack matrix variables method [87]. Basically, these delay-dependent methods include two perspectives, i.e. conservatism and complexity. The conservatism concerns about the development of delay-dependent conditions so as to allow a maximal delay, while the complexity turns to develop delay-dependent conditions not only allowing the maximal delay but also utilizing minimum decision variables.

Making a comparison between the aforementioned delay-dependent methods, it is observed that the complexity and conservatism, in fact, serve as performance criteria of the desired controller/filter, and thus a tradeoff between both criteria should be found when designing the controller/filter. Since it is often substantially difficult (*if not impossible*) to devise a globally optimal delay-dependent method with least conservatism yet least computation complexity, one may often need to weight between different delay-dependent methods according to different design requirements.

It should be noted that in contrast to other delay-dependent methods, the delay-fractioning method is capable of efficiently reducing the conservatism at the cost of increasing the computation complexity. Moreover, owing to the rapid development of computer science, the problem of the computation complexity can be easily handled by adopting the state-of-the-art data processing and computation unit. As such, a great deal of research effort has been devoted to control/filtering problems for various time-delay systems based on the delay-fractioning method with some representative works presented in [88–91].

1.3.2 FADING MEASUREMENTS

In a networked environment, when traveling along wireless communication links, signals are often susceptible to refraction, diffraction and reflection because of obstacles hindering signal propagation. Under such a circumstance, signals are therefore subject to the so-called multi-path propagation phenomenon [67, 92], resulting in fading signals. Hitherto, a lot of models have been put forward in the literature to characterize signal variations in both amplitudes and phases under the context of fading signals – for instance, the Rayleigh, Lth-order Rice and analog erasure fading models [93–95]. Based on such models, various control/filtering problems have been broadly studied in recent decades, with some typical works presented in [96–100].

To be specific, the linear-quadratic-Gaussian control over multi-erasure channels and the H_∞ output feedback-control over Lth-order Rice fading channels have been explored in [101] and [100], respectively. The mean-squared stabilization problem over various fading channels has been comprehensively studied in [102, 103]. When it comes to multiple-input multiple-output channels, a statistical discrete-time model has been proposed in [104] to characterize the Rayleigh fading channels that are triply selective due to the angle spread, delay spread and Doppler spread. Based on such a Rayleigh fading channel model, the emulator design and simulator modeling problems have been addressed in [105] and [106], respectively. It is worth mentioning that in contrast to the Rayleigh and analog erasure fading models, the Lth-order Rice fading model provides an intuitive characterization of multiple network-induced phenomena including packet dropouts and communication delays.

On the other hand, measurement signals transmitted through network channels may fade in a probabilistic way, and the frequently encountered missing measurement could be reviewed as an extreme case of the resultant fading measurement [107]. Similar to the missing measurement, the occurrence of the fading measurement can now be governed by a sequence of random variables obeying certain PDFs (e.g. the Bernoulli distribution) over some known intervals. In this sense, a Kalman-like filter has been developed in [108] to solve the filtering problem with multiple fading measurements by recurring to the forward Riccati difference equation. In addition, the Lth-order Rice fading channel has been modeled in [44] by a group of independently and identically distributed Gaussian variables and later employed to tackle control/filtering with multi-path fading measurements.

After being transmitted through the analog erasure, Rayleigh, Lth-order Rice fading channels are presented, the measurements of system (1.1) are, respectively, given by the following models.

- Analog Erasure fading measurement model [102].

$$\bar{y}_{m,k} = \xi_{m,k} y_{m,k},\tag{1.52}$$

where $y_{m,k}$ and $\bar{y}_k$ are, respectively, the measurement vectors before and after network transmission. $\xi_{m,k}$ is the Bernoulli random variable governing the erasure fading phenomenon with the following PDF:

$$\begin{cases} \text{Prob}\{\xi_{m,k} = 1\} = \bar{\xi}_{m,k}, \\ \text{Prob}\{\xi_{m,k} = 0\} = 1 - \bar{\xi}_{m,k}, \end{cases}\tag{1.53}$$

where $\bar{\gamma}_{m,k}$ is the known erasure probability. It is supposed that $\xi_{m,k}$ is uncorrelated with the initial system state and noise signals.

- Triply selective Rayleigh fading model [104].

$$\bar{y}_{m,k} = \sum_{l_m = -L_m^1}^{L_m^2} C_{m,k-l_m} x_{,k-l_m} + \upsilon_{m,k},\tag{1.54}$$

where $y_{m,k}$ and $\bar{y}_k$ are, respectively, the measurement vectors before and after network transmission; l_m is the fading coefficient index with its range being $[-L_m^1, L_m^2]$.

- Lth-order Rice fading model [101].

$$\bar{y}_{m,k} = \sum_{l_m=0}^{L_{m,k}} \vartheta_{m,k}^{l_m} C_{m,k-l_m} x_{k-l_m} + \upsilon_{m,k},\tag{1.55}$$

where $y_{m,k}$ and $\bar{y}_k$ are, respectively, the measurement vectors before and after network transmission. $L_{m,k} = \min\{L,k\}$ and L is the channel order. $\vartheta_{m,k}^{l_m}$ is the Gaussian distributed channel coefficient of mean $\bar{\vartheta}_{m,k}^{l_m}$ and variance $\tilde{\vartheta}_{m,k}^{l_m}$. It is assumed that $\vartheta_{m,k}^{l_m}$ is uncorrelated with the initial system state and noise signals.

1.3.3 NONLINEAR DISTURBANCES

Nonlinearities occur universally in a wide range of engineering practices ranging from wireless communication to battery modeling [109–113]. In most cases, the non-linearity is typically described as certain additive nonlinear disturbances coming from environmental circumstances. Note that in networked systems, e.g. the well-known internet-based three-tank system, such nonlinear disturbances may appear in a probabilistic way because of the random-occurrence nature of many network-induced phenomenon [114–118].

Taking into consideration limited communication bandwidth and perturbed network environments, the network transmission could face severe network congestions and unexpected network disturbances. Such disturbances can be characterized by some randomly occurring nonlinearities, with their occurrence probabilities obtained by statistical tests. Accordingly, a lot of research effort has been directed to the filtering/control problems subject to randomly occurring nonlinearities, see e.g. [119–126].

Focusing on state estimation for uncertain nonlinear stochastic systems, the linear matrix inequality method has been designed in [127], where randomly occurring nonlinearities have been described by statistical means. This type of description has been later utilized to design controllers/filters in the concurrence of both random nonlinearities and other network phenomena, such as communication delays, fading measurements and sensor saturations [108, 128, 129]. When it comes to the multi-sensor case, the distributed filtering problem has been addressed in [128] under random nonlinearities and sensor degradations, and a distributed robust variance-constrained filter has been proposed in [130] for time-varying stochastic systems with missing measurements and random nonlinearities. Sufficient conditions have also been achieved in both works to ascertain the mean-square boundedness of the resultant filtering errors.

Additionally, in case of gain variations, a resilient H_∞ filter has been developed in [131] for stochastically uncertain systems with the simultaneous presence of random nonlinearities and fading measurements. A resilient state estimator has been devised in [132] for uncertain recurrent neural networks with missing measurements and random nonlinearities, where both prescribed H_∞ performance and variance constraint have been sufficiently ensured. Recently, an enhanced TKF has been constructed in [47] for fractional-order systems with censoring measurements and random nonlinearities under the Round-Robin protocol and has further been applied to the spring-mass-damper system in order to optimally estimate parameters of the viscoelastically damped structure.

1.3.4 QUANTIZED MEASUREMENTS

It is known that quantization exists pervasively in computer-based scenarios where finite-precision arithmetic is extensively used. In these scenarios, owing to limited network bandwidth, network performance is inevitably affected by quantization errors imposed possibly by intense signal attenuation and unexpected perturbation [59, 60, 133, 134]. Thus, the quantization problem in a networked environment has long been researched, and abundant results have been reported in [135, 136] and references therein. Specifically, the time-varying quantization mechanism has been initially proposed in [135], where the number of quantization levels has been assumed to be fixed and finite, while the quantization resolution has been allowed to be manipulated arbitrarily over time. The problem of input-to-state stability with respect to the quantization

$$\Upsilon_m = \left\{\pm \upsilon_m^j, \upsilon_m^j = (\rho_m)^j \upsilon_m^0, j = 0, \pm 1, \pm 2, ...\right\} \cup \{0\}, 0 < \rho_m \left\langle 1, \upsilon_m^0 \right\rangle 0, \qquad (1.59)$$

error has been handled in [136] for nonlinear systems where the quantization effect has been treated as an additional disturbance.

Regarding the specific quantization strategy, two main quantization models, i.e. uniform quantization and logarithmic quantization models, have been developed in the literature. Letting the mapping of the quantization process be

$$q(y_k) = \left[q_1(y_{1,k}) \quad q_2(y_{2,k}) \quad \cdots \quad q_p y_{p,k}) \right]^T, \qquad (1.56)$$

the following equations present the uniformly and logarithmically quantized measurement models.

- Uniformly quantized measurement model [137].

For each $q_m (y_{m,k})$, the set of uniform quantization levels is

$$U_m = \left\{ \pm u_m^j, u_m^j = j[y_{m,k} / \delta_m], j = 0, \pm 1, \pm 2, ...\right\}, \qquad (1.57)$$

where δ_m is the quantization density with respect to $y_{m,k}$ and $[\cdot]$ is the integral function that rounds "$\cdot$" to its nearest integer. As a result, we have $q(y_{m,k}) = j[y_{m,k} / \delta_m]$, and the quantized signal is

$$\overline{y}_{m,k} = q(y_{m,k}) + \tilde{q}(y_{m,k}), \qquad (1.58)$$

where $\tilde{q}(y_{m,k})$ is the so-called quantization error satisfying certain norm or statistic conditions.

- Logarithmically quantized measurement model [138].

For each $q_m (y_{m,k})$, the set of logarithmic quantization levels is given by (1.59)

$$\Upsilon_m = \left\{ \pm \upsilon_m^j, \upsilon_m^j = (\rho_m)^j \upsilon_m^0, j = 0, \pm 1, \pm 2, \cdots \right\} \cup \{0\}, 0 < \rho_m < 1, \upsilon_m^0 > 0, \qquad (1.59)$$

where ρ_m is the quantization density and υ_m^0 is a scaling constant. The mapping from the whole segment to different quantization levels is accomplished through

$$q_m(y_{m,k}) = \begin{cases} v_m^j, & \dfrac{v_m^j}{1+\theta_m} < y_{m,k} \le \dfrac{v_m^j}{1-\theta_m}, \\ 0, & y_{m,k} = 0, \\ -q_m(-y_{m,k}), & y_{m,k} < 0, \end{cases} \qquad (1.60)$$

where $\theta_m = (1-\rho_m)/(1+\rho_m)$. As a result, we have $q_m(y_{m,k}) = (1+\Delta_{m,k})y_{m,k}$ with $|\Delta_{m,k}| \le \theta_m$, and the logarithmically quantized measurement signal is

$$\bar{y}_{m,k} = (1+\Delta_{m,k})C_{m,k}x_k + (1+\Delta_{m,k})v_{m,k}. \qquad (1.61)$$

Paying attention to these two quantization models, it has been verified that compared to the uniform quantized measurement model, the logarithmic quantized measurement model is much more preferable, as fewer bits are required to be propagated. A sector-bounded scheme has been formulated in [139] to resolve the logarithmic quantization influence by converting resultant quantization errors into certain sector-bounded uncertainties. Such an elegant yet efficient scheme has quickly gained widespread popularity among researchers, and this leads to vast applications in a substantial number of practical situations [25, 28, 140–144].

1.3.5 Disordered Measurements

In digital communication, data is generally transmitted through the form of "data packets" in the communication network, as data communication is unavoidably subject to delays which may occur in a random way due to limited network bandwidth. Such randomly occurring communication delays could give rise to the situation where data packets might not follow the principle of "first send first arrive". That is to say, data packets sent earlier (later) may arrive at the remote estimator later (earlier), which results in the so-called "packet disorder" or "disordered measurements".

Denote $\tau_{m,k}$ as the possible communication delay variable between the m-th sensor and local estimator. The PDF of the delay variable $\tau_{m,k}$ is

$$\text{Prob}\{\tau_{m,k} = i_m\} = p_{i_m}, \, i_m = 1, 2, \ldots, s_m, \qquad (1.62)$$

where s_m is the maximum delay value that $\tau_{m,k}$ can take. Consequently, taking into account the communication delay, the actual measurement arriving at the remote estimation is [145]

$$\bar{y}_{m,k} = \begin{cases} y_{m,k-\tau_{m,k}}, & k > 0 \\ y_{m,t_m}, & t_m = -s_m, -s_m+1, \ldots, 0, \end{cases} \qquad (1.63)$$

where $y_{m,-s_m}, y_{m,-s_m+1}, \ldots, y_{m,0}$ are the initial conditions.

An intuitive outcome of the disorder measurement is that the latest sensor measurement received by the remote filter might not have the newest information of the interested system. This would unquestionably bring unexpected computation challenge and/or degrade estimator performance. Consequently, the control/filtering

TABLE 1.2

Communication Constraints and Its Corresponding References

Communication constraints	References
Disordered measurements	[145–149].
Quantized measurements	[133–138, 150, 151]
Nonlinear disturbances	[47, 108, 127–132]
Fading measurements	[44, 100–108]
Communication delays	[81, 82, 84–91, 152–154]

problems with disordered measurements have stirred initial research attention in the last few years, and some pioneering works have been presented for nonlinear systems, complex networks and networked control systems [146–148].

Specifically, a compensation strategy has been proposed in [148] for a class of close-loop systems in order to ease the negative effects from packet disorders. Based on such a compensation strategy, the close-loop control system has been modeled by a Markov jump system, with its stochastic stability clearly proved. When packet disorders and time delays are separately considered, a novel mechanism has been presented in [147], where the latest arrived signal has been opted out as the actual input of the remote estimator. In addition, when it comes to the simultaneous presence of several networked-induced phenomena, e.g. measurement quantization and measurement missing, initial research attention has been drawn with some meaningful results presented in [145, 149].

Table 1.2 summarizes some typical references with respect to the investigated five communication constraints.

1.4 OUTLINE

The framework of this book is as follows:

- In Chapter 1, the research background is firstly introduced, where the state estimation and MSFF problems with censored measurements under a constrained network environment are comprehensively reviewed. It is clearly pointed out that most existing MSFF literature within a networked environment has focused on centralized MSFF with network-induced phenomena (e.g. missing measurement and communication delays). The corresponding results for a more broader network constraint (e.g. channel fading and communication protocols) case have been much fewer, especially in the context of networked systems where the usage of massive low-cost commercial off-the-shelf sensors gives rise to a special kind of measurement nonlinearity called censored measurements. After presenting the motivation, the outline of this book is then listed.
- In Chapter 2, the optimal filtering problem is studied for a class of discrete time-varying systems with both censored and fading measurements.

The censored measurements are described by the Tobit measurement model, and the fading measurements are characterized by the Lth-order Rice fading channel model capable of accounting for not only the packet dropout but also the communication phenomenon. The measurement fading occurs in a random way where the fading probabilities are regulated by a set of mutually independent Gaussian random variables. By resorting to the state augmentation technique and the orthogonality projection principle, the Tobit Kalman filter is designed in the presence of fading measurements. In the course of filter design, several state-augmentation-induced terms are introduced, all of which can be calculated recursively or off-line.

- Chapter 3 studies the minimum-variance recursive filtering with censored data and time-correlated multiplicative sensor noises under redundant channel transmission, with stochastic sensor gain degradation. The Tobit regression model is first modified to take into account the complexities contributed by measurement noises and intermittent failures as well as redundant channels. Then, an optimal Tobit Kalman filter is designed based on the modified Tobit regression model. In the developed algorithm for the filter design, several new terms are introduced to reflect how such complexities are addressed, all of which can be calculated recursively or off-line.

- Chapter 4 concerns the optimal state estimation problem with censored data under non-Gaussian Lévy and time-correlated additive sensor noises. The time correlation of the measurement noises is transformed into the cross-correlation between the equivalent measurement noise and the process noise. Then, by resorting to the Lévy-Ito theorem, the non-Gaussian Lévy measurement noises are transformed into equivalent Gaussian noises with unknown covariances. Based on the transformed Gaussian measurement noises, a modified recursive TKF is designed in which the unknown noise covariances are carefully calculated.

- In Chapter 5, we deal with the Tobit Kalman filtering for networked systems with integral measurements and probabilistic sensor failures, where self-propagated upper and lower bounds are determined on the filtering error covariance. The censored observations are characterized by the Tobit observation model, the integral measurements are described as functions of system states over a certain time interval required for data acquisition, and the sensor failures are governed by a set of uncorrelated random variables. The Round-Robin protocol (RRP) is employed to decide the transmission sequence of sensors in order to alleviate undesirable data collisions. By resorting to the augmentation technique and the orthogonality projection principle, a protocol-based Tobit Kalman filter is developed with the coexistence of integral measurements and sensor failures that lead to a couple of augmentation-induced terms. Moreover, the performance of the proposed filter is analyzed through examining the statistical property of the error covariance of the state estimation. Further analysis shows the existence of self-propagating upper and lower bounds on the estimation error covariance.

- In Chapter 6, the distributed filtering fusion problem is investigated under both packet dropouts and transmission delays via a probabilistic view.

The censored measurement is characterized by the Tobit measurement model, and the packet delay obeys the Poisson distribution. At the input terminal of each local estimator, a censoring detection device and a buffer with finite length are exploited to tackle the censored/delayed measurements. The filtering fusion is carried out in two steps: in the first step, every sensor in the network collects and sends its measurements to the corresponding local estimator via an unreliable network with probabilistic packet delays, and in the second step, the estimates from local estimators are further collected by the fusion center to form a fused estimate. The local estimator runs a modified Tobit Kalman filtering algorithm, which is designed based on the modified Tobit regression model, while the fusion center performs a distributed federated modified Tobit Kalman filtering algorithm, which is devised on the basis of the federated Kalman fusion rule. Given the buffer length, the prescribed bound on the estimation error covariance and the probability distribution of the packet delay, the stability performance of the proposed filtering fusion algorithm is evaluated via a probabilistic approach.

- Chapter 7 solves the Federated Tobit Kalman filtering fusion issue with dead-zone-like censoring and dynamical bias under the Round-Robin protocol. The uncertainties originate from three sources, namely, censored observations, dynamical biases and additive white noises. To reflect the dead-zone-like censoring phenomenon, the measurement observation is described by the Tobit model, where the censored region is constrained by prescribed left- and right-censoring thresholds. The bias is modeled as a dynamical stochastic process driven by a white noise in order to reflect the random behavior of possible ambient disturbances. The RRP is employed to decide the transmission sequence of sensors so as to alleviate undesirable data collisions. The filtering fusion is conducted via two stages: (1) the sensor observations arriving at its corresponding estimator are first leveraged to generate a local estimate, and (2) the local estimates are then gathered together at the fusion center in order to form the fused estimate. The local estimator implements a Tobit Kalman filtering algorithm on the basis of an enhanced Tobit regression model, whilst the fusion center realizes a filtering fusion algorithm in accordance with the well-known federated fusion principle.

- In Chapter 8, a distributed fusion estimator is designed for a class of time-varying systems under parametric uncertainties and measurement censoring. The parametric uncertainty is characterized by the multiplicative noise appearing in both state and observation equations, and the measurement censoring is governed by the Tobit observation model. The fusion estimation is implemented via two stages: at the first stage, each sensor sends its observations to the local estimator, and at the second stage, the local estimates are then transmitted to the fusion center so as to generate the fused estimate. The local estimator realizes a Tobit Kalman filtering algorithm which is devised in accordance with a modified regression model, whilst the fusion center carries out the fusion estimation by resorting to the federated fusion rule. Furthermore, the monotonicity of the fused error covariance with respect to the censoring threshold is discussed, and subsequently, the boundedness property is also examined, where both lower and upper bounds are acquired for the fused error covariance.

- In Chapter 9, we cope with the design issue of a class of protocol-based MSFF schemes under state saturation and cyber-attacks. In order to curb data collisions and ease communication overheads of the shared networks, the weighted try-once-discard protocol is implemented on the sensor-to-filter channel to orchestrate multiple sensors with a prescribed dynamic transmission order. Bernoulli distributed stochastic variables are utilized to characterize deception attacks initiated by potential adversaries. The well-known Tobit model is leveraged to characterize the dead-zone-like censoring phenomenon, where censored regions are constrained by certain left- and right-censoring thresholds. The fusion estimation is implemented via two stages: at the first stage, each sensor sends its observations to the local estimator and, at the second stage, the local estimates are then transmitted to the fusion center so as to generate the fused estimate. The local estimators realize Tobit Kalman filtering algorithms such that certain upper bounds (on the local filtering error covariances) are guaranteed and then minimized via adequately determining filter gains, while the fusion center carries out the fusion estimation by resorting to the federated fusion criterion. Furthermore, the performance of the devised fusion estimator is examined through assessing the boundedness of the upper bound on the fused error covariance

- Chapter 10 is concerned with the variance-constrained MSFF for nonlinear cyber-physical systems under stochastic communication protocol. To prevent data collision and reduce communication cost, the stochastic communication protocol is adopted in the sensor-to-filter channels to regulate the transmission order of sensors. Each sensor is allowed to enter the network according to the transmission priority decided by a set of independent and identically distributed random variables. From the defenders' view, the occurrence of the denial-of-service attack is governed by the randomly Bernoulli-distributed sequence. At the local filtering stage, a set of variance-constrained local filters is designed, where the upper bounds (on the filtering error covariances) are first acquired and later minimized by appropriately designing filter parameters. At the fusion stage, all local estimates and error covariances are combined to develop a variance-constrained fusion estimator under the federated fusion rule. Furthermore, the performance of the fusion estimator is examined by studying the boundedness of the fused error covariance.

- In Chapter 11, the conclusions and some potential topics for future work are given.

2 Optimal Filtering for Networked Systems with Channel Fading and Measurement Censoring

The past decades have witnessed the widespread applications of filtering techniques in areas such as manoeuvering target tracking, inertial navigation, global position system (GPS) positioning and so forth. Among a variety of filters designed under different circumstances, the Kalman filter (KF) has proven to be the most popular one. It is generally recognized that the KF is optimal in the sense of minimum mean-squared error (MMSE) under the assumption that the systems are linear and the noises are Gaussian. This assumption is, unfortunately, not always true in practice due to the fact that the nonlinearity, which serves as an important factor contributing to the complexities of the system modeling, exists in almost all the practical systems. In networked control systems, a typical kind of nonlinearities (e.g. censoring, fading, quantization and saturation) is unavoidable in the system measurements due mainly to the sudden environment changes, intermittent transmission congestions, random failures and repairs of components. In particular, censored and fading measurements are arguably two of the most frequently encountered measurement nonlinearities in engineering practice. The measurement nonlinearities, if not properly handled, could cause severe performance degradation of the traditional KF. To deal with this issue, many alternative filtering schemes have been reported, e.g. the extended Kalman filtering (EKF), the unscented Kalman filtering (UKF), and the Tobit Kalman filtering.

Investigating the literature concerning censored measurements, fading measurements and Kalman filtering, we can find that (1) the TKF has gained increasing research attention due primarily to its capability of suitably processing censored measurements as well as its recursive/concise structure; (2) in networked systems, the channel fading modeled as fading/degrading measurements in both the amplitude and the phase is unavoidable because of the multipath propagation/shadowing caused by the hinderance of obstacles; and (3) the Lth-order Rice fading model, as compared with the analog erasure channel model and the Rayleigh channel model, has a conciser yet simpler structure and is suitable for simultaneously representing the channel fadings, packet dropouts and communication delays. As such, a seemingly natural question is that when the measurement transmission is subject to both censoring and fading, can we design a filter to achieve the optimal estimate of the

DOI: 10.1201/9781003461623-2

system? Nevertheless, this appears to be a non-trivial question for the following two reasons: (1) it is unclear which method should be chosen to deal with the time-delays coupled in the Lth-order Rice fading model; and (2) it is pretty hard to establish a modified Tobit regression model which can not only correctly handle the time-delay and fading phenomena but also help design the subsequent filter design in a recursive form. This motivates us to conduct the present research.

This chapter deals with the optimal filtering problem for a class of discrete networked systems with both censored and fading measurements. The censored measurements are described by the Tobit measurement model, and the fading measurements are characterized by the Lth-order Rice fading channel model, which is capable of accounting for not only the packet dropout but also the communication phenomenon. The measurement fading occurs in a random way where the fading probabilities are regulated by a set of mutually independent Gaussian random variables. By resorting to the state augmentation technique and the orthogonality projection principle, the Tobit Kalman filter (TKF) is designed in the presence of fading measurements. In the course of filter design, several state-augmentation-induced terms are introduced, all of which can be calculated recursively or offline. A numerical example concerning the estimation of ballistic roll rates is provided to illustrate the usefulness of the proposed filter.

2.1 PROBLEM FORMULATION

Consider the linear discrete time-varying system with fading measurements described by the Lth-order Rice fading model (see [99, 155]) as follows:

$$x_{k+1} = A_k x_k + B_k \omega_k, \tag{2.1}$$

$$y_k^* = \sum_{s=0}^{\ell_k} \vartheta_k(s) C_{k-s} x_{k-s} + \upsilon_k, \tag{2.2}$$

where $x_k \in \mathbb{R}^{n_x}$ and $y_k^* \in \mathbb{R}^{n_y}$ are, respectively, the state vector and the uncensored measurement vector. $A_k \in \mathbb{R}^{n_x \times n_x}$, $B_k \in \mathbb{R}^{n_x \times n_\omega}$ and $C_k \in \mathbb{R}^{n_y \times n_x}$ are known time-varying matrices with appropriate dimensions. $\omega_k \in \mathbb{R}^{n_\omega}$ and $\upsilon_k \in \mathbb{R}^{n_y}$ are zero-mean and white Gaussian noises with covariances Q_k and R_k, respectively. $\ell_k = \min\{L,k\}$ where L is the order of the Lth-order Rice fading channel and k stands for the time instant. $\vartheta_k(s) \in \mathbb{R}$ $(s = 0,1,\cdots,\ell_k)$ are the channel coefficients.

We are now in the position to introduce the Tobit measurement model (see [75]) as follows:

$$y_k = \begin{cases} y_k^*, & y_k^* > \mathcal{I}, \\ \mathcal{I}, & y_k^* \leq \mathcal{I}, \end{cases} \tag{2.3}$$

where $y_k \in \mathbb{R}^{n_y}$ is the censored measurement vector and $\mathcal{I} = \begin{bmatrix} \mathcal{I}^1 & \mathcal{I}^2 & \cdots & \mathcal{I}^{n_y} \end{bmatrix}^T \in \mathbb{R}^{n_y}$ is the threshold vector, and $\mathcal{I}^m$ $(m = 1,2,\ldots,n_y)$ is the constant threshold of y_k^m.

Depending on whether y_k is censored or not, it is assumed that (2.3) can be transformed into the following form:

$$y_k = \Upsilon_k y_k^* + (I - \Upsilon_k)\mathcal{I} , \tag{2.4}$$

where $\Upsilon_k = \text{diag}\{\gamma_k^1, \gamma_k^2, \ldots, \gamma_k^{n_y}\}$ and γ_k^m $(m = 1, 2, \ldots, n_y)$ are known Bernoulli random variables which regulate the censoring phenomena of y_k^m $(m = 1, 2, \ldots, n_y)$ with the following probability distribution:

$$\text{Prob}\{\gamma_k^m = 1\} = \overline{\gamma}_k^m, \qquad \text{Prob}\{\gamma_k^m = 0\} = 1 - \overline{\gamma}_k^m, \tag{2.5}$$

where $\overline{\gamma}_k^m$ $(m = 1, 2, \ldots, n_y)$ are known non-negative constants. It is assumed that all the random variables γ_k^m $(m = 1, 2, \ldots, n_y)$ are mutually independent and are also uncorrelated with the fading coefficients and other noise signals.

Assumption 2.1 The initial state x_0 has the mean $\overline{x}_0$ and covariance P_0, and x_0, $\vartheta_k(s)$, ω_k and υ_k are mutually independent.

Assumption 2.2 The coefficients $\vartheta_k(s) = diag\{\vartheta_k^1(s), \vartheta_k^2(s), \ldots, \vartheta_k^{n_y}(s)\}$ where $\vartheta_k^m(s)$ $(m = 1, 2, \ldots, n_y)$ are n_y uncorrelated Gaussian random variables in m and k and are also uncorrelated with other noise signals. $\vartheta_k(s)$ have known means $\overline{\vartheta}_k(s) = diag\{\overline{\vartheta}_k^1(s), \overline{\vartheta}_k^2(s), \ldots, \overline{\vartheta}_k^{n_y}(s)\}$ and variances $\tilde{\vartheta}_k(s) = diag\{\tilde{\vartheta}_k^1(s), \tilde{\vartheta}_k^2(s), \ldots, \tilde{\vartheta}_k^{n_y}(s)\}$.

Remark 2.1 As is well known, network-induced phenomena (e.g. communication delays, packet dropouts, quantization effects, sensor saturations) are ubiquitous due to the widespread applications of the networked control systems. However, the channel fading caused by the multipath propagation of the signal has not been given full consideration despite its practical significance in networked communications. To deal with the measurement fading, models such as the analog erasure channel model, the Rayleigh channel model and the Lth-order Rice fading channel model have been established [102]. It is worth mentioning that the Lth-order Rice fading model given by (2.2) is one of the most frequently used ones in areas of signal processing and remote control. With such a model, a few phenomena of imperfect transmission including the channel fading and the time delay could be easily incorporated. As such, the Lth-order Rice fading model is adopted in this chapter for the filter design.

Remark 2.2 The KF has become widely used in tracking and estimation. The KF is optimal in the sense of minimum mean-squared error when the systems are linear and the noises are Gaussian. However, in many estimation application areas, sensors are subject to a specific type of measurement nonlinearities called censoring or censored measurements because of saturated sensors, limited detection effects and image frame effects. In this case, the direct use of the KF would probably result in biased estimates because (1) the measurement noise is correlated with the state value in the censored region and (2) the measurement noise distribution becomes censored Gaussian (instead of Gaussian), which violates the assumption of the Gaussian noise in the KF. Therefore, the KF should be modified to cater for the censored measurements. In addition, it is worth pointing out that as two important forms of measurement nonlinearities, the censored measurements and fading measurements have not been considered simultaneously in the framework of the KF, and our main motivation is to shorten such a gap.

Remark 2.3 As described in (2.5), the random variables γ_k^m ($m = 1, 2, \ldots, n_y$) are utilized to characterize the censoring phenomena of y_k^m ($m = 1, 2, \ldots, n_y$). According to (2.4), if no censoring occurs for y_k^m, i.e. $\gamma_k^m = 1$, the measurement output is $y_k^m = (y_k^*)^m$, which implies that the output measurement is the same as the latent measurement. If the censoring phenomena occur for y_k^m, i.e. $\gamma_k^m = 0$, the measurement output is $y_k^m = \mathcal{I}^m$, which implies that the threshold is assigned to the output measurement. Here, the censoring probabilities $\bar{\gamma}_k^m$ ($m = 1, 2, \ldots, n_y$) are assumed to be known a priori via some statistical experiments. Alternatively, $\bar{\gamma}_k^m$ can also be approximated by

$$\bar{\gamma}_k^m \approx \Phi\left(\frac{\hat{\lambda}_{k|k-1}^m - \mathcal{I}^m}{\sqrt{R_k^{m,m}}} \right), \tag{2.6}$$

where

$$\hat{\lambda}_{k|k-1}^m = \sum_{s=0}^{\ell_k} \sum_{a=1}^{n_y} \sum_{b=1}^{n_x} \bar{\vartheta}_k^{m,a}(s) C_{k-s}^{a,b} \hat{x}_{k-s|k-s-1}^b,$$

$$\bar{\vartheta}_k^{m,a}(s)(m = 1, 2, \ldots, n_y, a = 1, 2, \ldots, n_y),$$

$C_{k-s}^{a,b}$ ($b = 1, 2, \ldots, n_x$) and $R_k^{m,m}$ are, respectively, the (m,a)th, (a,b)th and (m,m)th elements of $\bar{\vartheta}_k(s)$, C_{k-s} and R_k; $\hat{x}_{k-s|k-s-1}^b$ is the bth element of $\hat{x}_{k-s|k-s-1}$ which is the one-step prediction of x_{k-s}; and $\Phi(\cdot)$ is the cumulative distribution function of the random variable "$\cdot$" which obeys the standard normal distribution.

Remark 2.4 Nowadays, because of the advanced microprocessor technology and the fact that microprocessors can provide greater accuracy and computing power than analog computers, digital filtering methods have been more widely utilized than the classical analog filtering methods. This means that for modern filtering applications, the discrete-time filter would be more preferable. However, a thorough study of optimal filtering must include the continuous-time filter. Therefore, parallel to the discrete-time system in this chapter, the continuous-time system is formulated as

$$\frac{dx_t}{dt} = A_t x_t + B_t \omega_t,$$

$$y_t^* = \int_0^{\ell_t} \vartheta_t(s) C_{t-s} x_{t-s} ds + \upsilon_t,$$

$$y_t = \Upsilon_t y_t^* + (I - \Upsilon_t)\mathcal{I},$$

where x_t, y_t^*, y_t, ω_t, υ_t, ℓ_t, A_t, B_t, C_t, $\vartheta_t(s)$ and Υ_t are the continuous-time counterparts of x_k, y_k^*, y_k, ω_k, υ_k, ℓ_k, A_k, B_k, $_k$, $\vartheta_k(s)$ and Υ_k. Based on the given continuous-time system, it would be interesting to design a continuous-time TKF to cope with censored and fading measurements and, meanwhile, examine the impacts from the censoring and fading phenomena on the design of the filter in the near future.

The objective of this chapter is to design the optimal TKF for the discrete time-varying system (2.1)–(2.4) with censored and fading measurements.

2.2 TOBIT KALMAN FILTER WITH FADING MEASUREMENTS

In this section, we aim to establish a unified framework to solve the addressed Tobit Kalman filtering problem in the simultaneous presence of censored and fading measurements. The filter design procedure is unique from two aspects: (1) a modified Tobit regression model resulting from the fading measurements and (2) additional computations of the gain matrix and the covariance matrix, where the additional computations include the augmentation of the delayed states, calculations with respect to the augmented state and calculation of the expected values of the channel coefficients.

Rewrite system (2.1)–(2.4) in the following compact form:

$$\xi_{k+1} = \mathcal{A}_k \xi_k + \mathcal{B}_k \omega_k, \tag{2.7}$$

$$y_k^* = \theta_k C_k \xi_k + \upsilon_k, \tag{2.8}$$

$$y_k = \begin{cases} \theta_k C_k \xi_k + \upsilon_k, & \theta_k C_k \xi_k + \upsilon_k > \mathcal{I}, \\ \mathcal{I}, & \theta_k C_k \xi_k + \upsilon_k \leq \mathcal{I}, \end{cases} \tag{2.9}$$

$$y_k = \Upsilon_k(\theta_k C_k \xi_k + \upsilon_k) + (I - \Upsilon_k)\mathcal{I}, \tag{2.10}$$

where

$$\xi_k = \mathrm{col}\{x_k, x_{k-1}, \cdots, x_{k-\ell_k}\},$$

$$\theta_k = \begin{bmatrix} \vartheta_k(0) & \vartheta_k(1) & \cdots & \vartheta_k(\ell_k) \end{bmatrix},$$

$$C_k = \mathrm{diag}\{C_k, C_{k-1}, \cdots, C_{k-\ell_k}\},$$

$$\mathcal{A}_k = \begin{bmatrix} A_k & 0 & \cdots & 0 \\ I & 0 & \cdots & 0 \\ \vdots & \ddots & \ddots & \vdots \\ 0 & \cdots & I & 0 \end{bmatrix}, \mathcal{B}_k = \begin{bmatrix} B_k \\ 0 \\ \vdots \\ 0 \end{bmatrix},$$

and coefficients θ_k have variances $P_{\tilde{\theta}_k} = diag\{\tilde{\vartheta}_k(0), \tilde{\vartheta}_k(1), \ldots, \tilde{\vartheta}_k(\ell_k)\}$ and means $\bar{\theta}_k$. Through the above augmentation, the system with time delays in (2.1)–(2.4) is transformed into the one in (2.7)–(2.10) without time delays.

Remark 2.5 In fact, many efficient approaches have been presented in the literature concerning the time-delay phenomenon in the past two decades, such as the augmentation technique, the bounding technique, the descriptor system method, the slack matrix variables technique and the delay-fractioning approach. Generally speaking, the conservatism and complexity of the algorithm should be considered when handling the time-delay phenomenon. That is to say, there should be a trade-off between the conservatism and the complexity of the proposed algorithm. Here, the conservatism means that the proposed algorithm should provide a maximal allowable delay, and the complexity means that the algorithm includes as few decision variables as possible while achieving the same maximal allowable delay. In this chapter, to remove the time-delay effect, the augmentation technique as shown in

(2.7)–(2.10) is chosen to benefit the following filter design at the cost of some extra computational cost.

Denote y_k^m and $\mathcal{I}^m$ ($m = 1, 2, \ldots, n_y$), respectively, as the mth elements of y_k and $\mathcal{I}$, $\theta_k^{m,\alpha}$ as the (m,α)th ($\alpha = 1, 2, \ldots, n_y(\ell_k + 1)$) element of θ_k, $C_k^{\alpha,\beta}$ as the (α,β)th ($\beta = 1, 2, \ldots, n_x(\ell_k + 1)$) element of C_k, and $R_k^{m,m}$ as the (m,m)th element of $R_k^{m,m}$. Let $\chi_k^m = \sum_{\alpha=1}^{n_y(\ell_k+1)} \sum_{\beta=1}^{n_x(\ell_k+1)} \theta_k^{m,\alpha} C_k^{\alpha,\beta} \xi_k^\beta$. To facilitate the subsequent filter design, the Tobit regression is first modified to encompass the fading measurements as shown in the following lemma.

Lemma 2.1 The mathematical expectation and variance of y_k^m ($m = 1, 2, \ldots, n_y$) are derived as follows:

$$\mathbb{E}\{y_k^m \mid x_k, R_k^{m,m}\} = \Phi\left(\frac{\chi_k^m - \mathcal{I}^m}{\sqrt{R_k^{m,m}}}\right)\left[\chi_k^m + \sqrt{R_k^{m,m}}\,\lambda\left(\frac{\mathcal{I}^m - \chi_k^m}{\sqrt{R_k^{m,m}}}\right)\right] + \Phi\left(\frac{\mathcal{I}^m - \chi_k^m}{\sqrt{R_k^{m,m}}}\right)\mathcal{I}^m, \tag{2.11}$$

$$Var\{y_k^m \mid x_k, R_k^{m,m}\} = R_k^{m,m}\left[1 - \varphi\left(\frac{\mathcal{I}^m - \chi_k^m}{\sqrt{R_k^{m,m}}}\right)\right], \tag{2.12}$$

where

$$\varphi\left(\frac{\mathcal{I}^m - \chi_k^m}{\sqrt{R_k^{m,m}}}\right) = \lambda\left(\frac{\mathcal{I}^m - \chi_k^m}{\sqrt{R_k^{m,m}}}\right)\left[\lambda\left(\frac{\mathcal{I}^m - \chi_k^m}{\sqrt{R_k^{m,m}}}\right) - \frac{\mathcal{I}^m - \chi_k^m}{\sqrt{R_k^{m,m}}}\right], \tag{2.13}$$

$$\lambda\left(\frac{\mathcal{I}^m - \chi_k^m}{\sqrt{R_k^{m,m}}}\right) = \frac{\phi\left(\dfrac{\mathcal{I}^m - \chi_k^m}{\sqrt{R_k^{m,m}}}\right)}{1 - \Phi\left(\dfrac{\mathcal{I}^m - \chi_k^m}{\sqrt{R_k^{m,m}}}\right)}, \tag{2.14}$$

with the probability density function (PDF) and the cumulative distribution function (CDF) of the Gaussian random variable y_k^m given by the following equations:

$$\phi\left(\frac{y_k^m - \chi_k^m}{\sqrt{R_k^{m,m}}}\right) = \frac{1}{\sqrt{2\pi}}e^{-\frac{(y_k^m - \chi_k^m)^2}{2R_k^{m,m}}}, \tag{2.15}$$

$$\Phi\left(\frac{y_k^m - \chi_k^m}{\sqrt{R_k^{m,m}}}\right) = \int_{-\infty}^{y_k^m} \frac{1}{\sqrt{2\pi R_k^{m,m}}}e^{-\frac{(z_k^m - \chi_k^m)^2}{2R_k^{m,m}}}\,d_{z_k^m}. \tag{2.16}$$

Proof According to (2.4), the probability distribution of y_k^m with a normally distributed noise v_k^m is as follows:

$$f(y_k^m \mid x_k) = \frac{1}{\sqrt{R_k^{m,m}}} \phi\left(\frac{y_k^m - \chi_k^m}{\sqrt{R_k^{m,m}}}\right) u(y_k^m - \mathcal{I}^m) + \delta(\mathcal{I}^m - y_k^m)\Phi\left(\frac{\mathcal{I}^m - \chi_k^m}{\sqrt{R_k^{m,m}}}\right), \quad (2.17)$$

where $\phi\left(\dfrac{y_k^m - \chi_k^m}{\sqrt{R_k^{m,m}}}\right)$ and $\Phi\left(\dfrac{\mathcal{I}^m - \chi_k^m}{\sqrt{R_k^{m,m}}}\right)$ are given by (2.15)–(2.16), $u(y_k^m - \mathcal{I}^m)$ is the

unit step function and $\delta\left(\mathcal{I}^m - y_k^m\right)$ is the Dirac delta function. Based on (2.17), we

have the mathematical expectation and variance of y_k^m as follows:

$$\begin{aligned}
&\mathbb{E}\{y_k^m \mid x_k, R_k^{m,m}\} \\
&= Prob\{y_k^m > \mathcal{I}^m \mid x_k, R_k^{m,m}\}\mathbb{E}\{y_k^m \mid y_k^m > \mathcal{I}^m, x_k, R_k^{m,m}\} \\
&\quad + Prob\{y_k^m = \mathcal{I}^m \mid x_k, R_k^{m,m}\}\mathbb{E}\{y_k^m \mid y_k^m = \mathcal{I}^m, x_k, R_k^{m,m}\}
\end{aligned} \quad (2.18)$$

From (2.18), we know that to compute $\mathbb{E}\{y_k^m \mid x_k, R_k^{m,m}\}$, we need to calculate the probabilities and expected values in (2.18).

$$\begin{aligned}
Prob\{y_k^m > \mathcal{I}^m \mid x_k, R_k^{m,m}\} &= Prob\{y_k^m > \mathcal{I}^m \mid x_k, R_k^{m,m}\} \\
&= Prob\{(y_k^*)^m > \mathcal{I}^m \mid x_k, R_k^{m,m}\} \\
&= Prob\{\chi_k^m + v_k^m > \mathcal{I}^m \mid x_k, R_k^{m,m}\} \\
&= Prob\{v_k^m > \mathcal{I}^m - \chi_k^m \mid x_k, R_k^{m,m}\} \\
&= \Phi\left(\frac{\chi_k^m - \mathcal{I}^m}{\sqrt{R_k^{m,m}}}\right).
\end{aligned} \quad (2.19)$$

Hence, we can easily obtain that

$$\mathbb{E}\{y_k^m \mid y_k^m > \mathcal{I}^m, x_k, R_k^{m,m}\} = \frac{1}{\sqrt{R_k^{m,m}}} \int_{\mathcal{I}^m}^{+\infty} z_k^m \frac{\Phi\left(\dfrac{z_k^m - \chi_k^m}{\sqrt{R_k^{m,m}}}\right)}{1 - \Phi\left(\dfrac{\mathcal{I}^m - \chi_k^m}{\sqrt{R_k^{m,m}}}\right)} d z_k^m \quad (2.20)$$

$$= \chi_k^m + \sqrt{R_k^{m,m}}\, \lambda\left(\frac{\mathcal{I}^m - \chi_k^m}{\sqrt{R_k^{m,m}}}\right),$$

where $\lambda\left(\dfrac{\mathcal{I}^m - \chi_k^m}{\sqrt{R_k^{m,m}}}\right)$ is given by (2.14). Similar to (2.19)–(2.20), we can obtain

$$Prob\{y_k^m = \mathcal{I}^m \mid x_k, R_k^{m,m}\} = \Phi\left(\frac{\mathcal{I}^m - \chi_k^m}{\sqrt{R_k^{m,m}}}\right), \qquad (2.21)$$

$$\mathbb{E}\{y_k^m \mid y_k^m = \mathcal{I}^m, x_k, R_k^{m,m}\} = \mathcal{I}^m. \qquad (2.22)$$

Putting (2.19)–(2.22) into (2.18) results in (2.11). It follows easily from (2.17) that $Var\{y_k^m \mid x_k, y_k^m < \mathcal{I}^m, R_k^{m,m}\} = 0$, and hence we have

$$\begin{aligned}
Var\{y_k^m \mid x_k, R_k^{m,m}\} &= Var\{y_k^m \mid x_k, y_k^m > \mathcal{I}^m, R_k^{m,m}\} \\
&= \mathbb{E}\{(y_k^m)^2 \mid y_k^m > \mathcal{I}^m, x_k, R_k^{m,m}\} \\
&\quad - \left(\mathbb{E}\{y_k^m \mid y_k^m > \mathcal{I}^m, x_k, R_k^{m,m}\}\right)^2 \\
&= R_k^{m,m}\left[1 - \varphi\left(\frac{\mathcal{I}^m - \chi_k^m}{\sqrt{R_k^{m,m}}}\right)\right],
\end{aligned}$$

which is exactly the same as (2.12), where $\varphi\left(\dfrac{\mathcal{I}^m - \chi_k^m}{\sqrt{R_k^{m,m}}}\right)$ is given by (2.13). The proof is now complete.

Remark 2.6 Equations (2.11)–(2.12) in Lemma 2.1 give the expressions of the mathematic expectation and variance of the censored and fading measurement y_k^m. Compared with the standard TKF, where only the censoring phenomenon is considered, it can be observed that, due to the introduction of the fading measurements, the term $\sum_{n=1}^{n_x} C_k^{m,n} x_k^n$ (in the case that y_k is a vector instead of a scalar) is replaced by the term χ_k^m, which characterizes the effect of the fading measurements. If the fading phenomenon is not taken into account, (2.2) becomes $y_k^* = C_k x_k + \upsilon_k$. Then, the term χ_k^m reduces to $\sum_{n=1}^{n_x} C_k^{m,n} x_k^n$, resulting in the degradation of (2.11) and (2.12) to (8) and (11) in [75], respectively.

Remark 2.7 Using (2.11)–(2.16), the Tobit model in [75] with only censored measurements has been extended to the one subjecting to both censored and fading measurements as shown in (2.4). Note that the extra fading measurements result in (1) the replacement of $\sum_{n=0}^{n_x} C_k^{m,n} x_k^n$ by χ_k^m in all the equations relative to the measurement y_k^m and (2) several new terms that further complicate the filter design.

Let $\overline{\Upsilon}_k = \mathbb{E}\{\Upsilon_k\} = \text{diag}\{\overline{\gamma}_k^1, \overline{\gamma}_k^2, \dots, \overline{\gamma}_k^{n_y}\}$ and $\hat{\xi}_{k|k-1} = \mathbb{E}\{\xi_k \mid y_{1:k-1}\}$. Denote

$$\hat{x}_{k|k-1} = \mathbb{E}\{x_k \mid y_{1:k-1}\}, \quad \hat{y}_{k|k-1} = \mathbb{E}\{y_k \mid y_{1:k-1}\}, \quad \hat{\xi}_{k|k} = \mathbb{E}\{\xi_k \mid y_{1:k}\}, \quad \hat{x}_{k|k} = \mathbb{E}\{x_k \mid y_{1:k}\},$$

$$\tilde{\xi}_{k|k-1} = \xi_k - \hat{\xi}_{k|k-1}, \quad \tilde{x}_{k|k-1} = x_k - \hat{x}_{k|k-1}, \quad \tilde{y}_{k|k-1} = y_k - \hat{y}_{k|k-1}, \quad \tilde{\xi}_{k|k} = \xi_k - \hat{\xi}_{k|k}, \quad \tilde{x}_{k|k} = x_k - \hat{x}_{k|k},$$

$$\mathcal{P}_{k|k-1} = \mathbb{E}\{\tilde{\xi}_{k|k-1}\tilde{\xi}_{k|k-1}^T\}, \quad P_{k|k-1} = \mathbb{E}\{\tilde{x}_{k|k-1}\tilde{x}_{k|k-1}^T\}, \quad \mathcal{P}_{k|k} = \mathbb{E}\{\tilde{\xi}_{k|k}\tilde{\xi}_{k|k}^T\}, \quad P_{k|k} = \mathbb{E}\{\tilde{x}_{k|k}\tilde{x}_{k|k}^T\},$$

$$\Sigma_{\tilde{y}_{k|k-1}\tilde{y}_{k|k-1}} = \mathbb{E}\{\tilde{y}_{k|k-1}\tilde{y}_{k|k-1}^T\} \text{ and } \Sigma_{\tilde{\xi}_{k|k-1}\tilde{y}_{k|k-1}} = \mathbb{E}\{\tilde{\xi}_{k|k-1}\tilde{y}_{k|k-1}^T\}. \text{ Let } \hat{\chi}_{k|k-1} = \overline{\theta}_k C_k \hat{\xi}_{k|k-1} \text{ with its}$$

mth element being $\hat{\chi}_{k|k-1}^m = \sum_{\alpha=1}^{n_y(\ell_k+1)} \sum_{\beta=1}^{n_x(\ell_k+1)} \overline{\theta}_k^{m,\alpha} C_k^{\alpha,\beta} \hat{\xi}_{k|k-1}^\beta.$

Based on Lemma 2.1, the optimal TKF with fading measurements can be obtained as follows.

Theorem 2.1 The optimal TKF for the augmented system (2.7)–(2.10) is given by

$$\hat{\xi}_{k|k-1} = \mathcal{A}_{k-1}\hat{\xi}_{k-1|k-1}, \tag{2.23}$$

$$\mathcal{P}_{k|k-1} = \mathcal{A}_{k-1}\mathcal{P}_{k-1|k-1}\mathcal{A}_{k-1}^{T} + \mathcal{B}_{k-1}\mathcal{Q}_{k-1}\mathcal{B}_{k-1}^{T}, \tag{2.24}$$

$$\hat{\xi}_{k|k} = \hat{\xi}_{k|k-1} + K_k(y_k - \hat{y}_{k|k-1}), \tag{2.25}$$

$$\mathcal{P}_{k|k} = \mathcal{P}_{k|k-1} - \bar{\Upsilon}_k \Sigma_{\tilde{\xi}_{k|k-1}\tilde{y}_{k|k-1}} \Sigma^{-1}_{\tilde{y}_{k|k-1}\tilde{y}_{k|k-1}} \Sigma^{T}_{\tilde{\xi}_{k|k-1}\tilde{y}_{k|k-1}}, \tag{2.26}$$

where $\hat{\xi}_{k|k-1}$ is the one-step prediction of ξ_{k-1}, $\hat{\xi}_{k|k}$ is the estimate of ξ_k, $\mathcal{P}_{k|k-1}$ is the one-step prediction error covariance matrix and $\mathcal{P}_{k|k}$ is the estimation error covariance matrix. The one-step measurement prediction in (2.25) is given by

$$\hat{y}_{k|k-1} = \bar{\Upsilon}_k\left[\hat{\chi}_{k|k-1} + \lambda\left(\frac{\mathcal{I} - \hat{\chi}_{k|k-1}}{\sqrt{R_k}}\right)R_k\right] + (I - \bar{\Upsilon}_k)\mathcal{I}, \tag{2.27}$$

where

$$\lambda\left(\frac{\mathcal{I} - \hat{\chi}_{k|k-1}}{\sqrt{R_k}}\right) = diag\left\{\lambda\left(\frac{\mathcal{I}^{m} - \hat{\chi}^{m}_{k|k-1}}{\sqrt{R_k^{m,m}}}\right)\right\}, \tag{2.28}$$

$\mathcal{R}_k = col\left\{\sqrt{R_k^{1,1}}, \sqrt{R_k^{1,1}}, \ldots, \sqrt{R_k^{n_y,n_y}}\right\}$, and $\sqrt{R_k^{m,m}}$ is the (m,m)th element of R_k. The gain matrix K_k in (2.25) has the following form:

$$K_k = \Sigma_{\tilde{\xi}_{k|k-1}\tilde{y}_{k|k-1}} \Sigma^{-1}_{\tilde{y}_{k|k-1}\tilde{y}_{k|k-1}}, \tag{2.29}$$

where

$$\Sigma_{\tilde{\xi}_{k|k-1}\tilde{y}_{k|k-1}} = \mathcal{P}_{k|k-1}(\bar{\Upsilon}_k\bar{\theta}_k\mathcal{C}_k)^{T}, \tag{2.30}$$

$$\Sigma_{\tilde{y}_{k|k-1}\tilde{y}_{k|k-1}} = \bar{\Upsilon}_k\bar{\theta}_k\mathcal{C}_k\mathcal{P}_{k|k-1}(\bar{\Upsilon}_k\bar{\theta}_k\mathcal{C}_k)^{T}$$

$$+ (\bar{\Upsilon}_k\mathcal{P}_{\bar{\theta}_k}\bar{\Upsilon}_k^{T})\circ\mathcal{C}_k\mathcal{P}_{\xi}\mathcal{C}_k^{T} + R_k\left[I - \varphi\left(\frac{\mathcal{I} - \hat{\chi}_{k|k-1}}{\sqrt{R_k}}\right)\right], \tag{2.31}$$

$$\varphi\left(\frac{\mathcal{I} - \hat{\chi}_{k|k-1}}{\sqrt{R_k}}\right) = diag\left\{\varphi\left(\frac{\mathcal{I}^{m} - \hat{\chi}^{m}_{k|k-1}}{\sqrt{R_k^{m,m}}}\right)\right\}. \tag{2.32}$$

Here $\mathcal{P}_{\xi_k} = \mathcal{A}_{k-1}\mathcal{P}_{\xi_{k-1}}\mathcal{A}_{k-1}^{T} + \mathcal{B}_{k-1}\mathcal{Q}_{k-1}\mathcal{B}_{k-1}^{T}$ and $\circ$ is the Hadamard product.

Proof From the definition of $\hat{\xi}_{k|k-1}$, $\hat{\xi}_{k|k}$, $\mathcal{P}_{k|k-1}$ and $\mathcal{P}_{k|k}$, we can directly have (2.23)–(2.26) by applying the orthogonality projection principle to system (2.7)–(2.10) with the corresponding gain matrix given by (2.29). According to (2.11), we have

$$\hat{y}_{k|k-1}^{m} = \overline{\gamma}_{k}^{m}\left[\hat{\chi}_{k|k-1}^{m} + \sqrt{R_{k}^{m,m}}\lambda\left(\frac{\mathcal{I}^{m} - \hat{\chi}_{k|k-1}^{m}}{\sqrt{R_{k}^{m,m}}}\right)\right] + (1 - \overline{\gamma}_{k}^{m})\mathcal{I}^{m}. \tag{2.33}$$

Noting that $y_{k} = diag\{y_{k}^{1}, y_{k}^{2}, \ldots, y_{k}^{n_{y}}\}$, we certainly have (2.27) from (2.33), where $\lambda\left(\dfrac{\mathcal{I} - \hat{\chi}_{k|k-1}}{\sqrt{R_{k}}}\right)$ is given by (2.28). Subtracting (2.27) from (2.10) leads to

$$\tilde{y}_{k|k-1} = \Upsilon_{k}\tilde{\theta}_{k}C_{k}\xi_{k} + \Upsilon_{k}\overline{\theta}_{k}C_{k}\tilde{\xi}_{k|k-1} + \Upsilon_{k}\left(\upsilon_{k} - \lambda\left(\frac{\mathcal{I} - \hat{\chi}_{k|k-1}}{\sqrt{R_{k}}}\right)R_{k}\right). \tag{2.34}$$

Substituting (2.34) into the definitions of $\Sigma_{\tilde{\xi}_{k|k-1}\tilde{y}_{k|k-1}}$ and $\Sigma_{\tilde{y}_{k|k-1}\tilde{y}_{k|k-1}}$, we have

$$\begin{aligned}
\Sigma_{\tilde{\xi}_{k|k-1}\tilde{y}_{k|k-1}} &= \mathbb{E}\left\{\tilde{\xi}_{k|k-1}\tilde{y}_{k|k-1}^{T}\right\} \\
&= \mathbb{E}\left\{\tilde{\xi}_{k|k-1}(\Upsilon_{k}\tilde{\theta}_{k}C_{k}\xi_{k} + \Upsilon_{k}\overline{\theta}_{k}C_{k}\tilde{\xi}_{k|k-1}\right. \\
&\quad \left. + \Upsilon_{k}\left(\upsilon_{k} - \lambda\left(\frac{\mathcal{I} - \hat{\chi}_{k|k-1}}{\sqrt{R_{k}}}\right)R_{k}\right)\right)^{T}\right\} \\
&= \mathcal{P}_{k|k-1}\left(\overline{\Upsilon}_{k}\overline{\theta}_{k}C_{k}\right)^{T},
\end{aligned}$$

which is exactly the same as (2.30), and

$$\begin{aligned}
&\Sigma_{\tilde{y}_{k|k-1}\tilde{y}_{k|k-1}} \\
&= \mathbb{E}\left\{\tilde{y}_{k|k-1}\tilde{y}_{k|k-1}^{T}\right\} \\
&= \overline{\Upsilon}_{k}\overline{\theta}_{k}C_{k}\mathcal{P}_{k|k-1}\left(\overline{\Upsilon}_{k}\overline{\theta}_{k}C_{k}\right)^{T} + (\overline{\Upsilon}_{k}\mathcal{P}_{\tilde{\theta}_{k}}\overline{\Upsilon}_{k}^{T})\circ C_{k}\mathcal{P}_{\xi_{k}}C_{k}^{T} \\
&\quad + \mathbb{E}\left\{\Upsilon_{k}\left(\upsilon_{k} - \lambda\left(\frac{\mathcal{I} - \hat{\chi}_{k|k-1}}{\sqrt{R_{k}}}\right)R_{k}\right)\left(\upsilon_{k} - \lambda\left(\frac{\mathcal{I} - \hat{\chi}_{k|k-1}}{\sqrt{R_{k}}}\right)R_{k}\right)^{T}\Upsilon_{k}^{T}\right\}.
\end{aligned} \tag{2.35}$$

For simplicity of the deduction, we assume that $cov\{y_{k}^{m}, y_{k}^{t}\} = 0$ for $m \neq t$, $m, t = 1, 2, \ldots, n_{y}$. The generalization of the result to the case where $cov\{y_{k}^{m}, y_{k}^{t}\} \neq 0$ for $m \neq t$ is straightforward but will bring notational burdens. Recalling the definition of the variance of y_{k} and noting (2.12), we have

$$
\mathbb{E}\left\{\Upsilon_k\left(v_k - \lambda\left(\frac{\mathcal{I} - \hat{\chi}_{k|k-1}}{\sqrt{R_k}}\right)\mathcal{R}_k\right)\left(v_k - \lambda\left(\frac{\mathcal{I} - \hat{\chi}_{k|k-1}}{\sqrt{R_k}}\right)\mathcal{R}_k\right)^T \Upsilon_k^T\right\}
$$

$$
= Var\{y_k \mid x_k, R_k\}
$$

$$
= diag\{Var\{y_k^1 \mid x_k, R_k^{1,1}\}, \ldots, Var\{y_k^{n_y} \mid x_k, R_k^{n_y,n_y}\}\}
$$

$$
= R_k\left[I - \varphi\left(\frac{\mathcal{I} - \hat{\chi}_{k|k-1}}{\sqrt{R_k}}\right)\right],
$$

(2.36)

where the second equality holds from the fact that $cov\{y_k^m, y_k^t\} = 0$ for $m \neq t$ and $\varphi\left(\frac{\mathcal{I} - \hat{\chi}_{k|k-1}}{\sqrt{R_k}}\right)$ is given by (2.32). Substituting (2.36) into (2.35), we have (2.31). The proof is now complete.

Remark 2.8 Comparing the standard TKF with the established filter in Theorem 2.4, there are two main differences. The first difference is that the term $C_k\hat{x}_{k|k-1}$ in [75] (which is the product of the measurement matrix C_k and the one-step state prediction $\hat{x}_{k|k-1}$) is replaced by a new term $\hat{\chi}_{k|k-1} = \bar{\theta}_k C_k \hat{\xi}_{k|k-1}$, which is now the product of the mean of the channel coefficient, the augmented measurement matrix and the predicted value of the augmented state. The second difference is the appearance of $\bar{\theta}_k$ in the calculations of $\hat{y}_{k|k-1}$ and K_k. The first difference is caused by the time-delay effect, while the second difference stems from the fading-channel phenomenon.

Theorem 2.2 Based on the filter given in Theorem 2.1 for the augmented system (2.7)–(2.10), the optimal TKF for the original system (2.1)–(2.4) with both censored and fading measurements is as follows:

$$
\hat{x}_{k|k-1} = \Gamma\hat{\xi}_{k|k-1}, \tag{2.37}
$$

$$
P_{k|k-1} = \Gamma\mathcal{P}_{k|k-1}\Gamma^T, \tag{2.38}
$$

$$
\hat{x}_{k|k} = \Gamma\hat{\xi}_{k|k}, \tag{2.39}
$$

$$
P_{k|k} = \Gamma\mathcal{P}_{k|k}\Gamma^T, \tag{2.40}
$$

where

$$
\Gamma = \begin{bmatrix} I & \underbrace{0 \quad \cdots \quad 0}_{\ell_k \; blocks} \end{bmatrix}.
$$

Proof Noticing the relationship between system (2.1)–(2.4) and system (2.7)–(2.10), we can easily obtain the result in Theorem 2.2 on the basis of Theorem 2.1.

Combining Lemma 2.1 and Theorems 2.1–2.2, we have accomplished the design of the optimal TKF with both censored measurements and fading measurements. Table 2.1 demonstrates the implementation procedure of the proposed filtering algorithm.

TABLE 2.1

Implementation Procedure of the Proposed Filtering Algorithm

Input: $\bar{x}_0$ and P_0

Output: $\hat{x}_{k|k}$ and $P_{k|k}$

for $k = 1, 2, \ldots$, do

    Compute the predicted parameter $\hat{\xi}_{k|k-1}$ and its error covariance

    matrix $\mathcal{P}_{k|k-1}$ by (2.23)–(2.24);

    Compute the predicted state estimate $\hat{x}_{k|k-1}$ and its error

    covariance matrix $P_{k|k-1}$ by (2.37)–(2.38);

    Compute the predicted measurement $\hat{y}_{k|k-1}$ by (2.27);

    Compute the parameter $\Sigma_{\tilde{\xi}_{k|k-1}\tilde{y}_{k|k-1}}$ by (2.30);

    Compute the parameter $\Sigma_{\tilde{y}_{k|k-1}\tilde{y}_{k|k-1}}$ by (2.31);

    Compute the gain matrix K_k by (2.29);

    Compute the updated parameter $\hat{\xi}_{k|k}$ and its error covariance

    $\mathcal{P}_{k|k-1}$ by (2.25)–(2.26);

    Compute the updated state estimate $\hat{x}_{k|k}$ and its error covariance

    $P_{k|k}$ by (2.39)–(2.40);

    Return $\hat{x}_{k|k}$ and $P_{k|k}$.

end

Remark 2.9 According to Lemma 2.1 and Theorems 2.1–2.2, a recursive filtering scheme is established to solve the novel filtering problem for the discrete time-varying systems suffering simultaneously from censored measurements and fading measurements. System (2.1)–(2.4) under consideration is comprehensive that covers two important measurement nonlinearities, i.e. censored and fading measurements, which are often encountered in engineering applications including target tracking and networked control. The two measurement nonlinearities are dealt with in a unified yet effective framework. Compared with the work in [75], several new terms have emerged in the filter design, and these terms explicitly reflect the effect of the fading-channel phenomenon. To be specific, the term $\theta_k^{m,\alpha}$ in Lemma 2.1 and the term $\bar{\theta}_k$ in Theorem 2.1 characterize the effect of the channel fading. The term ξ_k^β in Lemma 2.1, the terms $\hat{\xi}_{k|k-1}$, $\hat{\xi}_{k|k}$, $\mathcal{P}_{k|k-1}$, $\mathcal{P}_{k|k}$ in Theorem 2.1 and the term Γ in Theorem 2.2 characterize the time-delay effect. The term $\hat{\chi}_{k|k-1}^m$ in Theorem 2.1 characterize the effects of both the time-delay and the channel fading.

Remark 2.10 It can be seen from Lemma 2.1 and Theorems 2.1–2.2 that when implementing the presented Tobit Kalman filtering algorithm with fading measurements, the computation complexity mainly involves the matrix multiplication and the matrix inversion. In the matrix multiplication, the multiplication of matrices with dimensions $n_x(\ell_k+1) \times n_x(\ell_k+1)$ needs to be calculated, where n_x is the dimension of the system state; $\ell_k = min\{L,k\}$ where L is the order of the Lth-order Rice fading channel; and k stands for the time instant. As to the matrix inversion, the inverse of matrices with dimensions $n_y \times n_y$ needs to be calculated, where n_y is the dimension of the measurement. Hence, the proposed filtering algorithm has the computational complexity of $O[(n_x(\ell_k+1))^3 + (n_y)^3]$. Generally, we have $\ell_k \geq 1$, and therefore our proposed algorithm has more computation burden than the Tobit Kalman filtering algorithm in [75], which has the computational complexity of $O[(n_x)^3 + (n_y)^3]$. It can be apparently seen that the extra computation burden in our algorithm is caused by the fading measurements.

2.3 ILLUSTRATIVE EXAMPLES

In this section, we will employ an oscillator example and a radar tracking example to demonstrate the effectiveness of the proposed filtering algorithm. Let RMS1 denote the root mean-squared error (RMSE) for the estimation of x_k^1, i.e.

$$\text{RMSE1} = \sqrt{(1/M)\sum_{j=1}^{M}\left(x_k^{1(j)} - \hat{x}_{k|k}^{1(j)}\right)^2}, \text{ where } M \text{ is the number of simulation tests.}$$

Similarly, RMSE2 is the RMSE for the estimation of x_k^2, i.e.

$$\text{RMSE2} = \sqrt{(1/M)\sum_{j=1}^{M}\left(x_k^{2(j)} - \hat{x}_{k|k}^{2(j)}\right)^2}.$$

2.3.1 OSCILLATOR EXAMPLE

Consider the following class of discrete time-varying systems given by (2.1)–(2.4) with parameters:

$$\tilde{\vartheta}(1) = 0.0064, \tilde{\vartheta}(2) = 0.0049, \mathcal{I} = 0,$$

$$Q_k = \text{diag}\{0.0025, 0.0025\}, R_k = 1, L = 2,$$

$$C_k = \begin{bmatrix} 1 & 0 \end{bmatrix}, \bar{x}_0 = \begin{bmatrix} 5 & 0 \end{bmatrix}^T, P_0 = I_2,$$

$$\bar{\vartheta}(0) = 0.95, \bar{\vartheta}(1) = 0.75, \bar{\vartheta}(2) = 0.85, \tilde{\vartheta}(0) = 0.0081,$$

$$A_k = \begin{bmatrix} \cos(w) & -\sin(w) \\ \sin(w) & \cos(w) \end{bmatrix}, B_k = I_2, w = 0.052\pi$$

where L = 2 is the order of the Rice fading-channel model. The censoring probability $\bar{\Upsilon}_k = \bar{\gamma}_k^1$ can be calculated according to (2.6). The presented oscillator example considers the estimation of ballistic roll rates from fading and censored magnetometer data. The simulation shows a robust tracking ability with a known model and unknown disturbance that enters the system through ω_k.

Figure 2.1 illustrates the true values of the states and the estimates given by the TKF in [75] and our proposed TKF with fading measurements, which is named TKF-FM. Figure 2.2 shows the comparison of the RMSE curves between the TKF and the TKF-FM after 1000 times of Monte Carlo simulations. From Figure 2.1, we can observe that when the measurements are subject to both censoring and fading, our proposed TKF-FM is able to track the true values of the

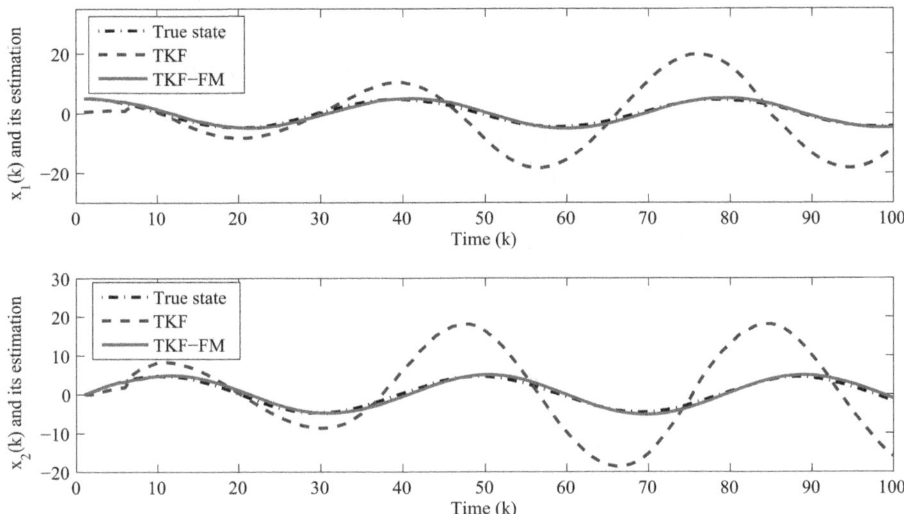

FIGURE 2.1 True values of the first and second dimensions of the state and their estimates.

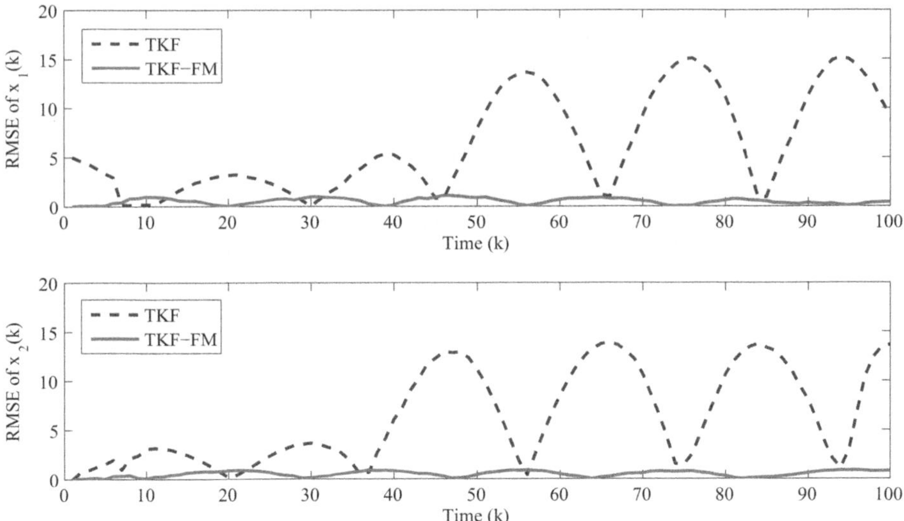

FIGURE 2.2 Method comparison with [75] in RMSE1 and RMSE2.

system states, while the TKF leads to significant deviations from the true state values. It can be seen from Figure 2.2 that the RMSE of our TKF-FM is always smaller than that of the TKF, which is simply due to the fact that the fading measurements are properly handled in our TKF-FM, while they are not addressed in the TKF.

2.3.2 Radar Tracking Example

Consider the radar tracking example with the following parameters:

$$A_k = \begin{bmatrix} 1 & T & 0 & 0 & 0 & 0 \\ 0 & 1 & 1 & 0 & 0 & 0 \\ 0 & 0 & \varrho_1 & 0 & 0 & 0 \\ 0 & 0 & 0 & 1 & T & 0 \\ 0 & 0 & 0 & 0 & 1 & 1 \\ 0 & 0 & 0 & 0 & 0 & \varrho_2 \end{bmatrix}, x_0 = \begin{bmatrix} 100 \\ 4 \\ 0 \\ 1 \\ 0.001 \\ 0 \end{bmatrix}^T,$$

$$C_k = \begin{bmatrix} 1 & 0 & 0 & 0 & 0 & 0 \\ 0 & 0 & 0 & 0 & 0 & 1 \end{bmatrix}, R_k = \mathrm{diag}\{100, 0.001\},$$

$$B_k = I_6, Q_k = 0.25, P_0 = \mathrm{diag}\{100, 10, 10, 1, 1, 1\}, \mathcal{I} = \begin{bmatrix} 100 & 1.01 \end{bmatrix}^T,$$

where $x_k = \begin{bmatrix} r_k & \dot{r}_k & u_{1,k} & \eta_k & \dot{\eta}_k & u_{2,k} \end{bmatrix}^T$ is the target state; r_k and $\dot{r}_k$ are, respectively, the range and range rate of the target at time k; η_k and $\dot{\eta}_k$ are, respectively, the bearing and bearing rate of the target at time k; $\varrho_1 = 0.4$ and $\varrho_2 = 0.5$ are the maneuvering correlated process noises; T = 1s is the sampling period. The remaining parameters are the same as those in the previous example.

The performance comparison between the Tobit Kalman filter (TKF) and the proposed Tobit Kalman filter with fading measurements (TKF-FM) has been made in RMSE after 1000 times of Monte Carlo simulations. Figure 2.3 illustrates the RMSE curves obtained by the two algorithms. From Figure 2.3, we find that the RMSE of our TKF-FM is always smaller than that of TKF, which indicates the superiority of the proposed filtering algorithm in handling censored and fading measurements. This is because our TKF-FM is capable of tackling the censoring and channel-fading phenomena at the same time, while the TKF cannot deal with fading measurements.

Theoretically, as time goes on, the RMSE of the TKF in Figure 2.3 should degrade to a certain value beyond zero because the TKF is unable to cope with the fading measurements. As to our TKF-FM, the RMSE should degrade to zero with the increase of k, since the proposed TKF-FM can properly deal with the fading phenomenon. Nevertheless, this is often not true due to the fact that part of the measurement information is lost because of the fading phenomenon, which makes it difficult

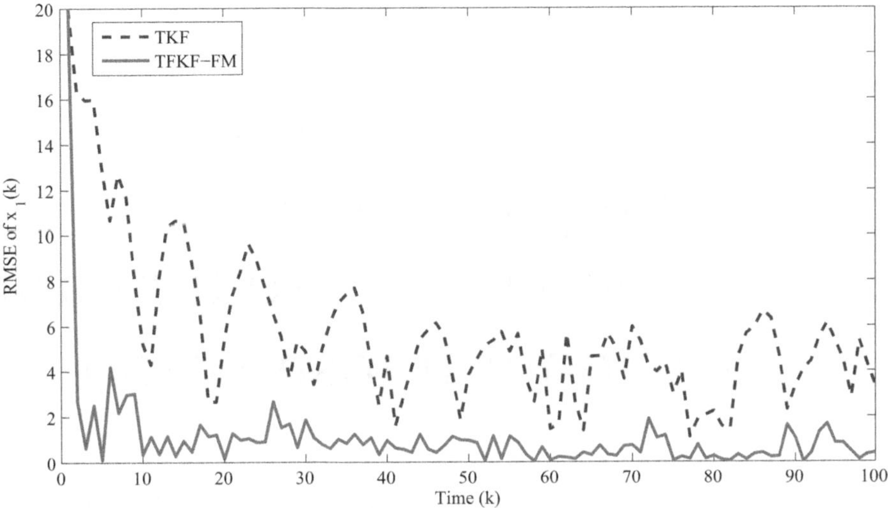

FIGURE 2.3 Method comparison with [75] in RMSE1.

to recover the true state from the obtained fading measurements. This results in the situation that there is a small deviation between the obtained state estimate and the true state, and hence, the RMSE of our TKF-FM in Figure 2.3 cannot degrade to zero as time goes on.

2.4 SUMMARY

In this chapter, we have investigated the recursive filtering problem for discrete time-varying systems in the coexistence of two important measurement nonlinearities, namely, censored and fading measurements. The fading measurements have been addressed under the assumption that the fading coefficients obey Gaussian distributions. The Gaussian distributions can be changed to other distributions, e.g. the uniform distribution on the interval $[\alpha_k, \beta_k]$ where $0 \leq \alpha_k < \beta_k \leq 1$. The adopted Lth-order Rice fading model not only covers the channel fading but also the time-delay phenomena. The two phenomena have been carefully handled by introducing the expected values of the channel coefficients and the augmented states in the filter design, which have led to additional computations in the calculations of the filter gain matrix and the estimation error covariance. Fortunately, these additional computations are recursive or can be performed offline. Accordingly, the designed filter is suitable for online applications. Finally, the effectiveness of the proposed filter has been illustrated by a numerical example.

3 Tobit Kalman Filter with Time-Correlated Multiplicative Sensor Noises under Redundant Channel Transmission

Since the seminal work in [156], the celebrated Kalman filter (KF) has proven to be a state estimation algorithm of central importance in a variety of engineering applications such as flight control, global position systems (GPS), target tracking and so forth. A standard assumption for the conventional KF is that the system model under consideration should be precisely known, and the process/measurement noises should be completely Gaussian. Such an assumption is, however, not always true because of unavoidable appearance of measurement nonlinearities, modeling uncertainties, networked-induced phenomena (e.g. communication delays, packet dropouts, quantization effects and channel fadings) etc., and all of these could lead to performance degradation of the standard KF. To solve such a problem, great efforts have been made to develop modified KFs, which include but are not limited to robust KF, extended Kalman filter (EKF), unscented Kalman filter (UKF) and Tobit Kalman filter (TKF), among which the newly developed TKF is a powerful tool in dealing with measurement censoring that is actually a special type of measurement nonlinearities. Censored measurements arise prevalently in various applications such as biological modeling, economics, computer vision tracking and distributed detection. The main reason for the occurrence of the censored measurements is that the measurements from sensors are often saturated as a result of dynamic changes or interferences.

It is quite common that the censored measurements take the forms of saturation, limits of the detection and occlusion regions. Such kinds of censoring, known as Tobit Type 1 censoring, can be described as a piecewise-linear transform of the output variable with a zero slope in the censored region. Despite its frequent presence in many application areas, the Tobit measurement model has not stirred adequate research attention in areas of estimation and tracking until recently. Nevertheless, it is often the case that the censored measurements are dependent heavily on the true states. In this case, the TKF has been developed for state estimation with censored measurements, and the proposed TKF has been further applied to the two-side censoring case.

The multiplicative noise, also called the state-dependent noise, is a special type of random perturbations on the system parameters that has gained a great deal of

DOI: 10.1201/9781003461623-3

attention in the past few decades. If not properly handled, the multiplicative noise could result in performance degradation of the proposed algorithm. By incorporating the multiplicative noise, more accurate and comprehensive models can be established for the estimation and control of physical processes from the viewpoints of engineering applications. So far, considerable efforts have been made to solve the filtering problems of discrete time-varying systems with multiplicative noises based on different performance criteria, e.g. the linear minimum mean-squared error (LMMSE) criterion, the H_∞ criterion, the minimum upper bound criterion and the Cesaro limit criterion.

As is well known, network-induced phenomena (e.g. communication delays, packet dropouts, quantization effects, sensor nonlinearities and channel fadings) are ubiquitous due to the limited bandwidth of the communication channel. If not properly handled, the network-induced phenomena, especially the packet dropouts (measurement missing), could render the performance of the filtering algorithm degraded or even divergent. To further ensure the quality and the reliability of the communication service, a new communication protocol, namely the redundant channel transmission protocol, has received much attention in recent years [157, 158]. The redundant channel transmission protocol has been widely utilized in industrial and public infrastructures such as ad hoc networked systems, wide-area control systems, wireless structural health monitoring systems and motor systems. In the redundant channel transmission protocol, when the transmission failure of a certain channel occurs and is detected by some detecting mechanisms, another redundant channel will be activated to prevent the data from being lost.

Summarizing these discussions, we can draw the conclusions that (1) sensor measurements are inclined to be corrupted by multiple time-correlated multiplicative noises, and ignorance of the multiplicative noises would result in a poor filtering performance and (2) the redundant channel transmission protocol is quite effective in improving the transmission quality/reliability in the case of transmission failures. As such, a seeming natural idea is to design a TKF to obtain the optimal state estimate for systems subject to censored measurements, intermittent transmission failures and time-correlated multiplicative measurement noises under the redundant channel transmission protocol. This appears to be a non-trivial question for the following three reasons: (1) it is unclear what redundant channel model should be chosen to handle the intermittent transmission failures; (2) it is pretty hard to determine the Tobit regression model that caters for the multiplicative measurement noises and the redundant channel transmission; and (3) it is fairly challenging to understand the impacts brought by the multiplicative measurement noise effect and the redundant channel transmission effect. This situation motivates us to conduct the present research in this chapter.

3.1 PROBLEM FORMULATION

Consider the following class of discrete time-varying systems:

$$x_{k+1} = A_k x_k + B_k \omega_k, \tag{3.1}$$

$$y_k^* = \alpha_{1,k}\left(C_{1,k} + \sum_{s=1}^{M} D_k^s \xi_k^s\right)x_k + \sum_{i=2}^{N}\prod_{j=1}^{i-1}\left(1-\alpha_{j,k}\right)$$
$$\times \alpha_{i,k}\left(C_{i,k} + \sum_{s=1}^{M} D_k^s \xi_k^s\right)x_k + \upsilon_k, \tag{3.2}$$

$$\xi_k^s = \sum_{t=1}^{M} h^{st}\xi_{k-1}^t + \zeta_{k-1}^s, \tag{3.3}$$

where $x_k \in \mathbb{R}^{n_x}$ and $y_k^* \in \mathbb{R}^{n_y}$ are, respectively, the state vector and the latent/uncensored measurement vector; $A_k \in \mathbb{R}^{n_x \times n_x}$, $B_k \in \mathbb{R}^{n_x \times n_\omega}$, $C_{1,k} \in \mathbb{R}^{n_y \times n_x}$, $C_{i,k} \in \mathbb{R}^{n_y \times n_x}$ $(i=2,3,\ldots,N)$ and $D_k^s \in \mathbb{R}^{n_y \times 1}$ are known matrices with appropriate dimensions; $\alpha_{i,k} \in \mathbb{R}$ $(i=1,2,\ldots,N)$ are random variables governing the transmission failure (packet dropout) over the ith channel; $h^{st} \in \mathbb{R}$ $(s,t=1,2,\ldots,M)$ are known constants; $\omega_k \in \mathbb{R}^{n_\omega}$ and $\upsilon_k \in \mathbb{R}^{n_y}$ are the additive process and measurement noises, respectively; $\xi_k^s \in \mathbb{R}$ and $\zeta_k^s \in \mathbb{R}$ are the multiplicative and the additive measurement noises, respectively.

We are now in the position to introduce the Tobit measurement model as follows:

$$y_k = \begin{cases} y_k^*, & y_k^* > \mathcal{I}, \\ \mathcal{I}, & y_k^* \le \mathcal{I}, \end{cases} \tag{3.4}$$

where $y_k \in \mathbb{R}^{n_y}$ is the censored measurement vector, $\mathcal{I} = \begin{bmatrix} \mathcal{I}^1 & \cdots & \mathcal{I}^{n_y} \end{bmatrix}^T \in \mathbb{R}^{n_y}$ is the threshold vector, and $\mathcal{I}^m$ $(m=1,2,\ldots,n_y)$ is the constant threshold of y_k^m. Depending on whether y_k is censored or not, (3.4) can be transformed into the following form:

$$y_k = \Upsilon_k y_k^* + (I - \Upsilon_k)\mathcal{I}, \tag{3.5}$$

where $\Upsilon_k = \text{diag}\{\gamma_k^1, \gamma_k^2, \ldots, \gamma_k^{n_y}\}$ and γ_k^m $(m=1,2,\ldots,n_y)$ are known Bernoulli random variables which regulate the censoring phenomena of y_k^m $(m=1,2,\ldots,n_y)$ with the following probability distributions:

$$\text{Prob}\{\gamma_k^m = 1\} = \overline{\gamma}_k^m, \quad \text{Prob}\{\gamma_k^m = 0\} = 1 - \overline{\gamma}_k^m. \tag{3.6}$$

Here, $\overline{\gamma}_k^m$ $(m=1,2,\ldots,n_y)$ are known non-negative constants. It is assumed that all the random variables γ_k^m $(m=1,2,\ldots,n_y)$ are mutually independent and are also uncorrelated with $\alpha_{i,k}$ $(i=1,2,\ldots,N)$ and other noise signals.

Denote $\overline{\Upsilon}_k \triangleq \text{diag}\{\overline{\gamma}_k^1, \overline{\gamma}_k^2, \ldots, \overline{\gamma}_k^{n_y}\}$ and $\pi_k^s \triangleq \xi_k^s x_k$ $(s=1,2,\ldots,M)$. We have the following assumptions.

Assumption 3.1 ω_k and υ_k are zero-mean white noise sequences with covariances Q_k and R_k, respectively. ζ_k^s is the zero-mean random variable with $\mathbb{E}\{\zeta_k^s \zeta_k^t\} = \sigma^{st}$ $(s,t=1,2,\ldots,M)$.

Assumption 3.2 The random variable $\alpha_{i,k}$ $(i=1,2,\ldots,N)$, which regulates the packet dropout of the ith channel, take values on 0 or 1 with $Prob\{\alpha_{i,k}=1\} = \overline{\alpha}_i$ and

$Prob\{\alpha_{i,k}=0\}=1-\bar{\alpha}_i$, where $\bar{\alpha}_i$ $(i=1,2,\ldots,N)$ are the known non-negative constants. $\alpha_{i,k}$ $(i=1,2,\ldots,N)$ are mutually independent.

Assumption 3.3 The initial state x_0 has the mean $\bar{x}_0$ and covariance P_0. The initial multiplicative measurement noises ξ_0^s $(s=1,2,\ldots,M)$ are zero-mean random variables uncorrelated with each other, and $\mathbb{E}\{\xi_0^s\xi_0^t\}=\rho^{st}$ $(s,t=1,2,\ldots,M)$.

Assumption 3.4 The random variables ω_k, υ_k, $\alpha_{i,k}$, ξ_k^s are uncorrelated with each other; x_0 and ξ_0^s are uncorrelated with ω_k, υ_k, $\alpha_{i,k}$, ξ_k^s; and x_0 and ξ_0^s are mutually uncorrelated.

Remark 3.1 Similar to [159], the N redundant channel transmission protocol described by the Bernoulli distributed random variables is adopted in (3.2). It can be found from (3.2) that $\alpha_{i,k}=1$ or 0 corresponds to whether the measurement transmitted in the ith channel is missing or not. It should be noted that at any time k, only one channel is activated for the data transmission, and other channels will not be activated unless the transmission failure of the working channel is detected by some detection devices. By utilizing such a redundant channel transmission protocol, it is possible to characterize the data transmission in each channel and improve the overall reliability of the transmitted information.

Remark 3.2 Equation (3.4) demonstrates the Tobit measurement model for censored measurements, which are often encountered in systems using low-cost commercial off-the-shelf sensors. It can be observed from (3.4) that, in the censored region, the distribution of the censored measurement noise conditioned on the censored measurement is censored Gaussian (rather than pure Gaussian), which violates the Gaussianity assumption in the standard KF and leads to the necessity of developing the TKF. Combining (3.1)–(3.4), it can be concluded that our system model covers three stochastic uncertainties, i.e. the intermittent failures which are handled by the redundant channel transmission protocol, the stochastic time-correlated multiplicative measurement noises and the stochastic measurement censoring. These stochastic phenomena, which often occur in the networked environments, have not been given full attention in the literature. The consideration of these stochastic phenomena is non-trivial since they bring substantial difficulties in the design of the desired TKF, which will be shown in the subsequent sections.

Remark 3.3 As described in (3.6), the random variables γ_k^m $(m=1,2,\ldots,n_y)$ are utilized to characterize the censoring phenomena of y_k^m $(m=1,2,\ldots,n_y)$. According to (3.5), if no censoring occurs for y_k^m, i.e. $\gamma_k^m=1$ $(m=1,2,\ldots,n_y)$ and $\Upsilon_k=I$, the measurement output is $y_k=y_k^*$, which implies that the output measurement is the same as the latent measurement. If the censoring phenomena occur for y_k^m, i.e., $\gamma_k^m=0$ and $\Upsilon_k=0$, the measurement output is $y_k=\mathcal{I}$, which implies that the threshold is assigned to the output measurement. Here, the censoring probability $\bar{\Upsilon}_k$ is assumed to be known a priori via some statistical experiments. Alternatively, $\bar{\Upsilon}_k$ can also be approximated by

$$\bar{\Upsilon}_k \approx diag\left\{\Phi\left(\frac{\hat{\chi}_{k|k-1}^{(m)}-\mathcal{I}^{(m)}}{\sqrt{R_k^{(m,m)}}}\right)\right\}, \tag{3.7}$$

where

$$\hat{\chi}_{k|k-1} = \bar{\alpha}_{1,k} C_{1,k} \hat{x}_{k|k-1} + \bar{\alpha}_{1,k} \sum_{s=1}^{M} D_k^s \hat{\pi}_{k|k-1}^s + \sum_{i=2}^{N} \bar{\beta}_{i,k} C_{i,k} \hat{x}_{k|k-1} + \sum_{i=2}^{N} \bar{\beta}_{i,k} \sum_{s=1}^{M} D_k^s \hat{\pi}_{k|k-1}^s,$$

$\bar{\beta}_{i,k} = \prod_{j=0}^{i-1} (1 - \bar{\alpha}_{j,k}) \bar{\alpha}_{i,k}$, $i = 2, 3, \ldots, N$, and $\Phi(\cdot)$ is the cumulative distribution function of the random variable "·", which obeys the standard normal distribution.

The objective of this chapter is to design the TKF for the discrete time-varying system (3.1)–(3.5) suffering from the censored measurements and time-correlated multiplicative measurement noises under the redundant channel transmission.

3.2 STATE-DEPENDENT TKF UNDER REDUNDANT CHANNELS

In this section, the TKF for discrete time-varying systems with redundant channel transmission and time-correlated multiplicative measurement noises is derived. The filter design procedure is along the similar lines to the standard TKF with two main differences: (1) a modified Tobit regression model resulting from the redundant channel transmission is introduced and (2) additional computations of the filtering gain matrix and the estimation error covariance matrix are included.

Denote $\mathcal{R}_k \triangleq \text{col}\left\{\sqrt{R_k^{(1,1)}}, \sqrt{R_k^{(2,2)}}, \ldots, \sqrt{R_k^{(n_y, n_y)}}\right\}$. For the convenience of the subsequent filter design, the following lemma is presented to modify the Tobit regression model to incorporate the redundant channel transmission phenomenon.

Lemma 3.1 The mathematical expectation and variance of the censored measurement y_k are derived as follows:

$$\mathbb{E}\{y_k | x_k, R_k\} = \Phi\left(\frac{\chi_k - \mathcal{I}}{\mathcal{R}_k}\right)\left[\chi_k + \lambda\left(\frac{\mathcal{I} - \chi_k}{\mathcal{R}_k}\right)\mathcal{R}_k\right] + \Phi\left(\frac{\mathcal{I} - \chi_k}{\mathcal{R}_k}\right)\mathcal{I}, \quad (3.8)$$

$$Var\{y_k | x_k, R_k\} = R_k\left[I - \varphi\left(\frac{\mathcal{I} - \chi_k}{\mathcal{R}_k}\right)\right], \quad (3.9)$$

where

$$\varsigma\left(\frac{\mathcal{I} - \chi_k}{\mathcal{R}_k}\right) = diag\left\{\varsigma\left(\frac{\mathcal{I}^{(m)} - \chi_k^{(m)}}{\sqrt{R_k^{(m,m)}}}\right)\right\}, \quad (3.10)$$

$\varsigma \in \{\varphi, \lambda, \Phi\}$, and

$$\varphi\left(\frac{\mathcal{I}^{(m)} - \chi_k^{(m)}}{\sqrt{R_k^{(m,m)}}}\right) = \lambda\left(\frac{\mathcal{I}^{(m)} - \chi_k^{(m)}}{\sqrt{R_k^{(m,m)}}}\right)\left[\lambda\left(\frac{\mathcal{I}^{(m)} - \chi_k^{(m)}}{\sqrt{R_k^{(m,m)}}}\right) - \frac{\mathcal{I}^{(m)} - \chi_k^{(m)}}{\sqrt{R_k^{(m,m)}}}\right], \quad (3.11)$$

$$\lambda\left(\frac{\mathcal{I}^{(m)} - \chi_k^{(m)}}{\sqrt{R_k^{(m,m)}}}\right) = \frac{\phi\left(\frac{\mathcal{I}^{(m)} - \chi_k^{(m)}}{\sqrt{R_k^{(m,m)}}}\right)}{1 - \Phi\left(\frac{\mathcal{I}^{(m)} - \chi_k^{(m)}}{\sqrt{R_k^{(m,m)}}}\right)}, \quad (3.12)$$

$$\phi\left(\frac{y_k^{(m)} - \chi_k^{(m)}}{\sqrt{R_k^{(m,m)}}}\right) = \frac{1}{\sqrt{2\pi}} e^{-\frac{\left(y_k^{(m)} - \chi_k^{(m)}\right)^2}{2R_k^{(m,m)}}}, \tag{3.13}$$

$$\Phi\left(\frac{y_k^{(m)} - \chi_k^{(m)}}{\sqrt{R_k^{(m,m)}}}\right) = \int_{-\infty}^{y_k^{(m)}} \frac{e^{-\frac{\left(z_k^{(m)} - \chi_k^{(m)}\right)^2}{2R_k^{(m,m)}}}}{\sqrt{2\pi R_k^{(m,m)}}} d_{z_k^{(m)}}. \tag{3.14}$$

Proof According to (3.5), the probability distribution of $y_k^{(m)}$ with a normally distributed noise $v_k^{(m)}$ is

$$f\left(y_k^{(m)} | x_k\right) = \frac{1}{\sqrt{R_k^{(m,m)}}} \phi\left(\frac{y_k^{(m)} - \chi_k^{(m)}}{\sqrt{R_k^{(m,m)}}}\right) u\left(y_k^{(m)} - \mathcal{I}^{(m)}\right)$$
$$+ 1_{\{\mathcal{I}^{(m)} - y_k^{(m)}\}} \Phi\left(\frac{\mathcal{I}^{(m)} - \chi_k^{(m)}}{\sqrt{R_k^{(m,m)}}}\right). \tag{3.15}$$

Based on (3.15), we have the mathematical expectation and variance of y_k^m as follows:

$$\mathbb{E}\{y_k^{(m)} | x_k\}$$
$$= Prob\{y_k^{(m)} > \mathcal{I}^{(m)} | x_k, R_k^{(m,m)}\} \mathbb{E}\{y_k^{(m)} | y_k^{(m)} > \mathcal{I}^{(m)}, x_k, R_k^{(m,m)}\} \tag{3.16}$$
$$+ Prob\{y_k^{(m)} = \mathcal{I}^{(m)} | x_k, R_k^{(m,m)}\} \mathbb{E}\{y_k^{(m)} | y_k^m = \mathcal{I}^{(m)}, x_k, R_k^{(m,m)}\}.$$

From (3.16), we know that, to compute $\mathbb{E}\{y_k^{(m)} | x_k, R_k^{(m,m)}\}$, we need to calculate probabilities and expected values in (3.18).

$$Prob\{y_k^{(m)} > \mathcal{I}^{(m)} | x_k, R_k^{(m,m)}\} = Prob\{v_k^{(m)} > \mathcal{I}^{(m)} - \chi_k^{(m)}\}$$
$$= \Phi\left(\frac{\chi_k^{(m)} - \mathcal{I}^{(m)}}{\sqrt{R_k^{(m,m)}}}\right). \tag{3.17}$$

Hence, we can easily obtain that

$$\mathbb{E}\{y_k^{(m)} | y_k^{(m)} > \mathcal{I}^{(m)}, x_k, R_k^{(m,m)}\} = \chi_k^{(m)} + \sqrt{R_k^{(m,m)}} \lambda\left(\frac{\mathcal{I}^{(m)} - \chi_k^{(m)}}{\sqrt{R_k^{(m,m)}}}\right), \tag{3.18}$$

where $\lambda\left(\frac{\mathcal{I}^{(m)} - \chi_k^{(m)}}{\sqrt{R_k^{(m,m)}}}\right)$ is given by (3.14). Similar to (3.17)–(3.18), we can obtain

$$Prob\{y_k^{(m)} = \mathcal{I}^{(m)} | x_k, R_k^{(m,m)}\} = \Phi\left(\frac{\mathcal{I}^{(m)} - \chi_k^{(m)}}{\sqrt{R_k^{(m,m)}}}\right),$$
$$\mathbb{E}\{y_k^{(m)} | y_k^{(m)} = \mathcal{I}^{(m)}, x_k, R_k^{(m,m)}\} = \mathcal{I}^{(m)}$$

Putting the above two equations into (3.16) results in

$$
\mathbb{E}\{y_k^{(m)}|x_k, R_k^{(m,m)}\} = \Phi\left(\frac{\mathcal{X}_k^{(m)} - \mathcal{I}^{(m)}}{\sqrt{R_k^{(m,m)}}}\right)\left[\mathcal{X}_k^{(m)} + \sqrt{R_k^{(m,m)}}\right.
$$

$$
\times \lambda\left(\frac{\mathcal{I}^{(m)} - \mathcal{X}_k^{(m)}}{\sqrt{R_k^{(m,m)}}}\right)\right] + \Phi\left(\frac{\mathcal{I}^{(m)} - \mathcal{X}_k^{(m)}}{\sqrt{R_k^{(m,m)}}}\right)\mathcal{I}^{(m)}. \tag{3.19}
$$

It follows easily from (3.15) that $Var\{y_k^{(m)}|x_k, y_k^{(m)} < \mathcal{I}^{(m)}, R_k^{(m,m)}\} = 0$, and hence we have

$$
Var\{y_k^{(m)}|x_k, R_k^{(m,m)}\} = R_k^{(m,m)}\left[1 - \varphi\left(\frac{\mathcal{I}^{(m)} - \mathcal{X}_k^{(m)}}{\sqrt{R_k^{(m,m)}}}\right)\right], \tag{3.20}
$$

where $\varphi\left(\dfrac{\mathcal{I}^{(m)} - \mathcal{X}_k^{(m)}}{\sqrt{R_k^{(m,m)}}}\right)$ is given by (3.13).

For simplicity of the deduction, we assume that $cov\{y_k^{(m)}, y_k^{(t)}\} = 0$ for $m \neq t$, $m, t = 1, 2, \ldots, n_y$. The generalization of the result to the case where $cov\{y_k^{(m)}, y_k^{(t)}\} \neq 0$ for $m \neq t$ is straightforward but will bring notational burdens. Therefore, combining (3.19)–(3.20) yields (3.8)–(3.9) where $\varphi\left(\dfrac{\mathcal{I} - \mathcal{X}_k}{R_k}\right)$, $\lambda\left(\dfrac{\mathcal{I} - \mathcal{X}_k}{R_k}\right)$ and $\Phi\left(\dfrac{y_k - \mathcal{X}_k}{R_k}\right)$ are given by (3.10). The proof is now complete.

Denote $\hat{y}_{k|k-1} \triangleq \mathbb{E}\{y_k|y_{1:k}\}$, $\hat{x}_{k|k} \triangleq \mathbb{E}\{x_k|y_{1:k}\}$, $\tilde{x}_{k|k-1} \triangleq x_k - \hat{x}_{k|k-1}$, $\tilde{x}_{k|k} \triangleq x_k - \hat{x}_{k|k}$, $\tilde{y}_{k|k-1} \triangleq y_k - \hat{y}_{k|k-1}$, $\tilde{\pi}_{k|k-1}^s \triangleq \pi_k - \hat{\pi}_{k|k-1}^s$, $P_{k|k-1} \triangleq \mathbb{E}\{\tilde{x}_{k|k-1}\tilde{x}_{k|k-1}^T\}$, $P_{k|k} \triangleq \mathbb{E}\{\tilde{x}_{k|k}\tilde{x}_{k|k}^T\}$, $\Sigma_{\tilde{y}_{k|k-1}\tilde{y}_{k|k-1}} \triangleq \mathbb{E}\{\tilde{y}_{k|k-1}\tilde{y}_{k|k-1}^T\}$, $\Sigma_{\tilde{x}_{k|k-1}\tilde{y}_{k|k-1}} \triangleq \mathbb{E}\{\tilde{x}_{k|k-1}\tilde{y}_{k|k-1}^T\}$, $\Sigma_{\tilde{x}_{k|k-1}\tilde{\pi}_{k|k-1}^s} \triangleq \mathbb{E}\{\tilde{x}_{k|k-1}(\tilde{\pi}_{k|k-1}^s)^T\}$ and $\Sigma_{\tilde{\pi}_{k|k-1}^s\tilde{\pi}_{k|k-1}^t} \triangleq \mathbb{E}\{\tilde{\pi}_{k|k-1}^s(\tilde{\pi}_{k|k-1}^t)^T\}$. Based on Lemma 3.1, the optimal TKF for system (3.1)–(3.5) is given as follows.

Theorem 3.1. For system (3.1)–(3.5), its optimal TKF has the following structure:

$$
\hat{x}_{k|k-1} = A_{k-1}\hat{x}_{k-1|k-1}, \tag{3.21}
$$

$$
P_{k|k-1} = A_{k-1}P_{k-1|k-1}A_{k-1}^T + B_{k-1}Q_{k-1}B_{k-1}^T, \tag{3.22}
$$

$$
\hat{x}_{k|k} = \hat{x}_{k|k-1} + K_k(y_k - \hat{y}_{k|k-1}), \tag{3.23}
$$

$$
P_{k|k} = P_{k|k-1} - \overline{\Upsilon}_k \Sigma_{\tilde{x}_{k|k-1}\tilde{y}_{k|k-1}} \Sigma_{\tilde{y}_{k|k-1}\tilde{y}_{k|k-1}}^{-1} \Sigma_{\tilde{x}_{k|k-1}\tilde{y}_{k|k-1}}^T, \tag{3.24}
$$

$$
\hat{y}_{k|k-1} = (I - \overline{\Upsilon}_k)\mathcal{I} + \overline{\Upsilon}_k[\bar{\alpha}_{1,k}C_{1,k}\hat{x}_{k|k-1} + \bar{\alpha}_{1,k}\sum_{s=1}^{M}D_k^s\hat{\pi}_{k|k-1}^s
$$

$$
+ \sum_{i=2}^{N}\bar{\beta}_{i,k}C_{i,k}\hat{x}_{k|k-1} + \sum_{i=2}^{N}\bar{\beta}_{i,k}\sum_{s=1}^{M}D_k^s\hat{\pi}_{k|k-1}^s + \lambda\left(\frac{\mathcal{I} - \hat{\mathcal{X}}_{k|k-1}}{R_k}\right)\mathcal{R}_k], \tag{3.25}
$$

where

$$\lambda\left(\frac{\mathcal{I}-\hat{\chi}_{k|k-1}}{R_k}\right) = diag\left\{\lambda\left(\frac{\mathcal{I}^{(m)}-\hat{\chi}_{k|k-1}^{(m)}}{\sqrt{R_k^{(m,m)}}}\right)\right\}. \tag{3.26}$$

The gain matrix K_k is

$$K_k = \Sigma_{\tilde{x}\tilde{y}_{k|k-1}}\Sigma_{\tilde{y}_{k|k-1}\tilde{y}_{k|k-1}}^{-1}, \tag{3.27}$$

where

$$\begin{aligned}
\Sigma_{\tilde{x}_{k|k-1}\tilde{y}_{k|k-1}} &= P_{k|k-1}(\overline{\Upsilon}_k\bar{\alpha}_{1,k}C_{1,k})^T \\
&+ \sum_{s=1}^{M}\Sigma_{\tilde{x}_{k|k-1}\tilde{\pi}_{k|k-1}^s}(D_k^s)^T(\overline{\Upsilon}_k\bar{\alpha}_{1,k})^T \\
&+ P_{k|k-1}\left(\overline{\Upsilon}_k\sum_{i=1}^{N}\bar{\beta}_{i,k}C_{i,k}\right)^T \\
&+ \sum_{s=1}^{M}\Sigma_{\tilde{x}_{k|k-1}\tilde{\pi}_{k|k-1}^s}(D_k^s)^T\left(\overline{\Upsilon}_k\sum_{i=2}^{N}\bar{\beta}_{i,k}\right)^T,
\end{aligned} \tag{3.28}$$

and

$$\begin{aligned}
\Sigma_{\tilde{y}_{k|k-1}\tilde{y}_{k|k-1}} &= \overline{\Upsilon}_k\bar{\alpha}_{1,k}C_{1,k}P_{k|k-1}(\overline{\Upsilon}_k\bar{\alpha}_{1,k}C_{1,k})^T \\
&+ \overline{\Upsilon}_k\bar{\alpha}_{1,k}C_{1,k}\sum_{t=1}^{M}\Sigma_{\tilde{x}_{k|k-1}\tilde{\pi}_{k|k-1}^t}(D_k^t)^T(\overline{\Upsilon}_k\bar{\alpha}_{1,k})^T \\
&+ \left[\overline{\Upsilon}_k\bar{\alpha}_{1,k}C_{1,k}\sum_{t=1}^{M}\Sigma_{\tilde{x}_{k|k-1}\tilde{\pi}_{k|k-1}^t}(D_k^t)^T(\overline{\Upsilon}_k\bar{\alpha}_{1,k})^T\right]^T \\
&+ R_k\left[I-\varphi\left(\frac{\mathcal{I}-\hat{\chi}_{k|k-1}}{R_k}\right)\right] \\
&+ \overline{\Upsilon}_k\bar{\alpha}_{1,k}C_{1,k}P_{k|k-1}\left(\overline{\Upsilon}_k\sum_{j=2}^{N}\bar{\beta}_{j,k}C_{j,k}\right)^T \\
&+ \overline{\Upsilon}_k\bar{\alpha}_{1,k}C_{1,k}\sum_{t=1}^{M}\Sigma_{\tilde{x}_{k|k-1}\tilde{\pi}_{k|k-1}^t}(D_k^t)^T\left(\overline{\Upsilon}_k\sum_{j=2}^{N}\bar{\beta}_{j,k}\right)^T \\
&+ \overline{\Upsilon}_k\bar{\alpha}_{1,k}\sum_{s=1}^{M}\sum_{t=1}^{M}D_k^s\Sigma_{\tilde{x}_{k|k-1}\tilde{\pi}_{k|k-1}^t}(D_k^t)^T\left(\overline{\Upsilon}_k\bar{\alpha}_{1,k}\right)^T \\
&+ \overline{\Upsilon}_k\bar{\alpha}_{1,k}\sum_{s=1}^{M}D_k^s\Sigma_{\tilde{\pi}_{k|k-1}^s\tilde{x}_{k|k-1}}\left(\overline{\Upsilon}_k\sum_{j=2}^{N}\bar{\beta}_{j,k}C_{j,k}\right)^T
\end{aligned}$$

$$
+\bar{\Upsilon}_k\bar{\alpha}_{1,k}\sum_{s=1}^{M}\sum_{t=1}^{M}D_k^s\Sigma_{\tilde{\pi}_{k|k-1}^s\tilde{\pi}_{k|k-1}^t}(D_k^t)^T\left(\bar{\Upsilon}_k\sum_{j=2}^{N}\bar{\beta}_{j,k}\right)^T
$$

$$
+\left[\bar{\Upsilon}_k\bar{\alpha}_{1,k}C_{1,k}P_{k|k-1}\left(\bar{\Upsilon}_k\sum_{j=2}^{N}\bar{\beta}_{j,k}C_{j,k}\right)^T\right]^T
$$

$$
+\bar{\Upsilon}_k\sum_{i=2}^{N}\bar{\beta}_{i,k}C_{i,k}\sum_{t=1}^{M}\Sigma_{\tilde{x}_{k|k-1}\tilde{\pi}_{k|k-1}^t}(D_k^t)^T(\bar{\Upsilon}_k\bar{\alpha}_{1,k})^T
$$

$$
+\bar{\Upsilon}_k\sum_{i=2}^{N}\bar{\beta}_{i,k}C_{i,k}P_{k|k-1}\left(\bar{\Upsilon}_k\sum_{j=2}^{N}\bar{\beta}_{j,k}C_{j,k}\right)^T
$$

$$
+\bar{\Upsilon}_k\sum_{i=2}^{N}\bar{\beta}_{i,k}C_{i,k}\sum_{t=1}^{M}\Sigma_{\tilde{x}_{k|k-1}\tilde{\pi}_{k|k-1}^t}(D_k^t)^T\left(\bar{\Upsilon}_k\sum_{j=2}^{N}\bar{\beta}_{j,k}\right)^T
$$

$$
+[\bar{\Upsilon}_k\bar{\alpha}_{1,k}C_{1,k}\sum_{t=1}^{M}\Sigma_{\tilde{x}_{k|k-1}\tilde{\pi}_{k|k-1}^t}\left((D_k^t)^T\left(\bar{\Upsilon}_k\sum_{j=2}^{N}\bar{\beta}_{j,k}\right)^T\right)^T]^T
$$

$$
+\bar{\Upsilon}_k\sum_{i=2}^{N}\bar{\beta}_{i,k}\sum_{s=1}^{M}\sum_{t=1}^{M}D_k^s\Sigma_{\tilde{\pi}_{k|k-1}^s\tilde{\pi}_{k|k-1}^t}(D_k^t)^T(\bar{\Upsilon}_k\bar{\alpha}_{1,k})^T \tag{3.29}
$$

$$
+\bar{\Upsilon}_k\sum_{i=2}^{N}\bar{\beta}_{i,k}\sum_{s=1}^{M}D_k^s\Sigma_{\tilde{\pi}_{k|k-1}^s\tilde{x}_{k|k-1}}\left(\bar{\Upsilon}_k\sum_{j=2}^{N}\bar{\beta}_{j,k}C_{j,k}\right)^T
$$

$$
+\bar{\Upsilon}_k\sum_{i=2}^{N}\bar{\beta}_{i,k}\sum_{s=1}^{M}\sum_{t=1}^{M}D_k^s\Sigma_{\tilde{\pi}_{k|k-1}^s\tilde{\pi}_{k|k-1}^t}(D_k^t)^T
$$

$$
\times\left(\bar{\Upsilon}_k\sum_{j=2}^{N}\bar{\beta}_{j,k}\right)^T,
$$

with

$$
\varphi\left(\frac{\mathcal{I}-\hat{\chi}_{k|k-1}}{R_k}\right)=diag\left\{\varphi\left(\frac{\mathcal{I}^{(m)}-\hat{\chi}_{k|k-1}^{(m)}}{\sqrt{R_k^{(m,m)}}}\right)\right\}. \tag{3.30}
$$

Proof Noticing the definitions of $\hat{x}_{k|k-1}$, $P_{k|k-1}$, $\hat{x}_{k|k}$ and $P_{k|k}$, we can directly obtain (3.21)–(3.25) and (3.27) by applying the orthogonality projection principle in to system (3.1)–(3.5). According to (3.8), we have (3.25) where $\lambda\left(\dfrac{\mathcal{I}-\hat{\chi}_{k|k-1}}{R_k}\right)$ is given by (3.26). Subtracting (3.25) from (3.5) leads to

$$
\tilde{y}_{k|k-1}=\Upsilon_k\Bigg[\alpha_{1,k}C_{1,k}\tilde{x}_{k|k-1}+\alpha_{1,k}\sum_{s=1}^{M}D_k^s\tilde{\pi}_{k|k-1}^s
$$

$$
+\sum_{i=2}^{N}\beta_{i,k}C_{i,k}\tilde{x}_{k|k-1}+\sum_{i=2}^{N}\beta_{i,k}\sum_{s=1}^{M}D_k^s\tilde{\pi}_{k|k-1}^s \tag{3.31}
$$

$$
+\left(\upsilon_k-\lambda\left(\frac{\mathcal{I}-\hat{\chi}_{k|k-1}}{R_k}\right)\mathcal{R}_k\right)\Bigg].
$$

Substituting (3.31) into the definition of $\Sigma_{\tilde{x}_{k|k-1}\tilde{y}_{k|k-1}}$, we have (3.28). Substituting (3.31) into the definition of $\Sigma_{\tilde{y}_{k|k-1}\tilde{y}_{k|k-1}}$, recalling the definition of the variance of y_k and noting (3.9), we have (3.29) where $\varphi\left(\dfrac{\mathcal{I}-\hat{\chi}_{k|k-1}}{R_k}\right)$ is given by (3.30). The proof is now complete.

Remark 3.4 In comparison with the standard TKF, one can observe that our TKF established in Theorem 3.1 has two differences in the calculations of the measurement prediction $\hat{y}_{k|k-1}$ and the gain matrix K_k. The first difference is that all the terms pertaining to the predicted values $\hat{x}_{k|k-1}$, $\hat{\pi}^s_{k|k-1}$ and their prediction covariances are multiplied by at least one of the means of the redundant channel coefficients. The second difference is that several new prediction and estimation cross-covariance terms appear in our filter structure. The two differences characterize the effect of the redundant channel transmission on the design of our filter. In addition, one can find that these newly appeared terms can all be computed offline or recursively.

It follows from Theorem 3.1 that in order to compute the estimate $\hat{x}_{k|k}$ in (3.23) and its covariance $P_{k|k}$ in (3.24), the innovation $\tilde{y}_{k|k-1}$ and the gain matrix K_k are required. The computation of $\tilde{y}_{k|k-1}$ needs $\hat{y}_{k|k-1}$ in (3.25), while the computation of K_k needs $\Sigma_{\tilde{x}_{k|k-1}\tilde{y}_{k|k-1}}$ in (3.28) and $\Sigma_{\tilde{y}_{k|k-1}\tilde{y}_{k|k-1}}$ in (3.32). It can be seen from (3.25) and (3.28)–(3.29) that $\hat{\pi}^s_{k|k-1}$, $\Sigma_{\tilde{x}_{k|k-1}\tilde{\pi}^s_{k|k-1}}$ and $\Sigma_{\tilde{\pi}^s_{k|k-1}\tilde{\pi}^t_{k|k-1}}$ are required in order to compute $\hat{y}_{k|k-1}$, $\Sigma_{\tilde{x}_{k|k-1}\tilde{y}_{k|k-1}}$ and $\Sigma_{\tilde{y}_{k|k-1}\tilde{y}_{k|k-1}}$. In the sequel, we will make specific efforts to solve the recursive computation of these terms.

Theorem 3.2 The predicted value $\hat{\pi}^s_{k|k-1}$ obeys the following recursion:

$$\hat{\pi}^s_{k|k-1} = A_{k-1}\sum_{t=1}^{M} h^{st}\hat{\pi}^t_{k-1|k-1}, \tag{3.32}$$

$$\hat{\pi}^s_{k|k} = \hat{\pi}^s_{k|k-1} + \mathbf{K}^s_k(y_k - \hat{y}_{k|k-1}). \tag{3.33}$$

The gain matrix $\mathbf{K}^s_k$ is as follows:

$$\mathbf{K}^s_k = \Sigma_{\tilde{\pi}^s_{k|k-1}\tilde{y}_{k|k-1}}\Sigma^{-1}_{\tilde{y}_{k|k-1}\tilde{y}_{k|k-1}}\Sigma^T_{\tilde{\pi}^s_{k|k-1}\tilde{y}_{k|k-1}}, \tag{3.34}$$

where

$$\begin{aligned}
\Sigma_{\tilde{\pi}^s_{k|k-1}\tilde{y}_{k|k-1}} &= \Sigma_{\tilde{\pi}^s_{k|k-1}\tilde{x}_{k|k-1}}(\overline{\Upsilon}_k\bar{\alpha}_{1,k}C_{1,k})^T \\
&\quad + \sum_{t=1}^{M}\Sigma_{\tilde{\pi}^s_{k|k-1}\tilde{\pi}^t_{k|k-1}}(D^t_k)^T\bar{\alpha}^T_{1,k} \\
&\quad + \Sigma_{\tilde{\pi}^s_{k|k-1}\tilde{x}_{k|k-1}}\left(\sum_{i=2}^{N}\bar{\beta}_{i,k}C_{i,k}\right)^T \\
&\quad + \sum_{t=1}^{M}\Sigma_{\tilde{\pi}^s_{k|k-1}\tilde{\pi}^t_{k|k-1}}(D^t_k)^T\sum_{i=2}^{N}\bar{\beta}^T_{i,k},
\end{aligned} \tag{3.35}$$

$$\Sigma_{\tilde{x}_{k|k-1}\tilde{\pi}^s_{k|k-1}} = A_{k-1}\sum_{t=1}^{M}h^{st}\Sigma_{\tilde{x}_{k-1|k-1}\tilde{\pi}^t_{k-1|k-1}}A^T_{k-1}, \tag{3.36}$$

$$
\Sigma_{\tilde{\pi}^s_{k|k-1}\tilde{\pi}^t_{k|k-1}} = \sum_{s=1}^{M}\sum_{t=1}^{M} h^{st}h^{ts}\Big(A_{k-1}\Sigma_{\tilde{\pi}^s_{k-1|k-1}\tilde{\pi}^t_{k-1|k-1}}A_{k-1}^T
$$

$$
+ B_{k-1}Q_{k-1}\Sigma_{\xi^t_{k-1}\xi^s_{k-1}}B_{k-1}^T\Big) \tag{3.37}
$$

with
$$
+ \sigma^{st}\Big(A_{k-1}\Sigma_{x_{k-1}x_{k-1}}A_{k-1}^T + B_{k-1}Q_{k-1}B_{k-1}^T\Big),
$$

$$
\Sigma_{\tilde{x}_{k|k}\tilde{\pi}^s_{k|k}} = \Sigma_{\tilde{x}_{k|k-1}\tilde{\pi}^s_{k|k-1}} - \Sigma_{\tilde{x}_{k|k-1}\tilde{y}_{k|k-1}}\Sigma^{-1}_{\tilde{y}_{k|k-1}\tilde{y}_{k|k-1}}\Sigma^T_{\tilde{\pi}^s_{k|k-1}\tilde{y}_{k|k-1}}. \tag{3.38}
$$

$$
\Sigma_{\tilde{\pi}^s_{k|k}\tilde{\pi}^t_{k|k}} = \Sigma_{\tilde{\pi}^s_{k|k-1}\tilde{\pi}^t_{k|k-1}} - \Sigma_{\tilde{\pi}^s_{k|k-1}\tilde{y}_{k|k-1}}\Sigma^{-1}_{\tilde{y}_{k|k-1}\tilde{y}_{k|k-1}}\Sigma^T_{\tilde{\pi}^t_{k|k-1}\tilde{y}_{k|k-1}}, \tag{3.39}
$$

$$
\Sigma_{\xi^s_k\xi^t_k} = \sum_{s=1}^{M}\sum_{t=1}^{M} h^{st}h^{ts}\Sigma_{\xi^t_{k-1}\xi^s_{k-1}} + \sigma^{st}, \tag{3.40}
$$

$$
\Sigma_{x_k x_k} = A_{k-1}\Sigma_{x_k x_k}A_{k-1}^T + B_{k-1}Q_{k-1}B_{k-1}^T. \tag{3.41}
$$

Proof Since $\pi^s_k = \xi^s_k x_k$, we have

$$
\pi^s_k = \left(\sum_{t=1}^{M} h^{st}\xi^t_{k-1} + \zeta^s_{k-1}\right)\big(A_{k-1}x_{k-1} + B_{k-1}\omega_{k-1}\big)
$$

$$
= A_{k-1}\sum_{t=1}^{M} h^{st}\pi^t_{k-1} + B_{k-1}\omega_{k-1}\sum_{t=1}^{M} h^{st}\xi^t_{k-1} \tag{3.42}
$$

$$
+ A_{k-1}x_{k-1}\zeta^s_{k-1} + B_{k-1}\omega_{k-1}\zeta^s_{k-1}.
$$

Based on (3.42) and noting $\mathbb{E}\{\omega_k\}=0$ and $\mathbb{E}\{\zeta^s_k\}=0$, we have (3.32). From (3.32), we can see that to get $\hat{\pi}^s_{k|k-1}$, $\hat{\pi}^t_{k-1|k-1}$ is required. Equations (3.33)–(3.34) follow easily from the orthogonality projection principle. From (3.33)–(3.34), we can see that, in order to get the recursive form of $\hat{\pi}^t_{k|k}$, the gain matrix $\mathbf{K}^s_k$ is required, which requires the computation of $\Sigma_{\tilde{\pi}^s_{k|k-1}\tilde{y}_{k|k-1}}$.

$$
\Sigma_{\tilde{\pi}^s_{k|k-1}\tilde{y}_{k|k-1}} = \Sigma_{\tilde{\pi}^s_{k|k-1}\tilde{x}_{k|k-1}}\big(\overline{\Upsilon}_k\bar{\alpha}_{1,k}C_{1,k}\big)^T + \sum_{t=1}^{M}\Sigma_{\tilde{\pi}^s_{k|k-1}\tilde{\pi}^t_{k|k-1}}\big(D^t_k\big)^T\bar{\alpha}^T_{1,k}
$$

$$
+ \Sigma_{\tilde{\pi}^s_{k|k-1}\tilde{x}_{k|k-1}}\left(\sum_{i=2}^{N}\beta_{i,k}C_{i,k}\right)^T + \sum_{t=1}^{M}\Sigma_{\tilde{\pi}^s_{k|k-1}\tilde{\pi}^t_{k|k-1}}\big(D^t_k\big)^T\sum_{i=2}^{N}\bar{\beta}^T_{i,k}. \tag{3.43}
$$

Equation (3.43) demonstrates that to compute $\Sigma_{\tilde{\pi}^s_{k|k-1}\tilde{y}_{k|k-1}}$, the recursive computations of $\Sigma_{\tilde{\pi}^s_{k|k-1}\tilde{x}_{k|k-1}}$ and $\Sigma_{\tilde{\pi}^s_{k|k-1}\tilde{\pi}^t_{k|k-1}}$ are required.

$$
\tilde{\pi}^s_{k|k-1} = A_{k-1}\sum_{t=1}^{M} h^{st}\tilde{\pi}^t_{k-1|k-1} + B_{k-1}\omega_{k-1}\sum_{t=1}^{M} h^{st}\xi^t_{k-1}
$$

$$
+ A_{k-1}x_{k-1}\zeta^s_{k-1} + B_{k-1}\omega_{k-1}\zeta^s_{k-1}. \tag{3.44}
$$

It follows from (3.44) that

$$\Sigma_{\tilde{x}_{k|k-1}\tilde{\pi}^s_{k|k-1}} = A_{k-1}\sum_{t=1}^{M} h^{st}\Sigma_{\tilde{x}_{k-1|k-1}\tilde{\pi}^t_{k-1|k-1}} A^T_{k-1}, \tag{3.45}$$

and

$$\Sigma_{\tilde{\pi}^s_{k|k-1}\tilde{\pi}^t_{k|k-1}} = \sum_{s=1}^{M}\sum_{t=1}^{M} h^{st}h^{ts}\left(A_{k-1}\Sigma_{\tilde{\pi}^s_{k-1|k-1}\tilde{\pi}^t_{k-1|k-1}}A^T_{k-1} + B_{k-1}Q_{k-1}\Sigma_{\xi^t_{k-1}\xi^s_{k-1}}B^T_{k-1}\right)$$
$$+ \sigma^{st}\left(A_{k-1}\Sigma_{x_{k-1}x_{k-1}}A^T_{k-1} + B_{k-1}Q_{k-1}B^T_{k-1}\right). \tag{3.46}$$

Then, to compute $\Sigma_{\tilde{x}_{k|k-1}\tilde{\pi}^s_{k|k-1}}$ and $\Sigma_{\tilde{\pi}^s_{k|k-1}\tilde{\pi}^t_{k|k-1}}$, the recursive forms of $\Sigma_{\tilde{x}_{k|k}\tilde{\pi}^s_{k|k}}$, $\Sigma_{\tilde{\pi}^s_{k|k}\tilde{\pi}^t_{k|k}}$, $\Sigma_{\xi^s_k\xi^t_k}$ and $\Sigma_{x_kx_k}$ are required.

$$\Sigma_{\tilde{x}_{k|k}\tilde{\pi}^s_{k|k}} = \Sigma_{\tilde{x}_{k|k-1}\tilde{\pi}^s_{k|k-1}}$$
$$- \Sigma_{\tilde{x}_{k|k-1}\tilde{y}_{k|k-1}}\mathbf{K}^T_k - K_k\Sigma^T_{\tilde{\pi}^s_{k|k-1}\tilde{y}_{k|k-1}} + \Sigma_{\tilde{x}_{k|k-1}\tilde{y}_{k|k-1}}\mathbf{K}^T_k$$
$$= \Sigma_{\tilde{x}_{k|k-1}\tilde{\pi}^s_{k|k-1}} - K_k\Sigma^T_{\tilde{\pi}^s_{k|k-1}\tilde{y}_{k|k-1}}.$$

which is exactly the same as (3.38). Similar to the derivation of (3.38), we have (3.39). From the expressions of ξ^s_k and x_k, we can directly have (3.40)–(3.41). The proof is now complete.

Remark 3.5 Theorem 3.2 presents the recursive forms of $\hat{\pi}^s_{k|k-1}$, $\hat{\pi}^s_{k|k}$, $\Sigma_{\tilde{\pi}^s_{k|k-1}\tilde{y}_{k|k-1}}$, $\Sigma_{\tilde{x}_{k|k-1}\tilde{\pi}^s_{k|k-1}}$, $\Sigma_{\tilde{x}_{k|k}\tilde{\pi}^s_{k|k}}$, $\Sigma_{\tilde{\pi}^s_{k|k-1}\tilde{\pi}^t_{k|k-1}}$, $\Sigma_{\tilde{\pi}^s_{k|k}\tilde{\pi}^t_{k|k}}$, $\Sigma_{\xi^s_k\xi^t_k}$ and $\Sigma_{x_kx_k}$. Substituting these recursive terms into Theorem 3.1, we can get the optimal estimate $\hat{x}^s_{k|k}$ in the LMMSE sense. We can also confirm the two differences explained in the previous remark. To be specific, for the first difference, the first and second terms in (3.35) contain $\bar{\alpha}_{1,k}$, while the third and fourth terms in (3.35) contain $\sum_{i=2}^{N}\bar{\beta}_{i,k}$. For the second difference, the third and fourth terms in (3.35) are not required in existing literature. The two differences can be explicitly observed from (3.35), while they can only be implicitly observed from other equations in Theorem 3.2 by combining Theorem 3.1.

Remark 3.6 According to Lemma 3.1 and Theorems 3.1–3.2, a recursive filtering scheme is presented to solve the novel filtering problem for the discrete time-varying systems subject to multiple modeling and communication uncertainties. To be specific, these uncertainties include the measurement nonlinearity which takes the form of censored measurements, the networked uncertainty which takes the form of the intermittent failures and are tackled by the redundant channel transmission protocol and the stochastic uncertainty which takes the form of the time-correlated multiplicative measurement noises. These uncertainties are often encountered in engineering applications such as target tracking and networked control. To deal with these uncertainties, a unified yet effective Tobit Kalman filtering framework is proposed. Compared with the work in [49], one can observe from our main results that the effect of the redundant channel transmission is explicitly reflected in the filter structure via the appearances of $\bar{\alpha}_{1,k}$, $\sum_{i=2}^{N} \bar{\beta}_{i,k}$ and some additional cross-covariance terms.

3.3 AN ILLUSTRATIVE EXAMPLE

In this section, we will use the oscillator example to demonstrate the effectiveness of the proposed filter design approach and the filtering performance.

Consider the following class of discrete time-varying systems given by (3.1)–(3.5) with te following parameters:

$$\Sigma_{x_0 \pi_0^1} = \Sigma_{x_0 \pi_0^2} = 0.01 I_2, \hat{\pi}_{0|0}^1 = \hat{\pi}_{0|0}^2 = \begin{bmatrix} 1 & 0 \end{bmatrix}^T,$$

$$\Sigma_{\xi_0^1 \xi_0^1} = \Sigma_{\xi_0^1 \xi_0^2} = \Sigma_{\xi_0^2 \xi_0^1} = \Sigma_{\xi_0^2 \xi_0^2} = 0.1,$$

$$\Sigma_{\pi_0^1 \pi_0^1} = \Sigma_{\pi_0^1 \pi_0^2} = \Sigma_{\pi_0^2 \pi_0^1} = \Sigma_{\pi_0^2 \pi_0^2} = 0.01 I_2,$$

$$h^{11} = h^{22} = 0.02, h^{12} = h^{21} = 0, T = 1, \mathcal{I} = 0,$$

$$\sigma^{11} = \sigma^{12} = \sigma^{21} = \sigma^{22} = 0.4, \Sigma_{x_0 x_0} = 0.01 I_2,$$

$$\bar{x}_0 = \begin{bmatrix} 5 & 0 \end{bmatrix}^T, P_0 = I_2, \rho^{11} = \rho^{12} = \rho^{21} = \rho^{22} = 0.01,$$

$$A_k = \begin{bmatrix} \cos(wT) & -\sin(wT) \\ \sin(wT) & \cos(wT) \end{bmatrix}, B_k = I_2, w = 0.052\pi,$$

$$C_1 = C_2 = \begin{bmatrix} 1 & 0 \end{bmatrix}, D_k^1 = D_k^2 = \begin{bmatrix} 0.2 & 0 \end{bmatrix}, M = 2, N = 3,$$

$$Q_k = 0.0025 I_2, R_k = 1, \bar{\alpha}_{1,k} = 0.6, \bar{\alpha}_{2,k} = 0.8, \bar{\alpha}_{3,k} = 0.9,$$

and the censoring probability $\bar{\Upsilon}_k = \bar{\gamma}_k^1$ can be calculated according to (3.7).

One thousand times of Monte Carlo simulations are conducted and the performance comparison are made under three different communication cases, that is, the case with only Channel 1, the case with Channels 1 and 2 and the case with all channels. Figure 3.1 describes the phenomena of packet dropouts in three channels with the aforementioned probabilities $\bar{\alpha}_{1,k}$, $\bar{\alpha}_{2,k}$ and $\bar{\alpha}_{3,k}$. Figure 3.2 illustrates the true values of the states and the estimates under different communication cases. From Figure 3.2, we can easily see that estimates obtained by the TKF utilizing measurements from Channels 1, 2 and 3 track the true state values very well, while estimates obtained by TKFs utilizing measurements only from Channel 1 or Channels 1 and 2 lead to deviations from true state values. Furthermore, Figure 3.3 shows the comparison of the RMSE curves under three different communication cases. It is observed in Figure 3.3 that the RMSE curve given by the TKF utilizing measurements from all channels is always lower than those given by TKFs utilizing measurements from one or two channels. In a word, the more channels are employed, the better performance will be achieved. This example validates the effectiveness of the proposed TKF with redundant channel transmission in improving the estimation performance.

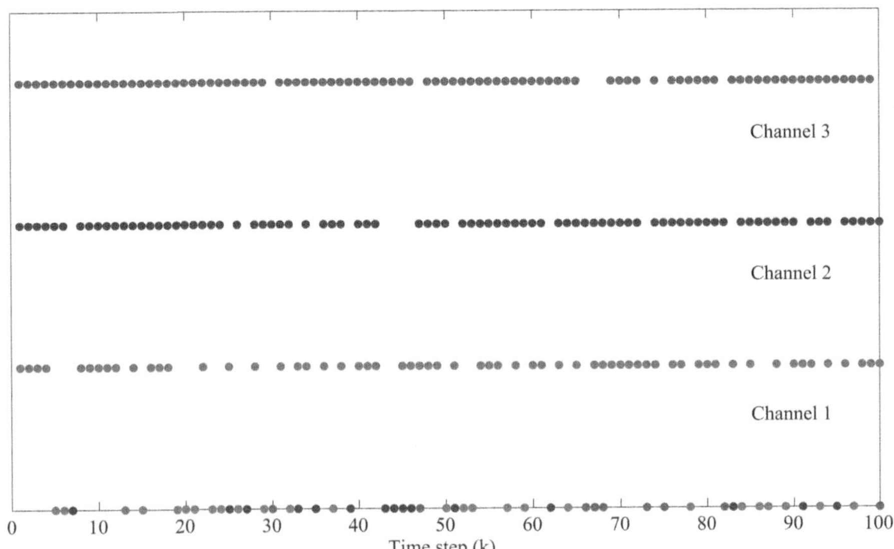

FIGURE 3.1 Random packet dropouts in three channels with probabilities $\bar{\alpha}_{1,k} = 0.6$, $\bar{\alpha}_{2,k} = 0.8$, $\bar{\alpha}_{3,k} = 0.9$. The dots above Channel 1, Channel 2 and Channel 3 or on the horizontal axis denote successful or unsuccessful packet transmission in Channels 1, 2 and 3, respectively. Take the dot above Channel 1 as an example. If a dot above Channel 1 is located above or on the horizontal axis at time k, it means that the packet at time k is successfully transmitted or dropped in Channel 1.

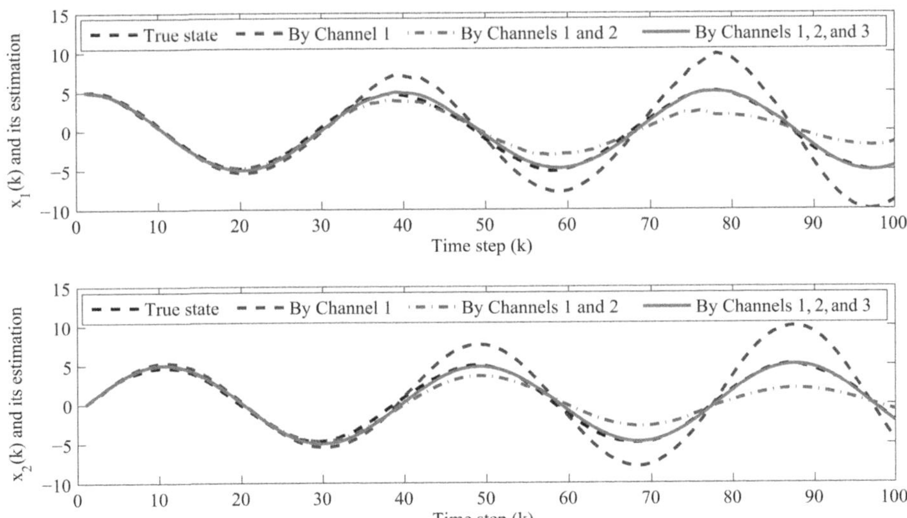

FIGURE 3.2 True state values and estimates under different communication cases where by Channel 1, by Channels 1 and 2, and by Channels 1, 2 and 3, respectively, mean that when calculating state estimates, the proposed TKF uses measurements from Channel 1, Channels 1, 2, and Channels 1, 2, 3.

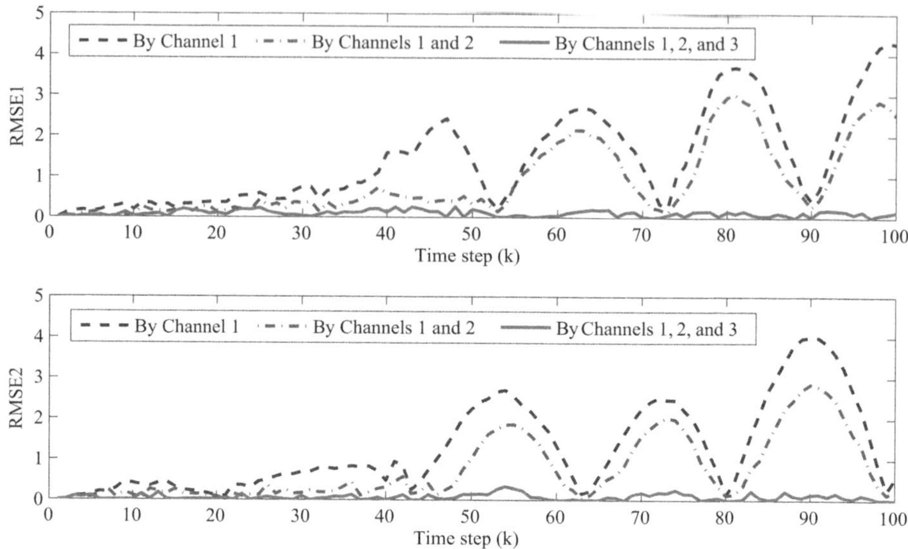

FIGURE 3.3 RMSE curves under different communication cases.

3.4 SUMMARY

In this chapter, we have investigated the recursive filtering problem for discrete time-varying systems in the presence of censored measurements, intermittent failures and time-correlated multiplicative measurement noises under the redundant channel transmission protocol. The Tobit regression model in [49] has been modified to account for the packet dropouts and the redundant channel transmission. Several new recursive terms have been introduced in the modified Tobit regression model as well as the design of the filtering algorithm. This gives rise to additional computations in the algorithm implementation. Fortunately, these additional computations are all recursive or offline. Accordingly, the designed filter is suitable for online applications. Finally, the effectiveness of the proposed filter has been illustrated by a numerical example.

4 State Estimation under Non-Gaussian Lévy and Time-Correlated Additive Sensor Noises

A Modified Tobit Kalman Filtering Approach

Non-Gaussian Lévy noises are typically used to model some commonly encountered noises including but not limited to noises on telephone lines, noises in the extremely low-frequency electromagnetic communication, tracking errors in code tracking loops and random walks in stock-market price models. The non-Gaussian Lévy noise is often approximated by the increments of the corresponding Lévy process per time step [160]. The main challenge in handling the Lévy noise lies in its infinite variance, which makes it difficult to use the standard KF in estimating the system state. Assuming that the contributions of the jumps are exactly known, the KF with a nonlinear and recursive form has been derived in [161] for systems with infinite variances. In [162], a nonlinear recursive filter, which minimizes the difference between the true state and the filtered observation in the L^2-norm sense, has been designed in the case of Lévy measurement noises.

As pointed out in [163], the filter in [162] concerning the Lévy noises has not actually addressed the Kalman filtering problem consisting of forecasts and observations. To solve this problem, the Kalman-Lévy filter with a robust performance has been developed by focusing on the large errors in [163] and has been adopted to solve the target tracking problems in [164]. Nevertheless, the implementation of the matrix diagonalization and the calculation of the fractional power in each step are needed in running the filter in [163], resulting in a high computational cost. To reduce the computation burden, a fractional KF has been proposed in [165] for nonlinear fractional order systems with non-Gaussian Lévy noises.

The Kalman filtering problems have recently stirred much research attention with respect to *time-correlated additive measurement noises* modeled mainly via two approaches. In the first approach, the measurement noises are assumed to be Gaussian and time-correlated with known covariances, while the second approach assumes the measurement noises to be the outputs of a linear recursive model with white noises. Regarding the first description approach, the state estimation problem has been studied in [166] for discrete-time systems with correlated measurement

DOI: 10.1201/9781003461623-4

noises, where the lifting technique has been adopted to augment the original state and the measurement noise to a new state in order to facilitate the application of the standard Kalman filtering method. To avoid the high computation load resulting from the augmentation, the globally optimal KF with finite-time correlated noises has been designed in [167] by decomposing the Kalman gain in [166] to two recursive factors and deriving recursive forms of the multi-step prediction error covariances of the state and the measurement. To further reduce the computational complexity and the storage load, a finite-dimensionally computable suboptimal algorithm has been put forward by using the truncated approximation strategy in [168]. In such a truncated approximation strategy, the state estimate has been calculated by utilizing only a finite set of the last several measurements rather than the whole measurement sequence.

Regarding the second description approach, the measurement differencing approach has been employed in [166] to remove the time-correlated portion of the measurement noises. With such an approach, an equivalent measurement, which is the linear combination of two sequential measurements, is constructed to eliminate the time-correlated noise effect. As a result, the obtained measurement contains only the purely random noise sequences instead of the sequentially correlated noise sequences. However, the dimension of the augmented system is sometimes too high, and consequently, the error covariance matrix of the state estimation could become singular, thereby invalidating the inverse operation in the Kalman filtering algorithm. To tackle this challenge, the measurement differencing over time has been utilized in [169] as an auxiliary signal to eliminate the time-correlation of the measurement noises.

Summarizing these discussions, it can be concluded that (1) the TKF has attracted increasing research attention due to its Kalman-type filter structure and its capability in handling censored measurements and (2) it is quite common in engineering practice that the measurement noises are non-Gaussian Lévy and/or time-correlated, in which case the traditional KF is no longer applicable. As such, a seeming natural idea is to design a TKF to obtain the optimal state estimate for systems subject to censored measurements, non-Gaussian Lévy and time-correlated measurement noises. This appears to be a non-trivial job for the following three reasons: (1) it is unclear how to calculate the covariances of the measurement noises when they are non-Gaussian Lévy and time-correlated; (2) it is quite hard to determine the Tobit regression model when both the non-Gaussian Lévy and the time-correlated measurement noises are present in the system model; and (3) it is theoretically difficult to quantify the impacts from the non-Gaussian Lévy and the time-correlated measurement noises on the design of the filtering algorithm. It is, therefore, the motivation of this chapter to address the identified challenges.

In this chapter, we consider the Tobit Kalman filtering problem in the presence of non-Gaussian Lévy and time-correlated measurement noises. Because of the infinite covariances and time correlation of the measurement noises, the direct utilization of the Tobit Kalman filtering method becomes infeasible, and there is a practical need to look for alternative methodologies. In this chapter, we first turn to the measurement differencing method, which makes it possible to transform the correlation of the measurement noises into the cross-correlation between the process noise and the equivalent measurement noise. Then, we use the Lévy-Ito theorem to transform the infinite measurement noise covariance into an unknown but finite covariance. After these two transformations, a modified TKF is then designed with its unknown noise covariance carefully calculated. Simulation experiments are conducted to illustrate the effectiveness of the proposed filtering algorithm.

4.1 PROBLEM FORMULATION

Consider the following class of discrete time-varying systems with time-correlated measurement noises:

$$x_{k+1} = A_k x_k + \omega_k, \tag{4.1}$$

$$y_k^* = C_k x_k + \upsilon_k, \tag{4.2}$$

$$\upsilon_{k+1} = H_k \upsilon_k + \varepsilon_k, \tag{4.3}$$

where $x_k \in \mathbb{R}^{n_x}$ and $y_k^* \in \mathbb{R}^{n_y}$ are, respectively, the state vector and the latent/uncensored measurement vector; $A_k \in \mathbb{R}^{n_x \times n_x}$, $C_k \in \mathbb{R}^{n_y \times n_x}$ and $H_k \in \mathbb{R}^{n_y \times n_y}$ are known matrices with appropriate dimensions; $\omega_k \in \mathbb{R}^{n_\omega}$, $\upsilon_k \in \mathbb{R}^{n_y}$ and $\varepsilon_k \in \mathbb{R}^{n_y}$ are, respectively, the Gaussian process noise, the time-correlated additive measurement noise and the non-Gaussian Lévy noise.

According to the Lévy-Ito theorem, the non-Gaussian Lévy noise ε_k in (6.3) can be approximated by a Gaussian white noise ε_k via clipping off its extremely large values. In practice, we usually do not know the exact value of the noise ε_k, and hence, we prefer to clip the latent measurement y_k^* instead of ε_k. Let $\acute{y}_k^*$ be the clipped latent measurement corresponding to y_k^* and $_k$ be the clipped Lévy noise (i.e. the approximated Gaussian noise) corresponding to ε_k. We now have the following clipped latent measurement model:

$$\begin{aligned}
\acute{y}_k^* &= C_k x_k + H_{k-1}\upsilon_{k-1} + \acute{\varepsilon}_{\ k-1} \\
&= \begin{cases} C_k \hat{x}_{k|k-1} + \mathcal{T} \cdot \mathrm{sign}(\aleph_k), & |\aleph_k| \geq \mathcal{T}, \\ y_k^*, & |\aleph_k| < \mathcal{T}, \end{cases}
\end{aligned} \tag{4.4}$$

where $\aleph_k = y_k^* - C_k \hat{x}_{k|k-1}$; $\hat{x}_{k|k-1}$ is the one-step prediction of the state x_{k-1}; $\mathcal{T} = \begin{bmatrix} \mathcal{T}^1 & \mathcal{T}^2 & \cdots & \mathcal{T}^{n_y} \end{bmatrix}^T \in \mathbb{R}^{n_y}$ is the clipping threshold vector of y_k^*; and $\mathcal{T}^m$ $(m = 1, 2, \ldots, n_y)$ is the constant clipping threshold of y_k^{*m}.

We are now in the position to introduce the Tobit measurement model as follows:

$$\acute{y}_k = \begin{cases} \acute{y}_k^*, & \acute{y}_k^* > \mathcal{I}, \\ \mathcal{I}, & \acute{y}_k^* \leq \mathcal{I}, \end{cases} \tag{4.5}$$

where $\acute{y}_k \in \mathbb{R}^{n_y}$ is the censored measurement vector, $\mathfrak{I} = \begin{bmatrix} \mathfrak{I}^1 & \mathfrak{I}^2 & \cdots & \mathfrak{I}^{n_y} \end{bmatrix}^T \in \mathbb{R}^{n_y}$ is the censoring threshold vector of $\acute{y}_k$ and $\mathcal{I}^m$ $(m = 1, 2, \ldots, n_y)$ is the constant censoring threshold of $\acute{y}_k^m$.

Depending on whether $\acute{y}_k$ is censored or not, it is assumed that (4.5) can be transformed into the following form:

$$\acute{y}_k = \Upsilon_k \acute{y}_k^* + (I - \Upsilon_k)\mathbf{I}, \tag{4.6}$$

where $\Upsilon_k = \mathrm{diag}\{\gamma_k^1, \gamma_k^2, \ldots, \gamma_k^{n_y}\}$ and γ_k^m $(m = 1, 2, \ldots, n_y)$ are known Bernoulli random variables regulating the censoring phenomena of $\acute{y}_k^m$ $(m = 1, 2, \ldots, n_y)$ with the following probability distribution:

$$\mathrm{Prob}\{\gamma_k^m = 1\} = \overline{\gamma}_k^m, \qquad \mathrm{Prob}\{\gamma_k^m = 0\} = 1 - \overline{\gamma}_k^m, \tag{4.7}$$

where $\overline{\gamma}_k^m$ $(m = 1, 2, \ldots, n_y)$ are known non-negative constants. It is assumed that all the random variables γ_k^m $(m = 1, 2, \ldots, n_y)$ are mutually independent and are also uncorrelated with other noise signals.

Assumption 4.1 ω_k is a zero-mean white noise sequence with covariance Q_k. ε_k is the symmetric α-stable Lévy noise with stability index μ and scale parameter σ, where $\mu = \begin{bmatrix} \mu^1 & \mu^2 & \cdots & \mu^{n_y} \end{bmatrix}^T$ and $\sigma = \begin{bmatrix} \sigma^1 & \sigma^2 & \cdots & \sigma^{n_y} \end{bmatrix}^T$ with μ^m and σ^m $(m = 1, 2, \ldots, n_y)$ being known constants.

Assumption 4.2 The initial state x_0 has the mean $\overline{x}_0$ and covariance P_0. x_0, ω_k and ε_k are mutually independent.

Remark 4.1 In (4.2), v_k stands for the time-correlated additive measurement noise, which is represented by the output of a linear recursive model with a non-Gaussian Lévy noise ε_{k-1} as shown in (4.3). In engineering practice, the time-correlation of sensor noises often stems from the fact that sensors work in a common time-correlated noisy environment or sensor noises are dependent on the system state. On the other hand, the non-Gaussian Lévy noise is better to model those heavy-tailed noises with infinite covariances (rather than the Gaussian noises), with examples including noises on telephone lines, noises in extremely low-frequency electromagnetic communication, tracking errors in code tracking loops and random-walk models of stock-market prices. The main challenge in handling the noise model given by (4.3)–(4.4) is that v_k is time-correlated not only with v_{k-1} but also with the non-Gaussian Lévy noise ε_{k-1} having an infinite covariance. In this case, the covariance of v_k does not exist, and hence, it is impossible to minimize the covariance of the estimation error to obtain the optimal estimate.

Remark 4.2 As described in (4.7), the random variables γ_k^m $(m = 1, 2, \ldots, n_y)$ are utilized to characterize the censoring phenomena of $\acute{y}_k^m$ $(m = 1, 2, \ldots, n_y)$. According to (4.6), if no censoring occurs for $\acute{y}_k^m$, i.e. $\gamma_k^m = 1$, the measurement output is $\acute{y}_k^m = \acute{y}_k^{*m}$, which implies that the output measurement is the same as the latent measurement. If the censoring phenomena occur for $\acute{y}_k^m$, i.e., $\gamma_k^m = 0$, the measurement output is $\acute{y}_k^m = \mathcal{I}^m$, which implies that the threshold is assigned to the output measurement. Here, the censoring probabilities $\overline{\gamma}_k^m$ $(m = 1, 2, \ldots, n_y)$ are assumed to be known a priori via some statistical experiments. Alternatively, $\overline{\gamma}_k^m$ can also be approximated by

$$\overline{\gamma}_k^m \approx \Phi\left(\frac{\sum_{n=1}^{n_x} C_k^{mn} \hat{x}_{k|k-1}^n - \mathcal{I}^m}{\sqrt{R_k^{mm}}}\right), \tag{4.8}$$

where C_k^{mn} $(n = 1, 2, \ldots, n_x)$ and R_k^{mm} are, respectively, the (m, n)th and (m, m)th elements of C_k and R_k; $\hat{x}_{k|k-1}^n$ is the nth element of $\hat{x}_{k|k-1}$, which is the one-step prediction of x_{k-1}; and $\Phi(\cdot)$ is the cumulative distribution function of the random variable "." which obeys the standard normal distribution.

The objective of this chapter is to design the TKF for the discrete time-varying system (4.1)–(4.6) subject to the censored measurements, the non-Gaussian Lévy noise and the time-correlated additive measurement noise.

4.2   A MODIFIED TOBIT KALMAN FILTER

In this section, the modified TKF for system (4.1)–(4.6) is designed. Due to the presence of the non-Gaussian Lévy noise and the time-correlated additive measurement noise, the measurement noise distribution is no longer the known Gaussian distribution. To solve this problem, the measurement differencing method and the Lévy-Ito theorem are used. Compared with the standard TKF, several new terms appear in the calculation of the filter gain matrix, state estimate as well as the state estimation covariance matrix. Fortunately, all these newly appeared terms can be computed offline or recursively and, therefore, won't add much computational burden to the filter design algorithm.

According to the measurement differencing method, we define an equivalent latent measurement $z_k^* = y_k^* - H_{k-1}y_{k-1}^*$ and transform system (4.1)–(4.3) into the following equivalent form:

$$x_{k+1} = A_k x_k + \omega_k, \tag{4.9}$$

$$z_k^* = \chi_k + \eta_k, \tag{4.10}$$

where

$$\chi_k = F_k x_k, F_k = C_k - \Gamma_{k-1},$$
$$\eta_k = \Gamma_{k-1}\omega_{k-1} + \varepsilon_{k-1}, \Gamma_{k-1} = H_{k-1}C_{k-1}A_{k-1}^{-1},$$

It can be observed from (4.9)–(4.10) that, with the adoption of the measurement differencing method, the original latent measurement y_k^* in (4.2) is transformed into the equivalent latent measurement z_k^* in (4.10). On the other hand, the time-correlated measurement noise υ_k in (4.2) is converted into an equivalent measurement noise η_{k-1}, which is the linear combination of the Gaussian noise ω_{k-1} and the Lévy noise ε_{k-1}.

On the basis of (4.9)–(4.10), the clipped latent measurement model, the Tobit measurement model and the equivalent Tobit measurement model corresponding to (4.4)–(4.6) are given as follows:

$$\check{z}_k^* = \chi_k + \eta_k$$
$$= \begin{cases} \hat{\chi}_{k|k-1} + \mathcal{T} \cdot \text{sign}(\Im_k), & |\Im_k| \geq \mathcal{T}, \\ z_k^*, & |\Im_k| < \mathcal{T}, \end{cases} \tag{4.11}$$

$$\acute{z}_k = \begin{cases} \check{z}_k^*, & \check{z}_k^* > \mathcal{I}, \\ \Im, & \check{z}_k^* \leq \mathcal{I}, \end{cases} \tag{4.12}$$

$$\acute{z}_k = \Upsilon_k \check{z}_k^* + (I - \Upsilon_k)\mathcal{I}, \tag{4.13}$$

where $\hat{\chi}_{k|k-1}$ is the predicted value of χ_k; $\Im_k = z_k^* - \hat{\chi}_{k|k-1}$ is the clipped measurement noise; $\acute{z}_k^*$ and $\acute{z}_k$ are, respectively, the clipped equivalent latent measurement and the clipped equivalent censored measurement corresponding to $\acute{y}_k^*$ and $\acute{y}_k$; $\acute{\eta}_k = \Gamma_{k-1}\omega_{k-1} + \acute{\varepsilon}_{k-1}$ is the equivalent clipped measurement noise and $\acute{\varepsilon}_{k-1}$ is the Gaussian noise; $\mathcal{T} = (I - H_{k-1})\Im$ is the equivalent clipping threshold vector corresponding to $\acute{z}_k^*$; $\mathcal{I} = (I - H_{k-1})\mathcal{J}$ is the equivalent censoring threshold vector corresponding to $\acute{z}_k$; $\mathcal{T} = \begin{bmatrix} \mathcal{T}^1 & \mathcal{T}^2 & \cdots & \mathcal{T}^{n_z} \end{bmatrix}^T \in \mathbb{R}^{n_z}$ where $\mathcal{T}^m$ $(m = 1,2,\ldots,n_y)$ is the constant clipping threshold of $\acute{z}_k^{*m}$; and $\mathcal{I} = \begin{bmatrix} \mathcal{I}^1 & \mathcal{I}^2 & \cdots & \mathcal{I}^{n_z} \end{bmatrix}^T \in \mathbb{R}^{n_z}$ where $\mathcal{I}^m$ $(m = 1,2,\ldots,n_y)$ is the constant censoring threshold of $\acute{z}_k^m$. Correspondingly, the censoring probability in (6.8) is converted to

$$\overline{\gamma}_k^m \approx \Phi\left(\frac{\hat{\chi}_{k|k-1}^m - \mathcal{I}^m}{\sqrt{R_k^{mm}}} \right), \tag{4.14}$$

Where $\hat{\chi}_{k|k-1}^m$ is the mth element of $\hat{\chi}_{k|k-1}$.

Remark 4.3 In (4.11)–(4.13), by approximating the Lévy noise ε_{k-1} with a Gaussian noise $\acute{\varepsilon}_{k-1}$, the noise η_{k-1} can be further transformed into a Gaussian noise $\acute{\eta}_{k-1}$. Hence, we can conclude that the distribution of the measurement noise $\acute{\eta}_{k-1}$ conditional on the clipped latent measurement $\acute{z}_k^*$ and the clipped measurement $\acute{z}_k$ are, respectively, Gaussian and censored Gaussian. This satisfies the assumption of the noise distribution in the standard TFK and results in the feasibility of deriving the TKF for system (4.9)–(4.13).

Remark 4.4 Equation (4.11) shows that if the absolute value of the clipped measurement noise $\Im_k$ is smaller than the prescribed threshold $\mathcal{T}$, the value of the clipped latent measurement $\acute{z}_k^*$ will be equal to the value of the latent measurement value z_k^* under the non-Gaussian Lévy noise. Otherwise, $\acute{z}_k^*$ will be evaluated by the value of $\hat{\chi}_{k|k-1}$ plus or minus the prescribed threshold $\mathcal{T}$. It should be noted that when adopting the measurement differencing method, the latent measurement y_k^* is transformed into the equivalent latent measurement z_k^*; correspondingly, the clipping threshold $\mathcal{T}$ and the censoring threshold $\Im$ should be, respectively, transformed into the equivalent clipping threshold $\mathcal{T} = (I - H_{k-1})\mathcal{T}$ and the equivalent censoring threshold $\mathcal{I} = (I - H_{k-1})\mathcal{I}$.

Denote $\quad R_k \triangleq \mathbb{E}\{\acute{\eta}_k\acute{\eta}_k^T\}$, $\quad S_k \triangleq \mathbb{E}\{\omega_{k-1}\acute{\eta}_k^T\}$ $\quad$ and $\quad \Xi_k \triangleq \mathbb{E}\{\varepsilon_k\varepsilon_k^T\}$. $\quad$ Since $\acute{\eta}_k = \Gamma_{k-1}\omega_{k-1} + \acute{\varepsilon}_{k-1}$, we have

$$R_k = \Gamma_{k-1}Q_{k-1}\Gamma_{k-1}^T + \Xi_{k-1}, \mathrm{X} \tag{4.15}$$

$$S_k = Q_{k-1}\Gamma_{k-1}^T. \tag{4.16}$$

Remark 4.5 It can be found from this analysis that compared with the original measurement noise υ_k in (4.2)–(4.3), the equivalent clipped measurement noise $\acute{\eta}_k$ in (4.11) is no longer time-correlated but correlated with ω_{k-1} as shown in (4.16). It can be also concluded from (4.15) that the computation of R_k requires Q_{k-1} and Ξ_{k-1} where Q_{k-1} is the known process noise covariance and Ξ_{k-1} is the unknown covariance of the clipped Lévy noise ε_{k-1}. Since ε_{k-1} is a Gaussian noise, we can compute Ξ_{k-1} according to the definition of the Gaussian noise covariance. As a result, we are able to obtain the measurement noise covariance R_k, which is essential in the following filter design.

To facilitate the subsequent filter design, the Tobit regression for system (4.9)–(4.13) is first given in the following lemma.

Lemma 4.1 The mathematical expectation and variance of z_k are derived as follows:

$$\mathbb{E}\{\acute{z}_k \mid x_k, R_k\} = \Phi\left(\frac{\chi_k - \mathcal{I}}{R_k}\right)\left[F_k x_k + \lambda\left(\frac{\mathcal{I} - \chi_k}{R_k}\right)R_k\right] + \Phi\left(\frac{\mathcal{I} - \chi_k}{R_k}\right)\mathcal{I}, \quad (4.17)$$

$$Var\{\acute{z}_k \mid x_k, R_k\} = R_k\left[1 - \varphi\left(\frac{\mathcal{I} - \chi_k}{R_k}\right)\right], \quad (4.18)$$

where

$$\varphi\left(\frac{\mathcal{I} - \chi_k}{R_k}\right) = diag\left\{\varphi\left(\frac{\mathcal{I}^m - \chi_k^m}{\sqrt{R_k^{mm}}}\right)\right\}, \quad (4.19)$$

$$\lambda\left(\frac{\mathcal{I} - \chi_k}{R_k}\right) = diag\left\{\lambda\left(\frac{\mathcal{I}^m - \chi_k^m}{\sqrt{R_k^{mm}}}\right)\right\}, \quad (4.20)$$

$$\Phi\left(\frac{\acute{z}_k - \chi_k}{R_k}\right) = diag\left\{\Phi\left(\frac{\acute{z}_k^m - \chi_k^m}{\sqrt{R_k^{mm}}}\right)\right\}, \quad (4.21)$$

with

$$\varphi\left(\frac{\mathcal{I}^m - \chi_k^m}{\sqrt{R_k^{mm}}}\right) = \lambda\left(\frac{\mathcal{I}^m - \chi_k^m}{\sqrt{R_k^{mm}}}\right)\left[\lambda\left(\frac{\mathcal{I}^m - \chi_k^m}{\sqrt{R_k^{mm}}}\right) - \frac{\mathcal{I}^m - \chi_k^m}{\sqrt{R_k^{mm}}}\right], \quad (4.22)$$

$$\lambda\left(\frac{\mathcal{I}^m - \chi_k^m}{\sqrt{R_k^{mm}}}\right) = \frac{\phi\left(\dfrac{\mathcal{I}^m - \chi_k^m}{\sqrt{R_k^{mm}}}\right)}{1 - \Phi\left(\dfrac{\mathcal{I}^m - \chi_k^m}{\sqrt{R_k^{mm}}}\right)}. \quad (4.23)$$

Here, $\phi\left(\dfrac{z_k^m - \chi_k^m}{\sqrt{R_k^{mm}}}\right)$ and $\Phi\left(\dfrac{z_k^m - \chi_k^m}{\sqrt{R_k^{mm}}}\right)$ $(m = 1, 2, \ldots, n_y)$ are, respectively, the probability density function (pdf) and the cumulative distribution function (CDF) of the

Gaussian random variable $\acute{z}_k^m$ with mean χ_k^m and covariance R_k^{mm}. The pdf and CDF of $\acute{z}_k^m$ have the following form:

$$\phi\left(\frac{\acute{z}_k^m - \chi_k^m}{\sqrt{R_k^{mm}}}\right) = \frac{1}{\sqrt{2\pi}} e^{-\frac{\left(\acute{z}_k^m - \chi_k^m\right)^2}{2R_k^{mm}}}, \tag{4.24}$$

$$\Phi\left(\frac{\acute{z}_k^m - \chi_k^m}{\sqrt{R_k^{mm}}}\right) = \int_{-\infty}^{\acute{z}_k^m} \frac{1}{\sqrt{2\pi R_k^{mm}}} e^{-\frac{\left(\xi_k^m - \chi_k^m\right)^2}{2R_k^{mm}}} d_{\xi_k^m}. \tag{4.25}$$

Proof According to (4.13), the probability distribution of $\acute{z}_k^m$ with a normally distributed noise ε_k^m is as follows:

$$f(\acute{z}_k^m \mid x_k) = \frac{1}{\sqrt{R_k^{mm}}} \phi\left(\frac{\acute{z}_k^m - \chi_k^m}{\sqrt{R_k^{mm}}}\right) u(\acute{z}_k^m - \mathcal{I}^m) + 1_{\{\mathcal{I}^m - \acute{z}_k^m\}} \Phi\left(\frac{\mathcal{I}^m - \chi_k^m}{\sqrt{R_k^{mm}}}\right), \tag{4.26}$$

where $u(\acute{z}_k^m - \mathcal{I}^m)$ is the unit step function, $1_{\{\mathcal{I}^m - \acute{z}_k^m\}}$ is the Dirac delta function, $\phi\left(\frac{\acute{z}_k^m - \chi_k^m}{\sqrt{R_k^{mm}}}\right)$ and $\Phi\left(\frac{\acute{z}_k^m - \chi_k^m}{\sqrt{R_k^{mm}}}\right)$ are, respectively, given by (4.32) and (4.33). Based on (4.20), we have the mathematical expectation and variance of $\acute{z}_k^m$ as follows:

$$\mathbb{E}\{\acute{z}_k^m \mid x_k, R_k^{mm}\} = Prob\{\acute{z}_k^m > \mathcal{I}^m \mid x_k, R_k^{mm}\} \mathbb{E}\{\acute{z}_k^m \mid \acute{z}_k^m > \mathcal{I}^m, x_k, R_k^{mm}\}$$
$$+ Prob\{\acute{z}_k^m = \mathcal{I}^m \mid x_k, R_k^{mm}\} \mathbb{E}\{\acute{z}_k^m \mid \acute{z}_k^m = \mathcal{I}^m, x_k, R_k^{mm}\}. \tag{4.27}$$

It is known from (4.21) that to compute $\mathbb{E}\{\acute{z}_k^m | x_k, R_k^{mm}\}$*, we need to calculate the probabilities and expected values in (4.21).*

$$\begin{aligned} &Prob\{\acute{z}_k^m > \mathcal{I}^m | x_k, R_k^{mm}\} \\ &= Prob\{z_k^{*m} > \mathcal{I}^m | x_k, R_k^{mm}\} \\ &= Prob\{\chi_k^m + \upsilon_k^m > \mathcal{I}^m | x_k, R_k^{mm}\} \\ &= Prob\{\upsilon_k^m > \mathcal{I}^m - \chi_k^m | x_k, R_k^{mm}\} \\ &= \Phi\left(\frac{\chi_k^m - \mathcal{I}^m}{\sqrt{R_k^{mm}}}\right). \end{aligned} \tag{4.28}$$

Hence, we can easily obtain that

$$\mathbb{E}\{\acute{z}_k^m \mid z_k^m > \mathcal{I}^{\ m}, x_k, R_k^{mm}\}$$

$$= \frac{1}{\sqrt{R_k^{m,m}}} \int_{\mathcal{I}^{\ m}}^{+\infty} \xi_k^m \frac{\Phi\left(\dfrac{\xi_k^m - \chi_k^m}{\sqrt{R_k^{m,m}}}\right)}{1 - \Phi\left(\dfrac{\mathcal{I}^{\ m} - \chi_k^m}{\sqrt{R_k^{m,m}}}\right)} d_{\xi_k^m} \tag{4.29}$$

$$= \sum_{n=1}^{n_x} \acute{z}_k^{mn} x_k^n + \sqrt{R_k^{m,m}} \lambda\left(\frac{\mathcal{I}^{\ m} - \chi_k^m}{\sqrt{R_k^{m,m}}}\right),$$

where $\lambda\left(\dfrac{\mathcal{I}^{\ m} - \chi_k^m}{\sqrt{R_k^{mm}}}\right)$ is given by (4.31).

Similar to (4.22)–(4.23), we can obtain

$$Prob\{\acute{z}_k^m = \mathcal{I}^{\ m} \mid x_k, R_k^{mm}\} = \Phi\left(\frac{\mathcal{I}^{\ m} - \chi_k^m}{\sqrt{R_k^{mm}}}\right), \tag{4.30}$$

$$\mathbb{E}\{\acute{z}_k^m \mid \acute{z}_k^m = \mathcal{I}^{\ m}, x_k, R_k^{mm}\} = \mathcal{I}^{\ m}. \tag{4.31}$$

Putting (4.22)–(4.38) into (4.21) results in

$$\mathbb{E}\{\acute{z}_k^m \mid x_k, R_k^{mm}\} = \Phi\left(\frac{\mathcal{I}^{\ m} - \chi_k^m}{\sqrt{R_k^{mm}}}\right) \mathcal{I}^{\ m}$$

$$+ \Phi\left(\frac{\chi_k^m - \mathcal{I}^{\ m}}{\sqrt{R_k^{mm}}}\right)\left[\chi_k^m + \sqrt{R_k^{mm}} \lambda\left(\frac{\mathcal{I}^{\ m} - \chi_k^m}{\sqrt{R_k^{mm}}}\right)\right]. \tag{4.32}$$

From (4.20), we know that $Var\{z_k'^m \mid x_k, z_k'^m < \mathcal{I}^{\ m}, R_k^{mm}\} = 0$, and hence we have

$$Var\{\acute{z}_k^m \mid x_k, R_k^{mm}\} = Var\{\acute{z}_k^m \mid x_k, \acute{z}_k^m > \mathcal{I}^{\ m}, R_k^{mm}\}$$

$$= \mathbb{E}\{(\acute{z}_k^m)^2 \mid \acute{z}_k^m > \mathcal{I}^{\ m}, x_k, R_k^{mm}\}$$

$$- \left(\mathbb{E}\{\acute{z}_k^m \mid \acute{z}_k^m > \mathcal{I}^{\ m}, x_k, R_k^{mm}\}\right)^2 \tag{4.33}$$

$$= R_k^{mm}\left[1 - \varphi\left(\frac{\mathcal{I}^{\ m} - \chi_k^m}{\sqrt{R_k^{mm}}}\right)\right],$$

where $\varphi\left(\dfrac{\mathcal{I}^{\ m} - \chi_k^m}{\sqrt{R_k^{mm}}}\right)$ is given by (4.30).

For simplicity of the deduction, we assume that $cov\{\acute{z}_k^m, \acute{z}_k^t\} = 0$ for $m \neq t$, $m, t = 1, 2, \ldots, n_y$. The generalization of the result to the case where $cov\{\acute{z}_k^m, \acute{z}_k^t\} \neq 0$ for $m \neq t$ is straightforward but will bring notational burdens. Therefore, we have

$$\mathbb{E}\{\acute{z}_k \mid x_k, R_k\}$$
$$= col\{\mathbb{E}\{\acute{z}_k^1 \mid x_k, R_k^{11}\}, \mathbb{E}\{\acute{z}_k^2 \mid x_k, R_k^{22}\}, \cdots, \mathbb{E}\{\acute{z}_k^{n_y} \mid x_k, R_k^{n_y n_y}\}\}, \tag{4.34}$$

$$Var\{\acute{z}_k \mid x_k, R_k\}$$
$$= diag\{Var\{\acute{z}_k^1 \mid x_k, R_k^{11}\}, Var\{\acute{z}_k^2 \mid x_k, R_k^{22}\}, \cdots, Var\{\acute{z}_k^{n_y} \mid x_k, R_k^{n_y n_y}\}\}. \tag{4.35}$$

Combining (4.39)–(4.47) yields (4.25)–(4.26) where $\varphi\left(\dfrac{\mathcal{I} - \chi_k}{R_k}\right)$, $\lambda\left(\dfrac{\mathcal{I} - \chi_k}{R_k}\right)$ and $\Phi\left(\dfrac{\acute{z}_k - \chi_k}{R_k}\right)$ are given by (4.27)–(4.29). The proof is now complete.

Remark 4.6 Compared with the expectation and variance expressions in the standard TKF where only the censoring phenomenon is considered, it can be observed from Lemma 4.1 that, our obtained expectation and variance of the censored measurement $\acute{z}_k$ are different in two aspects due to the introduction of the non-Gaussian Lévy noise and the time-correlated measurement noise. Firstly, the computation of the expectation and variance of $\acute{z}_k$ (which is conditioned on the originally known Gaussian measurement noise covariance) is now conditioned on the unknown noise covariance R_k which is dependent on the known process noise covariance Q_{k-1} and the unknown clipped Lévy noise covariance Ξ_{k-1}. Secondly, the term $C_k x_k$ (which is the product of the measurement matrix C_k and the state x_k in the standard TKF) is now replaced by the term χ_k. The first difference is caused by the non-Gaussian Lévy noise, while the second difference arises from the time-correlated additive measurement noise.

Let $\hat{z}_{k|k-1} \triangleq \mathbb{E}\{\acute{z}_k \mid \acute{z}_{1:k}\}$, $\hat{x}_{k|k} \triangleq \mathbb{E}\{x_k \mid \acute{z}_{1:k}\}$, $\tilde{x}_{k|k-1} \triangleq x_k - \hat{x}_{k|k-1}$, $\tilde{x}_{k|k} \triangleq x_k - \hat{x}_{k|k}$, $\tilde{z}_{k|k-1} \triangleq \acute{z}_k - \hat{z}_{k|k-1}$, $P_{k|k-1} \triangleq \mathbb{E}\{\tilde{x}_{k|k-1}\tilde{x}_{k|k-1}^T\}$, $P_{k|k} \triangleq \mathbb{E}\{\tilde{x}_{k|k}\tilde{x}_{k|k}^T\}$, $\Sigma_{\tilde{z}_{k|k-1}\tilde{z}_{k|k-1}} \triangleq \mathbb{E}\{\tilde{z}_{k|k-1}\tilde{z}_{k|k-1}^T\}$ and $\Sigma_{\tilde{x}_{k|k-1}\tilde{z}_{k|k-1}} \triangleq \mathbb{E}\{\tilde{x}_{k|k-1}\tilde{z}_{k|k-1}^T\}$. Based on Lemma 6.1, the TKF with both non-Gaussian Lévy noises and time-correlated additive measurement noises can be derived using the following theorem.

Theorem 4.1 Given system (4.1)–(4.6), its optimal TKF is given by

$$\hat{x}_{k|k-1} = A_{k-1}\hat{x}_{k-1|k-1}, \tag{4.36}$$

$$P_{k|k-1} = A_{k-1}P_{k-1|k-1}A_{k-1}^T + Q_{k-1}, \tag{4.37}$$

$$\hat{x}_{k|k} = \hat{x}_{k|k-1} + K_k(\acute{z}_k - \hat{z}_{k|k-1}), \tag{4.38}$$

$$P_{k|k} = P_{k|k-1} - \Sigma_{\tilde{x}_{k|k-1}\tilde{z}_{k|k-1}} \Sigma_{\tilde{z}_{k|k-1}\tilde{z}_{k|k-1}}^{-1} \Sigma_{\tilde{x}_{k|k-1}\tilde{z}_{k|k-1}}^T. \tag{4.39}$$

The one-step measurement prediction $\hat{z}_{k|k-1}$ in (4.36) is given by

$$\hat{z}_{k|k-1} = \bar{\Upsilon}_k\left[\hat{\chi}_{k|k-1} + \lambda\left(\dfrac{\mathcal{I} - \hat{\chi}_{k|k-1}}{R_k}\right)R_k\right] + (I - \bar{\Upsilon}_k)\mathcal{I}, \tag{4.40}$$

where

$$\lambda\left(\frac{\mathcal{I}-\hat{\mathcal{X}}_{k|k-1}}{R_k}\right)=diag\left\{\lambda\left(\frac{\mathcal{I}^m-\hat{\mathcal{X}}_{k|k-1}^m}{\sqrt{R_k^{mm}}}\right)\right\}, \tag{4.41}$$

and $\mathcal{R}_k=col\{\sqrt{R_k^{11}},\sqrt{R_k^{22}},\ldots,\sqrt{R_k^{n_y n_y}}\}$ with R_k^{mm} $(m=1,2,\ldots,n_y)$ being the (m,m)th element of R_k. The gain matrix K_k in (4.36) has the following form:

$$K_k=\Sigma_{\tilde{x}_{k|k-1}\tilde{z}_{k|k-1}}\Sigma^{-1}_{\tilde{z}_{k|k-1}\tilde{z}_{k|k-1}}, \tag{4.42}$$

where

$$\Sigma_{\tilde{x}_{k|k-1}\tilde{z}_{k|k-1}}=P_{k|k-1}(\overline{\Upsilon}_k F_k)^T+S_k\overline{\Upsilon}_k^T, \tag{4.43}$$

$$\begin{aligned}\Sigma_{\tilde{y}_{k|k-1}\tilde{y}_{k|k-1}}&=\overline{\Upsilon}_k F_k P_{k|k-1}(\overline{\Upsilon}_k F_k)^T+\overline{\Upsilon}_k F_k S_k\overline{\Upsilon}_k^T\\&\quad+(\overline{\Upsilon}_k F_k S_k\overline{\Upsilon}_k^T)^T+R_k\left[I-\varphi\left(\frac{\mathcal{I}-\hat{\mathcal{X}}_{k|k-1}}{R_k}\right)\right],\end{aligned} \tag{4.44}$$

$$\varphi\left(\frac{\mathcal{I}-\hat{\mathcal{X}}_{k|k-1}}{R_k}\right)=diag\left\{\varphi\left(\frac{\mathcal{I}^m-\hat{\mathcal{X}}_{k|k-1}^m}{\sqrt{R_k^{mm}}}\right)\right\}, \tag{4.45}$$

$$R_k=\Gamma_{k-1}S_k+\Xi_{k-1}, \tag{4.46}$$

$$\Xi_{k-1}=(\acute{z}_k^*-\hat{\mathcal{X}}_{k|k-1})(\acute{z}_k^*-\hat{\mathcal{X}}_{k|k-1})^T+F_k P_{k|k}F_k^T. \tag{4.47}$$

Proof Noticing the definitions of $\hat{x}_{k|k-1}$, $P_{k|k-1}$, $\hat{x}_{k|k}$ and $P_{k|k}$, we can easily obtain (4.34)–(4.37) and (4.43) by applying the orthogonality projection principle to system (4.9)–(4.13). According to (4.25), we have

$$\hat{z}_{k|k-1}=\overline{\Upsilon}_k\left[\hat{\mathcal{X}}_{k|k-1}+\lambda\left(\frac{\mathcal{I}-\hat{\mathcal{X}}_{k|k-1}}{R_k}\right)\mathcal{R}_k\right]+(I-\overline{\Upsilon}_k)\mathcal{I}, \tag{4.48}$$

which is exactly the same as (4.41) where $\lambda\left(\dfrac{\mathcal{I}-\hat{\mathcal{X}}_{k|k-1}}{R_k}\right)$ given by (4.42). Subtracting (4.41) from (4.13) yields

$$\tilde{z}_{k|k-1}=\Upsilon_k\left[\tilde{\mathcal{X}}_{k|k-1}+\left(\acute{\eta}_k-\lambda\left(\frac{\mathcal{I}-\hat{\mathcal{X}}_{k|k-1}}{R_k}\right)\mathcal{R}_k\right)\right], \tag{4.49}$$

where $\tilde{\mathcal{X}}_{k|k-1}=z_k\tilde{x}_{k|k-1}$. Substituting (4.48) into the definition of $\Sigma_{\tilde{x}_{k|k-1}\tilde{z}_{k|k-1}}$, we have

$$\begin{aligned}\Sigma_{\tilde{x}_{k|k-1}\tilde{z}_{k|k-1}}&=\mathbb{E}\{\tilde{x}_{k|k-1}\tilde{z}_{k|k-1}^T\}\\&=\mathbb{E}\left\{\tilde{x}_{k|k-1}\left(\tilde{\mathcal{X}}_{k|k-1}+\left(\eta_k-\lambda\left(\frac{\mathcal{I}-\hat{\mathcal{X}}_{k|k-1}}{R_k}\right)\mathcal{R}_k\right)\right)^T\Upsilon_k^T\right\}\\&=P_{k|k-1}(\overline{\Upsilon}_k z_k)^T+S_k\overline{\Upsilon}_k^T,\end{aligned} \tag{4.50}$$

where the third equality holds from the fact that $\tilde{x}_{k|k-1} = A_{k-1}\tilde{x}_{k-1|k-1} + \omega_{k-1}$ is correlated with $\acute{\eta}_k = \Gamma_{k-1}\omega_{k-1} + \acute{\varepsilon}_{k-1}$. (4.50) is exactly the same as (4.45).

Substituting (4.49) into the definition of $\Sigma_{\tilde{z}_{k|k-1}\tilde{z}_{k|k-1}}$, we have

$$
\begin{aligned}
\Sigma_{\tilde{z}_{k|k-1}\tilde{z}_{k|k-1}} &= \mathbb{E}\{\tilde{z}_{k|k-1}\tilde{z}_{k|k-1}^T\} \\
&= \overline{\Upsilon}_k F_k P_{k|k-1}(\overline{\Upsilon}_k F_k)^T + \overline{\Upsilon}_k F_k S_k \overline{\Upsilon}_k^T + (\overline{\Upsilon}_k F_k S_k \overline{\Upsilon}_k^T)^T \\
&\quad + \mathbb{E}\left\{\Upsilon_k\left(\acute{\eta}_k - \lambda\left(\frac{\mathcal{I} - \hat{\chi}_{k|k-1}}{R_k}\right)\mathcal{R}_k\right)\left(\acute{\eta}_k - \lambda\left(\frac{\mathcal{I} - \hat{\chi}_{k|k-1}}{R_k}\right)\mathcal{R}_k\right)^T \Upsilon_k^T\right\},
\end{aligned}
\tag{4.51}
$$

where the last equality holds from the fact that $\tilde{x}_{k|k-1} = A_{k-1}\tilde{x}_{k-1|k-1} + \omega_{k-1}$ is orthogonal to $\hat{x}_{k|k-1} = A_{k-1}\hat{x}_{k-1|k-1}$ and $\tilde{x}_{k|k-1} = A_{k-1}\tilde{x}_{k-1|k-1} + \omega_{k-1}$ is correlated with $\acute{\eta}_k = \Gamma_{k-1}\omega_{k-1} + \acute{\varepsilon}_{k-1}$. Recalling the definition of the variance of z_k and noting (4.26), we have

$$
\begin{aligned}
\mathbb{E}&\left\{\Upsilon_k\left(\acute{\eta}_k - \lambda\left(\frac{\mathcal{I} - \hat{\chi}_{k|k-1}}{R_k}\right)\mathcal{R}_k\right)\left(\acute{\eta}_k - \lambda\left(\frac{\mathcal{I} - \hat{\chi}_{k|k-1}}{R_k}\right)\mathcal{R}_k\right)^T \Upsilon_k^T\right\} \\
&= Var\{\acute{z}_k \mid x_k, R_k\} \\
&= R_{k-1}\left[I - \varphi\left(\frac{\mathcal{I} - \hat{\chi}_{k|k-1}}{R_k}\right)\right],
\end{aligned}
\tag{4.52}
$$

where the second equality holds from the fact that $cov\{\acute{z}_k^m, \acute{z}_k^t\} = 0$ for $m \neq t$; and $\varphi\left(\dfrac{\mathcal{I} - \hat{\chi}_{k|k-1}}{R_k}\right)$ is given by (4.17). Substituting (4.52) into (4.51) results in (4.45).

Finally, we arrive at the computation of R_k in (4.15), which needs to solve the computation of Ξ_{k-1}. It follows from (4.11) that $\acute{\varepsilon}_{k-1} = \acute{z}_k^* - \Phi_k x_k$, and hence we have

$$
\begin{aligned}
\Xi_{k-1} &= \mathbb{E}\{\acute{\varepsilon}_{k-1}\acute{\varepsilon}_{k-1}^T\} \\
&= \mathbb{E}\{(\acute{z}_k^* - \chi_k)(\acute{z}_k^* - \chi_k)^T\} \\
&= \mathbb{E}\{(\acute{z}_k^* - \tilde{\chi}_{k|k-1} - \hat{\chi}_{k|k-1})(\acute{z}_k^* - \tilde{\chi}_{k|k-1} - \hat{\chi}_{k|k-1})^T\} \\
&= (\acute{z}_k^* - \hat{\chi}_{k|k-1})(\acute{z}_k^* - \hat{\chi}_{k|k-1})^T + F_k P_{k|k-1} F_k^T,
\end{aligned}
\tag{4.53}
$$

where the fourth equality holds from the fact that $\tilde{x}_{k|k-1}$ is orthogonal to $\hat{x}_{k|k-1}$, and z_k^* and $\hat{x}_{k|k-1}$ are known values. (4.53) is exactly the same as (4.19) and (4.18) is the same as (4.15).

Combining Lemma 4.1 and Theorem 4.1, we have accomplished the design of the optimal TKF with the censored measurements, the non-Gaussian Lévy noise and the time-correlated measurement noise. Table 4.1 demonstrates the implementation procedure of the proposed filtering algorithm.

Remark 4.7 The main difficulty in designing the TKF for systems with non-Gaussian Lévy noises and time-correlated measurement noises lies in the handling of infinite covariances of the Lévy noises and the time-correlation of the

---

TABLE 4.1

Implementation Procedure of the Proposed Filtering Algorithm

Algorithm 1: The proposed modified TKF

Input: $\bar{x}_0$ and P_0

Output: $\hat{x}_{k|k}$ and $P_{k|k}$

for $k = 1, 2, \ldots$, do

    Compute the predicted estimate $\hat{x}_{k|k-1}$ and its error covariance

    $P_{k|k-1}$ by (4.34) and (4.35), respectively;

    Compute the measurement noise covariance R_k by (4.18)–(4.19);

    Compute the predicted measurement $\hat{z}_{k|k-1}$ by (4.41)–(4.42);

    Compute the predicted covariance $\pounds_{\tilde{x}_{k|k-1}\tilde{z}_{k|k-1}}$ by (4.44);

    Compute the predicted covariance $\pounds_{\tilde{z}_{k|k-1}\tilde{z}_{k|k-1}}$ by (4.45)–(4.17);

    Compute the gain matrix K_k by (4.43);

    Compute the updated estimate $\hat{x}_{k|k}$ and error covariance $P_{k|k}$

    by (4.36) and (4.37), respectively;

    Return $\hat{x}_{k|k}$ and $P_{k|k}$;

end

---

measurement noises. By resorting to the measurement differencing method and the Lévy-Ito theorem, the non-Gaussian Lévy noise ε_k is approximated by a Gaussian noise $\acute{\varepsilon}_k$, while the time-correlation of the measurement noises υ_k is replaced by the cross-correlation between the equivalent measurement noise $\acute{\eta}_k$ and the process noise ω_k. In this case, the equivalent measurement noise $\acute{\eta}_k = \Gamma_{k-1}\omega_{k-1} + \acute{\varepsilon}_{k-1}$, which is the linear combination of the Gaussian process noise ω_{k-1} and the approximated Gaussian noise $\acute{\varepsilon}_{k-1}$, becomes a Gaussian measurement noise. Hence, the noise distributions of $\acute{\eta}_k$ conditional on the latent measurement $\acute{z}_k^*$ and the censored measurement $\acute{z}_k$ are, respectively, Gaussian and censored Gaussian. This satisfies the assumption the standard TKF and makes the derivation of the modified TKF presented in Theorem 4.1 possible.

Remark 4.8 Compared with the standard TKF, one can observe that our established TKF in Theorem 4.1 has two distinct differences. The first one is the appearance of S_k (which is the cross-covariance of the process noise and the equivalent measurement noise) in the calculation of the gain matrix, the state estimate as well as the state estimation error covariance. The second one is the appearance of the clipped Lévy noise covariance Ξ_{k-1} in the calculation of the unknown equivalent measurement noise covariance R_k. The two differences are, respectively, caused by the time-correlated measurement noise and the non-Gaussian Lévy noise. To be specific, the second term in (4.44), the second and third terms in (4.45) and the first term in (4.18) are originated from the time-correlated measurement noise. The appearance of (4.18)–(4.19) stems from the non-Gaussian Lévy noise. In addition, one can find that these terms can all be computed off-line or recursively.

Remark 4.9 It can be seen from Lemma 4.1 and Theorem 4.1 that when implementing the presented Tobit Kalman filtering algorithm with non-Gaussian Lévy and time-correlated sensor noises, the computational complexity mainly involves the matrix multiplication and matrix inversion. In the matrix multiplication, the multiplication of matrices with dimensions $n_x \times n_x$ needs to be calculated, where n_x is the dimension of the system state. As to the matrix inversion, the inverse of matrices with dimensions $n_y \times n_y$ needs to be calculated, where n_y is the dimension of the sensor measurement. Hence, the proposed filtering algorithm has the computational complexity of $\mathcal{O}[(n_x)^3 + (n_y)^3]$.

4.3 AN ILLUSTRATIVE EXAMPLE

In this section, we will use the oscillator example to demonstrate the effectiveness of the proposed filter design approach and the filtering performance. Consider the following class of discrete time-varying systems given by (4.1)–(4.6) with the following parameters:

$$A_k = \begin{bmatrix} \cos(wT) & -\sin(wT) \\ \sin(wT) & \cos(wT) \end{bmatrix}, C_k = \begin{bmatrix} 1 & 0 \end{bmatrix}, H_k = 1,$$

$$\bar{x}_0 = \begin{bmatrix} 5 & 0 \end{bmatrix}^T, P_0 = I_2, Q_k = 0.0025 I_2, w = 0.052\pi,$$

$$\alpha = 1, \mu = 1.3, \sigma = 10, T = 1, \mathcal{I} = 0, \mathcal{T} = 40, r = 1,$$

where r is the known covariance of the measurement noise υ_k in the standard TKF. In the simulation, we compare the traditional TKF with the proposed TKF with Lévy and time-correlated additive measurement noises (named TKF-LV-TC). For the traditional TKF, the measurement noise υ_k is time-uncorrelated with a known covariance $r = 1$. Nevertheless, in our proposed TKF-LV-TC, the measurement noise υ_k is time-correlated as described by (4.3), and its covariance R_k is unknown, which needs to be recursively computed using (4.18).

One thousand times of Monte Carlo simulations are conducted, and the result of the performance comparison is shown in Figures 4.1–4.2. Figure 4.1 illustrates the true values of the states and the estimates given by the two methods, while Figure 4.2 shows the comparison of the RMSE (root mean-squared error) curves between the two methods. It can be observed from Figure 4.1 that our proposed TKF-LV-TC is able to track the true values of the system states, while the standard TKF deviates from the true state values. Figure 4.2 indicates that the RMSE of our TKF-LV-TC is always smaller than that of the TKF. This is because that in our TKF-LV-TC, the non-Gaussian Lévy and time-correlated measurement noises are handled through computing the unknown noise covariance and using the measurement differencing

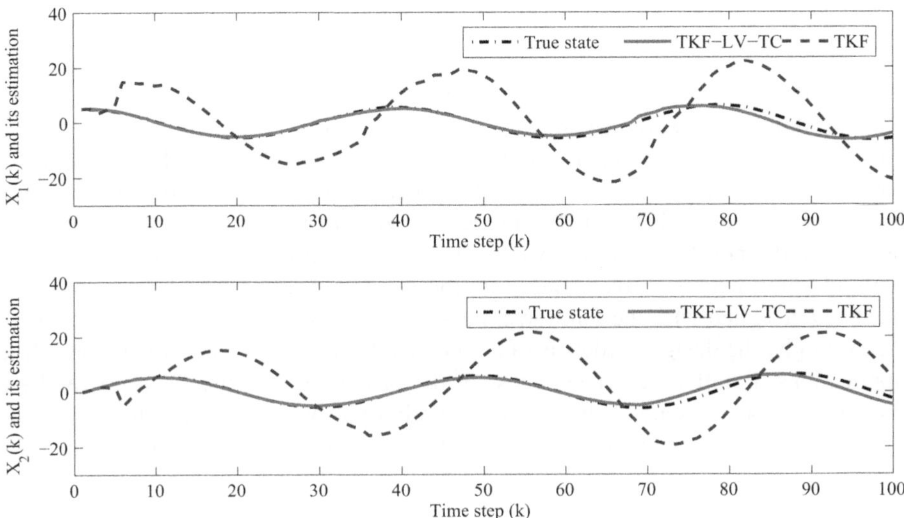

FIGURE 4.1 True values of the first and second dimensions of the state and the corresponding estimates.

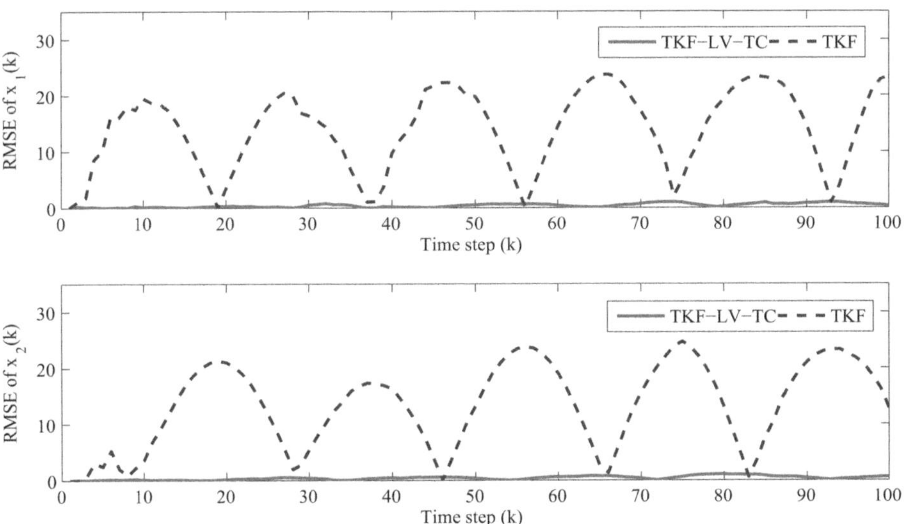

FIGURE 4.2 Method comparison with [35] in RMSE.

method, while they are still regarded as time uncorrelated Gaussian noises with known covariances in the TKF.

4.4 SUMMARY

In this chapter, we have investigated the Tobit Kalman filtering problem for discrete time-varying systems in the presence of non-Gaussian Lévy and time-correlated additive measurement noises. The problem has been addressed by taking advantage of the measurement differencing method and the Lévy-Ito theorem. As a result, the time correlation of the additive measurement noises is transformed into the cross-correlation between the process noise and the equivalent measurement noise, and the non-Gaussian Lévy noise is transformed into a Gaussian white noise with an unknown covariance. Based on the two transformations, the optimal TKF has been designed with the unknown measurement covariance carefully calculated. Accordingly, the designed filter is of a recursive form and thus is suitable for online applications. Finally, the effectiveness of the proposed filter has been proven by a numerical example. One of the future research topics would be the extension of the main results to more complex systems with different filtering performance indices.

5 Protocol-Based Filter Design under Integral Measurements and Probabilistic Sensor Failures

The Censored Data Case

In the context of state estimation, the overwhelming majority of the existing work has implicitly assumed that the sensor observation relies only on the *current* system state. This assumption is, unfortunately, sometimes unreasonable in certain applications such as chemical reactions, nuclear fusion and synchrotron radiation. In these applications, the sensor observation is actually proportional to so-called *integral measurements* [170], that is, the integral of system states within a prescribed time interval due presumably to delayed data acquisition and analysis. Another underlying assumption behind the conventional estimation schemes is that the sensors are equipped with the capability to provide accurate observations to the designed estimator. This assumption, however, does not always hold in reality. More often than not, sensors working in real-world circumstances are ineluctably confronted with all sorts of failures due mainly to abrupt environment changes, unexpected exogenous disturbances, internal component ageing and so forth.

In the past few decades, the networked system has gained a surge of research attention due largely to its broad applications in industrial fabrication, environmental monitoring and target localization. In an ideal situation, *all* system components (e.g. actuators, controllers, filters and sensors) are supposed to have privileges for information propagation via shared communication networks. This supposition, however, is often impractical, as limited-bandwidth-induced data collisions are likely to happen when the information exchanges take place simultaneously by more than one component. In this respect, communication protocols have been leveraged to orchestrate the transmission sequence of system components by giving the transmission permission to a single component at each time instant so as to avoid possible data collisions. Among the various communication protocols, the Round-Robin protocol (RRP) has drawn particular research attention because of its succinct execution manner, where the information propagation among system components is conducted in a *fixed circular* order.

DOI: 10.1201/9781003461623-5

To conclude these discussions, we make the following observations: (1) although much work has been done hitherto on the design of TKFs under different circumstances, there has been a lack of analysis results on the corresponding filtering performance within a holistic Tobit Kalman filtering framework; (2) due to unreliable working conditions and delayed data acquisition and analysis, sensors are inclined to experience probabilistic failures and integral measurements, and the negligence of such phenomena would result in deteriorated filtering performances; (3) communication protocols have proven to be beneficial in boosting transmission scheduling and circumventing data collisions, but the investigation into TKFs under communication protocols is still in its infancy due mainly to the difficulty of appropriately describing TKF-embedded protocol characteristics.

Following the observations made previously, a seemingly natural research topic is to devise a protocol-based TKF, in the presence of sensor outputs undergoing integral measurements and sensor failures, to achieve the optimal state estimation and also evaluate the associated filtering performance. This topic, though theoretically important and practically significant, is quite challenging for three reasons: (1) it is unclear how to derive a protocol-based Tobit regression model in conjunction with integral measurements and sensor failures; (2) it is fairly difficult to conduct the performance analysis on the developed filter due to its time-varying and stochastic nature; and (3) it is mathematically hard to examine the joint impacts from the communication protocol, integral measurements and sensor failures on the design and performance analysis of the filter. Therefore, the main purpose of this chapter is to overcome the identified challenges.

In this chapter, we endeavor to deal with the protocol-based Tobit Kalman filter under integral measurements and probabilistic sensor failures. To be more specific, a protocol-based Tobit regression model is first built that accommodates the integral measurements and randomly occurring sensor failures. By resorting to the orthogonality projection principle, an optimal protocol-based TKF is designed in the sense of linear minimum mean-squared error (LMMSE), where most of its computation can be carried out recursively or offline. In addition, the performance of the desired filter is statistically assessed, and sufficient conditions are established for the existence of self-propagating upper and lower bounds on the estimation error covariance.

5.1 PROBLEM FORMULATION

Consider the Tobit Kalman filtering problem for a networked system as shown in Figure 5.1. In this framework, the sensor is susceptible to probabilistic failures, the sampling is subject to delays, the signal transmission between the filter and the sensor is implemented through a communication network under the RRP, and the measurement arriving at the filter is inclined to censoring. In what follows, let us introduce the plant, the communication network and the degraded, integral and censored measurement in a mathematical way.

Consider the following linear discrete time-varying system subject to integral measurements and probabilistic sensor failures.

$$x_{k+1} = A_k x_k + \omega_k, \tag{5.1}$$

$$z_{m,k} = \Lambda_{m,k} C_{m,k} \sum_{s=0}^{\ell} x_{k-s} + \upsilon_{m,k}, \quad m = 1, 2, \ldots, p, \tag{5.2}$$

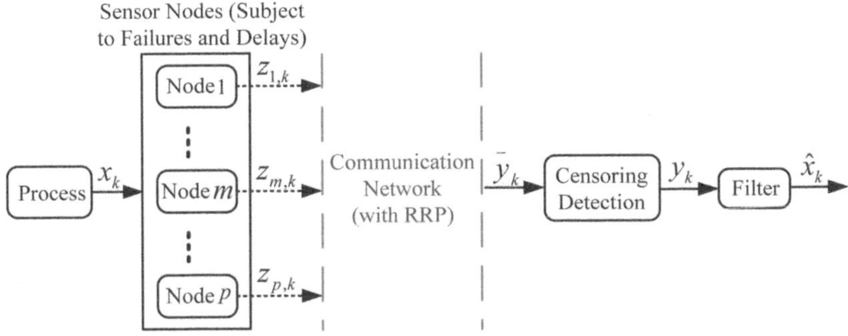

FIGURE 5.1 Schematic diagram for the concerned Tobit Kalman filtering problem.

where $x_k \in \mathbb{R}^{n_x}$ is the state vector and $z_{m,k} \in \mathbb{R}$ is the uncensored observation of the mth sensor. A_k and $C_{m,k}$ are known time-varying matrices with compatible dimensions. $\Lambda_{m,k} \in \mathbb{R}$ is the sensor failure coefficient, ℓ is the time length required for the data collection, and p is the number of sensors. $\omega_k \in \mathbb{R}^{n_x}$ and $v_{m,k} \in \mathbb{R}$ are zero-mean white Gaussian noises with covariances Q_k and $R_{m,k}$, respectively.

In the current investigation, the sensor measurements $z_{m,k}$ $(m = 1, 2, \ldots, p)$ are transmitted to the remote estimator via a shared communication network. Due to limited communication bandwidth, it is assumed that at each communication time instant, only one single sensor is granted the access to the shared channel to transmit its output through the network. Accordingly, the RRP is leveraged to orchestrate the transmission order of the sensors with a view to avoiding data collisions.

Define $\hbar_k \triangleq \mathrm{mod}(k-1, p) + 1 \in \{1, 2, \ldots, p\}$ as the selected sensor that has access to the network at time k where $\mathrm{mod}(k-1, p)$ is the unique non-negative remainder on division of $k-1$ by p, and $\Gamma_{m,\hbar_k} \triangleq \delta(\hbar_k - m)$ is the measurement update matrix that regulates the token-dependent scheduling of the mth sensor. Under the RRP and the zero-input strategy, the actual measurement that is sent to the estimator at time k is

$$\bar{y}_k = \sum_{m=1}^{p} \Gamma_{m,\hbar_k} z_{m,k}. \tag{5.3}$$

At the input terminal of the estimator, let an additional detection device be equipped to check whether the received $\bar{y}_k$ is censored or not, and this gives rise to the following Tobit observation model:

$$y_k = \begin{cases} \bar{y}_k, & \bar{y}_k > \tau, \\ \tau, & \bar{y}_k \leq \tau, \end{cases} \tag{5.4}$$

where $y_k \in \mathbb{R}$ is the censored observation with a constant threshold τ.

Based on that Tobit observation model (5.4), let us define a Bernoulli random variable γ_k to regulate the censoring phenomenon of y_k as follows:

$$\gamma_k = \begin{cases} 1, & \bar{y}_k > \tau, \\ 0 & \bar{y}_k \leq \tau, \end{cases} \tag{5.5}$$

with the following probability distribution:

$$\text{Prob}\{\gamma_k = 1\} = \overline{\gamma}_k, \quad \text{Prob}\{\gamma_k = 0\} = 1 - \overline{\gamma}_k. \tag{5.6}$$

Here, $\overline{\gamma}_k$ is a known non-negative constant. It is supposed that γ_k is known *a priori* and is uncorrelated with other noise signals. Taking advantage of γ_k, y_k in (5.4) can be rewritten as follows:

$$y_k = \gamma_k \overline{y}_k + (1 - \gamma_k)\tau. \tag{5.7}$$

Let

$$y_{1:k} \triangleq \{y_1, y_2, \ldots, y_k\}, \quad \gamma_{1:k} \triangleq \{\gamma_1, \gamma_2, \ldots, \gamma_k\}$$

be the measurement and censoring sequences up till time k, respectively. Furthermore, we denote

$$\hat{x}_k \triangleq \mathbb{E}\{x_k | y_{1:k}, \gamma_{1:k}\}, \qquad \tilde{x}_k \triangleq x_k - \hat{x}_k,$$

$$\hat{x}_k^- \triangleq \mathbb{E}\{x_k | y_{1:k-1}, \gamma_{1:k-1}\}, \tilde{x}_k^- \triangleq x_k - \hat{x}_k^-,$$

$$\hat{y}_k^- \triangleq \mathbb{E}\{y_k | y_{1:k-1}, \gamma_{1:k-1}\}, \tilde{y}_k^- \triangleq y_k - \hat{y}_k^-,$$

$$P_{\tilde{x}_k} \triangleq \mathbb{E}\{\tilde{x}_k \tilde{x}_k^T | y_{1:k}, \gamma_{1:k}\}, P_{\tilde{y}_k^-} \triangleq \mathbb{E}\{\tilde{y}_k^- (\tilde{y}_k^-)^T | y_{1:k-1}, \gamma_{1:k-1}\},$$

$$P_{\tilde{x}_k^-} \triangleq \mathbb{E}\{\tilde{x}_k^- (\tilde{x}_k^-)^T | y_{1:k-1}, \gamma_{1:k-1}\}, P_{\tilde{x}_k^- \tilde{y}_k^-} \triangleq \mathbb{E}\{\tilde{x}_k^- (\tilde{y}_k^-)^T | y_{1:k-1}, \gamma_{1:k-1}\}.$$

Assumption 5.1 The initial state x_0 has the mean $\overline{x}_0$ and covariance P_0. The random variables x_0, $\Lambda_{m,k}$, ω_k and $\upsilon_{m,k}$ are mutually independent.

Assumption 5.2 The sensor failure coefficients $\Lambda_{m,k}$ $(m = 1, 2, \ldots, p)$ are mutually independent random variables in m and k and are also uncorrelated with γ_k and other noise signals. $\Lambda_{m,k}$ regulate the probabilistic failure phenomena of the mth sensor at time k and take values on the interval $[0, 1]$ with certain probability density functions (PDFs) of means $\overline{\Lambda}_{m,k}$ and variances $\tilde{\Lambda}_{m,k}$.

Remark 5.1 It is noteworthy that the measurement sequence $y_{1:k}$ relies on the random censoring sequence $\gamma_{1:k}$, implying that $y_{1:k}$ contains information of $\gamma_{1:k}$, and $y_{1:k}$ will be different for different realizations of $\gamma_{1:k}$. As a result, all expectations defined here are conditional expectations in regard to $\gamma_{1:k}$. Thus, in this chapter, we are interested in the statistical property of the error covariance $P_{\tilde{x}_k^-}$.

When designing filtering algorithms, a widely accepted assumption is that the current sensor observation depends merely on the current state, whereas past states do exert influence on the current observation in the event that a time interval is required for data acquisition and analysis. Hence, the observation in (5.2) is modeled as the integral of states over a prescribed time slot to characterize such influence. A typical example of the integral measurement can be found in distillation columns [171], where lab analysis is often required for measurements of the distillate and bottom compositions as the use of online analyzers is often infeasible due to economic considerations or technological difficulties. In order to analyze the composition, a sufficient amount of samples must be collected, and the sample collection process cannot be completed instantaneously but in a time interval. Accordingly, the measurement model for the sampled distillate and bottom compositions can be written as (5.2), where x_k is the state of distillate and bottom compositions, ℓ is the time interval required to complete the sample collection, $C_{m,k}$ are the known measurement matrices of the lab analysis, and $v_{m,k}$ are the lab analysis errors which are assumed to be zero-mean white Gaussian noises with covariances $R_{m,k}$.

Remark 5.2 It is worth noting that model (5.2) is comprehensive, as it accounts for several frequently encountered measurement uncertainties (e.g. the packet dropout, time delay and measurement degradation). Specifically, at time k, if $\Lambda_{m,k} = 0$, it is implied that the mth sensor suffers from the entire failure and its output signa is completely missing; if $\Lambda_{m,k} = 1$ and $\ell = 0$, it is implied that the mth sensor works in a good condition and no delayed sample collections exist; if $\Lambda_{m,k} = 1$ and $\ell > 0$, it is implied that though the mth sensor functions well, its sample collection is delayed by a time interval ℓ; if $0 < \Lambda_{m,k} < 1$ and $\ell = 0$, it is implied that the mth sensor undergoes partial failures and its output signal is measured with reduced gains that lead to degraded measurements; if $0 < \Lambda_{m,k} < 1$ and $\ell > 0$, it is implied that the mth sensor is susceptible to both failures and delayed sample collections, yielding the integral and degraded observations.

As an efficient tool for handling censored observations, the TKF has stirred much research interest during the last few years. By bringing in new definitions (of the measurement expectation, residual as well as variance), the TKF is capable of formalizing a fully recursive state estimation paradigm to process the uncertainty caused by censored observations. Apart from the measurement censoring, sensor outputs are easily prone to uncertainties ranging from integral measurements to probabilistic sensor failures as shown in (5.2), and the corresponding Tobit Kalman filtering problem has not yet been fully investigated, let alone the case where the RRP is employed to reinforce the reliability of network communication. As such, there is a practical need to establish a holistic protocol-based Tobit Kalman filtering framework to fill in such a gap.

It is observed from (5.4) that the random variable γ_k is employed to describe the censoring phenomenon of y_k. In accordance with (5.4), if no censoring occurs for y_k, i.e. $\gamma_k = 1$, the measurement becomes $y_k = \bar{y}_k$, which means that the output observation is equivalent to the latent one. If the censoring occurs for y_k, i.e. $\gamma_k = 0$, the measurement becomes $y_k = \tau$, which means that the censoring threshold is allocated to the output observation. Here, we suppose that the censoring probability $\bar{\gamma}_k$ is known *a priori* via some statistical experiments. Alternatively, $\bar{\gamma}_k$ can also be approximated by

$$\bar{\gamma}_k \approx \Phi\left(\frac{\sum_{m=1}^{p}\sum_{s=0}^{\ell}\Gamma_{m,h_k}\bar{\Lambda}_{m,k}C_{m,k}\varsigma_{k-s} - \tau}{\sqrt{\sum_{m=1}^{p}\Gamma_{m,h_k}^2 R_{m,k}}}\right), \tag{5.8}$$

where $\varsigma_{k-s} = \hat{x}_{k-s}^-$ for $s = 0$ and $\varsigma_{k-s} = \hat{x}_{k-s}$ for $s = 1,2,\dots,\ell$. $\Phi(\cdot)$ is the cumulative distribution function (CDF) of the random variable "·" obeying the standard normal distribution.

The objectives of this chapter are to (i) design an optimal protocol-based TKF for system (5.1)–(5.6) in the LMMSE sense under the RRP and (ii) analyze the performance of the obtained filter via the evaluation index $\mathbb{E}\{P_{\tilde{x}_k}^{-}\}$.

5.2 PROTOCOL-BASED TOBIT KALMAN FILTER

In this section, we aim to formalize an ameliorated Tobit Kalman filtering paradigm to surmount the challenges brought by the coexistence of sensor failures, integral observations and measurement censoring under the RRP. The formulation procedure differentiates itself in the following aspects: (1) a protocol-based Tobit regression model where impacts from the integral measurements, sensor failures and RRP are taken into consideration and (2) extra computations of the gain and covariance matrices (resulting from the integral measurements and sensor failures) which comprise the augmentation of states, derivation with respect to the augmented state as well as calculations in terms of failure coefficients.

As integral measurements are often caused by the delayed sample collection and signal processing, the augmentation technique is first applied to system (5.1)–(5.6) to accommodate such integral effects. Letting

$$\xi_k \triangleq \begin{bmatrix} x_k^T & x_{k-1}^T & \cdots & x_{k-\ell}^T \end{bmatrix}^T,$$

we have

$$\xi_k = \mathcal{A}_k \xi_{k-1} + \mathcal{B}_k \omega_k, \tag{5.9}$$

$$y_k = \gamma_k \sum_{m=1}^{p} \Gamma_{m,h_k} \left(\Lambda_{m,k} \mathcal{C}_{m,k} \xi_k + \upsilon_{m,k} \right) + (1 - \gamma_k)\tau, \tag{5.10}$$

where

$$\mathcal{A}_k = \begin{bmatrix} A_k & 0 & \cdots & 0 \\ I & 0 & \cdots & 0 \\ \vdots & \ddots & \ddots & \vdots \\ 0 & \cdots & I & 0 \end{bmatrix}, \mathcal{B}_k = \begin{bmatrix} I \\ 0 \\ \vdots \\ 0 \end{bmatrix}, \mathcal{C}_{m,k} = \begin{bmatrix} C_{m,k} & C_{m,k} & \cdots & C_{m,k} \end{bmatrix}.$$

By means of augmentation, system (5.1)–(5.6) with integral measurements is converted into the integral-free one in (5.9)–(5.10) at the cost of extra computations pertinent to the augmented state.

Let

$$P_{\tilde{\xi}_k^{-}} \triangleq \mathbb{E}\{\tilde{\xi}_k^{-}(\tilde{\xi}_k^{-})^T | y_{1:k-1}, \gamma_{1:k-1}\}, P_{\tilde{\xi}_k} \triangleq \mathbb{E}\{\tilde{\xi}_k \tilde{\xi}_k^T | y_{1:k}, \gamma_{1:k}\},$$

$$\zeta_k \triangleq \sum_{m=1}^{p} \Gamma_{m,h_k} \Lambda_{m,k} \mathcal{C}_{m,k} \xi_k, \mathcal{R}_k \triangleq \sum_{m=1}^{p} \Gamma_{m,h_k}^2 R_{m,k}, \vartheta_k \triangleq \frac{\mathcal{I} - \zeta_k}{\mathcal{R}_k},$$

$$\hat{\xi}_k^{-} \triangleq \mathbb{E}\{\xi_k | y_{1:k-1}, \gamma_{1:k-1}\}, \tilde{\xi}_k^{-} \triangleq \xi_k - \hat{\xi}_k^{-}, \hat{\xi}_k \triangleq \mathbb{E}\{\xi_k | y_{1:k}, \gamma_{1:k}\}, \tilde{\xi}_k \triangleq \xi_k - \hat{\xi}_k.$$

Before embarking on the filter design, the protocol-based Tobit regression model entailing the integral measurements and sensor failures is first derived.

Lemma 5.1 The expectation and variance of y_k conditional on the observation sequence $y_{1:k-1}$ and censoring sequence $\gamma_{1:k}$ are

$$\mathbb{E}\{y_k \mid y_{1:k-1}, \gamma_{1:k}\} = \gamma_k \left[\zeta_k + \sqrt{\mathcal{R}_k}\, \lambda(\vartheta_k) \right] + (1-\gamma_k)\tau, \tag{5.11}$$

$$var\{y_k \mid y_{1:k-1}, \gamma_{1:k}\} = \mathcal{R}_k \left[1 - \varphi(\vartheta_k) \right], \tag{5.12}$$

where

$$\lambda(\vartheta_k) = \frac{\phi(\vartheta_k)}{1 - \Phi(\vartheta_k)}, \tag{5.13}$$

$$\varphi(\vartheta_k) = \lambda(\vartheta_k)\left[\lambda(\vartheta_k) - \vartheta_k \right]. \tag{5.14}$$

Here, $\phi(\vartheta_k)$ and $\Phi(\vartheta_k)$ are, respectively, the PDF and CDF of the Gaussian random variable ϑ_k of the following structures:

$$\phi(\vartheta_k) = \frac{1}{\sqrt{2\pi}} e^{-\frac{(\tau - \zeta_k)^2}{2\mathcal{R}_k}}, \tag{5.15}$$

$$\Phi(\vartheta_k) = \int_{-\infty}^{\tau} \frac{1}{\sqrt{2\pi\mathcal{R}_k}} e^{-\frac{(y_k - \zeta_k)^2}{2\mathcal{R}_k}} \, d_{y_k}. \tag{5.16}$$

Proof It follows from (5.2)–(5.3) that $\bar{y}_k$ is a Gaussian variable with the mean ζ_k and variance $\mathcal{R}_k$. Then, the PDF of y_k conditioned on $y_{1:k-1}$ and $\gamma_{1:k}$ can be expressed as

$$f(y_k \mid y_{1:k-1}, \gamma_{1:k}) = \frac{1}{\sqrt{\mathcal{R}_k}} \phi\left(\frac{y_k - \zeta_k}{\sqrt{\mathcal{R}_k}} \right) u(y_k - \tau) + \delta(\tau - y_k)\Phi(\vartheta_k), \tag{5.17}$$

where $\phi\left(\dfrac{y_k - \zeta_k}{\sqrt{\mathcal{R}_k}} \right)$ and $\Phi(\vartheta_k)$ are calculated via (5.15)–(5.16), and $u(y_k - \tau)$ is the unit step function. Taking advantage of (5.4), (5.17) is further translated into

$$f(y_k \mid y_{1:k-1}, \gamma_{1:k}) = \frac{\gamma_k}{\sqrt{\mathcal{R}_k}} \frac{\phi\left(\dfrac{y_k - \zeta_k}{\sqrt{\mathcal{R}_k}} \right)}{1 - \Phi(\vartheta_k)} + (1-\gamma_k). \tag{5.18}$$

In the light of (5.18), the conditional expectation of y_k *is*

$$\mathbb{E}\{y_k|y_{1:k-1},\gamma_{1:k}\} = \int_{-\infty}^{+\infty} \eta_k f\left(\eta_k|\eta_{1:k-1},\gamma_{1:k}\right) d_{\eta_k}$$

$$= \gamma_k \int_{\tau}^{+\infty} \frac{\eta_k}{\sqrt{\mathcal{R}_k}} \frac{\phi\left(\dfrac{\eta_k-\zeta_k}{\sqrt{\mathcal{R}_k}}\right)}{1-\Phi(\vartheta_k)} d_{\eta_k} + (1-\gamma_k)\tau$$

$$= \gamma_k \left[\zeta_k + \sqrt{\mathcal{R}_k}\lambda(\vartheta_k)\right] + (1-\gamma_k)\tau,$$

which is exactly the same as (5.11), where $\lambda(\vartheta_k)$ *is calculated via (5.13). In line with (5.11), we get*

$$\mathbb{E}\{y_k|y_{1:k-1},\gamma_{1:k-1},\gamma_k = 1\} = \zeta_k + \sqrt{\mathcal{R}_k}\lambda(\vartheta_k),$$

$$\mathbb{E}\{y_k|y_{1:k-1},\gamma_{1:k-1},\gamma_k = 0\} = \tau,$$

$$var\{y_k | y_{1:k-1},\gamma_{1:k-1},\gamma_k = 0\} = 0.$$

As a result, we have

$$var\{y_k|y_{1:k-1},\gamma_{1:k}\}$$
$$= var\{y_k|y_{1:k-1},\gamma_{1:k-1},\gamma_k = 1\}$$
$$= \mathbb{E}\{y_k^2|y_{1:k-1},\gamma_{1:k-1},\gamma_k = 1\} - \left(\mathbb{E}\{y_k | y_{1:k-1},\gamma_{1:k-1},\gamma_k = 1\}\right)^2$$
$$= \mathcal{R}_k\left[1-\varphi(\vartheta_k)\right],$$

which is exactly the same as (5.12), where $\varphi(\vartheta_k)$ *is calculated via (5.14). This completes the proof.*

The protocol-based Tobit regression model given by Lemma 5.1 embodies the expectation and variance of y_k conditional on sequences $y_{1:k-1}$ and $\gamma_{1:k}$. In contrast with its counterpart in the standard TKF which barely cares about the measurement censoring in case of unknown $\gamma_{1:k}$, two noteworthy distinctions of model (5.11)–(5.12) can be encapsulated. The first distinction is the substitution of the censoring probability $\bar{\gamma}_k$ by the true censoring variable γ_k owing to the exact acknowledgement of $\gamma_{1:k}$. The second distinction is the replacement of $C_k x_k$ (the product of the original measurement matrix C_k and state x_k) and $R_{m,k}$ (the original noise covariance), respectively, by $\zeta_k = \sum_{m=1}^{p}\Gamma_{m,h_k}\Lambda_{m,k}C_{m,k}\xi_k$ (the sum of p products with respect to the update coefficient Γ_{m,h_k}, failure coefficient $\Lambda_{m,k}$, augmented measurement matrix $\Lambda_{m,k}$ and augmented state ξ_k) and $\mathcal{R}_k = \sum_{m=1}^{p}\Gamma_{m,h_k}^2 R_{m,k}$ (the sum of p products with regard to the update coefficient Γ_{m,h_k} and original noise covariance $R_{m,k}$) due to the involvement of the RRP, integral measurements and sensor failures. If these phenomena are disregarded, (5.11) and (5.12) would degrade to (8) and (11) in [35], respectively.

Remark 5.3 Thanks to (5.11)–(5.16), the standard Tobit regression model has been modified to the protocol-based one with integral measurements, sensor failures and acknowledged measurement censoring. Be aware that such a modification brings on board (1) the substitution of $\bar{\gamma}_k$ by γ_k, $C_k x_k$ by ζ_k and $R_{m,k}$ by $\mathcal{R}_k$ throughout all measurement-related terms and (2) the emergence of a suite of new terms that further sophisticate the subsequent algorithm design.

Denote

$$\mathfrak{C}_{m,k} \triangleq \Gamma_{m,h_k} \overline{\Lambda}_{m,k} C_{m,k}, \qquad \hat{\zeta}_k^- \triangleq \sum_{m=1}^{p} \mathfrak{C}_{m,k} \hat{\xi}_k^-,$$

$$\overline{\vartheta}_k \triangleq \frac{\tau - \hat{\zeta}_k^-}{\mathcal{R}_k}, \tilde{\Lambda}_{m,k} \triangleq \overline{\Lambda}_{m,k}^2 + \breve{\Lambda}_{m,k}.$$

The following theorem presents the optimal protocol-based TKF in the LMMSE sense subject to integral measurements and sensor failures.

Theorem 5.1 The optimal protocol-based TKF for the augmented system (5.9)–(5.10) is

$$\hat{\xi}_k^- = \mathcal{A}_k \hat{\xi}_{k-1}, \tag{5.19}$$

$$P_{\xi_k^-} = \mathcal{A}_{k-1} P_{\xi_{k-1}} \mathcal{A}_{k-1}^T + \mathcal{B}_{k-1} \mathcal{Q}_{k-1} \mathcal{B}_{k-1}^T, \tag{5.20}$$

$$\hat{\xi}_k = \hat{\xi}_k^- + K_k (y_k - \hat{y}_k^-), \tag{5.21}$$

$$P_{\xi_k} = P_{\xi_k^-} - K_k P_{\xi_k^- \tilde{y}_k}^T. \tag{5.22}$$

The one-step measurement prediction and filtering gain are

$$\hat{y}_k^- = \gamma_k \left[\hat{\xi}_k^- + \sqrt{\mathcal{R}_k}\, \lambda(\overline{\vartheta}_k) \right] + (1-\gamma_k)\tau, \tag{5.23}$$

$$K_k = P_{\xi_k^- \tilde{y}_k} P_{\tilde{y}_k}^{-1}, \tag{5.24}$$

where

$$P_{\xi_k^- \tilde{y}_k} = P_{\xi_k^-} \left(\gamma_k \sum_{m=1}^{p} \mathfrak{C}_{m,k} \right)^T, \quad (5.25)$$

$$\begin{aligned}
P_{\tilde{y}_k} &= \gamma_k \sum_{m=1}^{p} \sum_{n=1}^{p} \mathfrak{C}_{m,k} P_{\xi_k^-} \left(\gamma_k \mathfrak{C}_{n,k} \right)^T \\
&\quad + \gamma_k \sum_{m=1}^{p} \sum_{n=1}^{p} \Gamma_{m,h_k} \tilde{\Lambda}_{m,k} C_{m,k} P_{\xi_k} \left(\gamma_k \Gamma_{n,h_k} C_{n,k} \right)^T + \mathcal{R}_k \left[1 - \varphi(\overline{\vartheta}_k) \right].
\end{aligned} \tag{5.26}$$

Here, $\lambda(\overline{\vartheta}_{h_k,k})$ and $\varphi(\overline{\vartheta}_{h_k,k})$ can be calculated via (5.13)–(5.14) by replacing ζ_k with $\hat{\zeta}_k^-$; and $P_{\xi_k} = \mathcal{A}_{k-1} P_{\xi_{k-1}} \mathcal{A}_{k-1}^T + \mathcal{B}_{k-1} \mathcal{Q}_{k-1} \mathcal{B}_{k-1}^T.$

Proof A straightforward exploitation of the orthogonality projection principle to system (5.9)–(5.10) yields (5.19)–(5.22) where the optimal gain matrix is computed by (5.24). The combination of (5.10) and (5.11) generates

$$
\begin{aligned}
\tilde{y}_k^- &= \gamma_k \sum_{m=1}^{p} \Gamma_{m,h_k} \left(\Lambda_{m,k} C_{m,k} \xi_k + \upsilon_{m,k} \right) \\
&\quad - \gamma_k \left[\sum_{m=1}^{p} \Gamma_{m,h_k} \overline{\Lambda}_{m,k} C_{m,k} \hat{\xi}_k^- + \sqrt{\mathcal{R}_k} \lambda(\vartheta_k) \right] \\
&= \gamma_k \sum_{m=1}^{p} \Gamma_{m,h_k} \left(\overline{\Lambda}_{m,k} C_{m,k} \tilde{\xi}_k^- + \left(\Lambda_{m,k} - \overline{\Lambda}_{m,k} \right) C_{m,k} \xi_k \right) \\
&\quad + \gamma_k \left(\sum_{m=1}^{p} \Gamma_{m,h_k} \upsilon_{m,k} - \sqrt{\mathcal{R}_k} \lambda(\vartheta_k) \right),
\end{aligned}
\tag{5.27}
$$

where $\lambda(\vartheta_k)$ can be calculated via (5.13) by replacing ζ_k with $\hat{\xi}_k^-$.

Putting (5.27) into the definitions of $P_{\tilde{\xi}_k^- \tilde{y}_k^-}$ and $P_{\tilde{y}_k^-}$, respectively, gives (5.25) and

$$
\begin{aligned}
P_{\tilde{y}_k^-} &= \mathbb{E}\{ \tilde{y}_k^- (\tilde{y}_k^-)^T \mid y_{1:k-1}, \gamma_{1:k} \} \\
&= \gamma_k \sum_{m=1}^{p} \sum_{n=1}^{p} \mathfrak{C}_{m,k} P_{\tilde{\xi}_k^-} \left(\gamma_k \mathfrak{C}_{n,k} \right)^T + var\{ y_{h_k,k} \mid y_{h_{1:k-1}}, \gamma_{h_{1:k}} \} \\
&\quad + \gamma_k \sum_{m=1}^{p} \sum_{n=1}^{p} \Gamma_{m,h_k} \tilde{\Lambda}_{m,k} C_{m,k} P_{\xi_k} \left(\gamma_k \Gamma_{n,h_k} C_{n,k} \right)^T \\
&= \gamma_k \sum_{m=1}^{p} \sum_{n=1}^{p} \mathfrak{C}_{m,k} P_{\tilde{\xi}_k^-} \left(\gamma_k \mathfrak{C}_{n,k} \right)^T + \mathcal{R}_k \left[1 - \varphi(\overline{\vartheta}_k) \right] \\
&\quad + \gamma_k \sum_{m=1}^{p} \sum_{n=1}^{p} \Gamma_{m,h_k} \tilde{\Lambda}_{m,k} C_{m,k} P_{\xi_k} \left(\gamma_k \Gamma_{n,h_k} C_{n,k} \right)^T,
\end{aligned}
$$

which is exactly the same as (5.26). This completes the proof.

Two exceptional features can be spotted when comparing the proposed protocol-based TKF in Theorem 5.1 with its standard counterpart. One is the substitution of the term $\overline{\gamma}_k C_k$ (which is the product of the censoring probability and the original measurement coefficient) by the term $\gamma_k \sum_{m=1}^{p} \mathfrak{C}_{m,k}$ (which is the sum of p products in relation to the known censoring variable γ_k and the equivalent measurement coefficient $\mathfrak{C}_{m,k}$) in all prediction-related equations. The other is the emergence of the term $\gamma_k \sum_{m=1}^{p} \sum_{n=1}^{p} \Gamma_{m,h_k} \tilde{\Lambda}_{m,k} C_{m,k} P_{\xi_k} \left(\gamma_k \Gamma_{n,h_k} C_{n,k} \right)^T$ in the calculation of $P_{\tilde{y}_k^-}$. The first feature originates from the concurrence of the RRP, sensor failures, integral measurements and exact acknowledgement of the censoring phenomenon, whilst the second feature stems from the presence of sensor failures.

Let $\Pi \triangleq \begin{bmatrix} I & 0 & \cdots & 0 \end{bmatrix}$. On the basis of Theorem 5.1, the following theorem presents the optimal protocol-based TKF for the original system (5.1)–(5.7).

Theorem 5.2 The optimal protocol-based TKF for system (5.1)–(5.7) is

$$\begin{cases} \hat{x}_k^- = \Pi \hat{\xi}_k^-, \\ \hat{x}_k = \Pi \hat{\xi}_k, \\ P_{\tilde{x}_k^-} = \Pi P_{\tilde{\xi}_k^-} \Pi^T, \\ P_{\tilde{x}_k} = \Pi P_{\tilde{\xi}_k} \Pi^T. \end{cases} \tag{5.28}$$

Proof Theorem 5.2 follows readily from Theorem 5.1 by noting the correlation between system (5.1)–(5.7) and system (5.9)–(5.10).

Theorems 5.1–5.2, together with Lemma 5.1, constitute the protocol-based Tobit Kalman filtering algorithm, with its pseudocode outlined in Table 5.1.

It can be seen from Lemma 5.1 and Theorems 5.1–5.2 that when implementing the presented protocol-based Tobit Kalman filtering algorithm with integral measurements and probabilistic sensor failures, the computational complexity mainly involves the matrix multiplication and inversion. In the matrix multiplication, the multiplication of matrices with dimensions $n_x(\ell+1) \times n_x(\ell+1)$ needs to be calculated, where n_x is the dimension of the system state and ℓ is the time length required for the data collection. As to the matrix inversion, the inverse of matrices with dimensions $n_y \times n_y$ needs to be calculated, where n_y is the dimension of the measurement y_k. Hence, the proposed filtering algorithm has the computational complexity of $O\left[\left(n_x(\ell+1)\right)^3 + (n_y)^3 \right]$.

Remark 5.4 The specifically tailored Tobit Kalman filtering architecture is composed of Lemma 5.1 and Theorems 5.1–5.2 and exhibits two extraordinary advantages. On one hand, system (5.1)–(5.7) under investigation is comprehensive for its

---

TABLE 5.1 The Pseudocode of the Protocol-Based TKF

Algorithm: Protocol-Based TKF

Input: $\overline{x}_0, P_0, y_{1:k}$

Output: $\hat{x}_k, P_{\tilde{x}_k}$

1: let $\hat{x}_0 = \overline{x}_0$, $P_{\tilde{x}_0} = P_0$.

2: **for** $k = 1 : N$ **do**

3: compute the predicted value $\hat{\xi}_k^-$ and associate covariance $P_{\tilde{\xi}_k^-}$ by (5.19)–(5.20);

4: compute the predicted value $\hat{x}_k^-$ and associate covariance $P_{\tilde{x}_k^-}$ by (5.28);

5: compute the gain matrix K_k by (5.24)–(5.26);

6: compute the updated estimate $\hat{\xi}_k$ and associate covariance $P_{\tilde{\xi}_k}$ by (5.21)–(5.22);

7: compute the updated estimate $\hat{x}_k$ and associate covariance $P_{\tilde{x}_k}$ by (5.28);

8: **end for**

---

inclusion of the RRP and multiple measurement uncertainties (e.g. measurement censoring, integral measurements and sensor failures), which are prevalently confronted in a myriad of application ranges and are elegantly settled in a holistic yet valid framework. On the other hand, the protocol-induced measurement update coefficient Γ_{m,h_k} and a bank of new terms arise in the development of our TKF, which transparently reveals the impacts from the RRP and measurement uncertainties. Explicitly, the term γ_k throughout the chapter manifests the influence of measurement censoring, the term $\mathcal{R}_k$ in Lemma 5.1 and Theorem 5.1 reveals the impact of the RRP, the term Π in Theorem 5.2 reflects the effect of the integral measurements, the term $\gamma_k \sum_{m=1}^{p} \sum_{n=1}^{p} \Gamma_{m,h_k} \tilde{\Lambda}_{m,k} C_{m,k} P_{\xi_k} \left(\gamma_k \Gamma_{n,h_k} C_{n,k} \right)^T$ in Theorem 5.1 sketches the influence of sensor failures, and terms ζ_k and $\hat{\zeta}_k^-$ characterize the simultaneous influence of the RRP, integral measurements and sensor failures.

It should also be noted that the protocol-based Tobit Kalman filtering paradigm proposed in Theorems 5.1–5.2 is stochastic due to its dependence on the random censoring variable γ_k. This indicates that the state estimates and associate error covariances are now functions of γ_k. Given such a stochastic filtering paradigm, a viable way to analyze its performance is to investigate the statistical property of the filtering error covariance $P_{\tilde{x}_k}$, as shown in the later section.

5.3   SELF-PROPAGATING LOWER AND UPPER BOUNDS

In this section, we aim to come up with a holistic analysis of the filtering performance in relation to the proposed optimal protocol-based TKF. The analysis is performed by taking advantage of the mean estimation error covariance $\mathbb{E}\left\{P_{\tilde{x}_k}\right\}$. Considering the time-varying nature of the filter, we show that there exist self-propagating lower and upper bounds on $\mathbb{E}\left\{P_{\tilde{x}_k}\right\}$. The pursuit of such bounds complies with two principles: (1) the optimality of the filter motivates us to construct a suboptimal filter whose mean estimation error covariance is envisioned to be the upper bound on $\mathbb{E}\left\{P_{\tilde{x}_k}\right\}$ and (2) the semi-positive definiteness of $P_{\tilde{\xi}_k^-}$, $P_{\tilde{\xi}_k}$, $\tilde{\Lambda}_{m,k}$ and $\mathcal{R}_k$ paves the way for us to envisage the lower bound on $\mathbb{E}\left\{P_{\tilde{x}_k}\right\}$ via some subtle matrix manipulations.

For the sake of notation brevity, we define $M_k \triangleq \mathbb{E}\left\{P_{\tilde{\xi}_k}\right\}$, $N_k \triangleq \mathbb{E}\left\{P_{\tilde{x}_k}\right\}$ and $\bar{\varphi}_k \triangleq \varphi\left(\bar{\vartheta}_k\right)$. To begin with, the error covariance $P_{\tilde{\xi}_{k+1}^-}$ in (5.19) is rearranged into

$$
\begin{aligned}
P_{\tilde{\xi}_{k+1}^-} = {} & \mathcal{B}_k Q_k \mathcal{B}_k^T - \mathcal{A}_k P_{\tilde{\xi}_k^-} \left(\gamma_k \sum_{s=1}^{p} \mathcal{C}_{s,k} \right)^T \left\{ \gamma_k \sum_{m=1}^{p} \sum_{n=1}^{p} \mathcal{C}_{m,k} P_{\tilde{\xi}_k^-} \left(\gamma_k \mathcal{C}_{n,k} \right)^T \right. \\
& \left. + \gamma_k \sum_{m=1}^{p} \sum_{n=1}^{p} \Gamma_{m,k} \tilde{\Lambda}_{m,k} C_{m,k} P_{\tilde{\xi}_k} \left(\gamma_k \Gamma_{n,h_k} C_{n,k} \right)^T + \mathcal{R}_k \left[1 - \bar{\varphi}_k \right] \right\}^{-1} \\
& \times \gamma_k \sum_{t=1}^{p} \mathcal{C}_{t,k} P_{\tilde{\xi}_k^-} \mathcal{A}_k^T + \mathcal{A}_k P_{\tilde{\xi}_k^-} \mathcal{A}_k^T .
\end{aligned}
\tag{5.29}
$$

Taking expectation on both sides of (5.29) leads to

$$
\begin{aligned}
M_{k+1} = {}& \mathcal{B}_k Q_k \mathcal{B}_k^T \mathcal{A}_k \mathbb{E}\Bigg\{ P_{\tilde{\xi}_k^-}\left(\gamma_k \sum_{s=1}^{p} \mathfrak{C}_{s,k}\right)^T \Bigg[\gamma_k \sum_{m=1}^{p}\sum_{n=1}^{p}\mathfrak{C}_{m,k} P_{\tilde{\xi}_k^-}\left(\gamma_k \mathfrak{C}_{n,k}\right)^T \\
& + \gamma_k \sum_{m=1}^{p}\sum_{n=1}^{p}\Gamma_{m,h_k}\tilde{\Lambda}_{m,k}C_{m,k}P_{\xi_k}\left(\gamma_k\Gamma_{n,h_k}C_{n,k}\right)^T + \mathcal{R}_k\left[1-\bar{\varphi}_k\right]\Bigg]^{-1} \\
& \times \gamma_k \sum_{t=1}^{p}\mathfrak{C}_{t,k}P_{\tilde{\xi}_k^-}\Bigg\}\mathcal{A}_k^T + \mathcal{A}_k M_k \mathcal{A}_k^T .
\end{aligned}
\tag{5.30}
$$

It is apparent that the complex structure of the third term on the right side of (5.30) prevents the self-propagation of M_{k+1}, that is to say, the acknowledgement of M_k is not sufficient for the determination of M_{k+1}, which further hinders us from finding the self-propagating bound on M_k. Nonetheless, it will be shown later that such an upper bound is well expected through some subtle matrix manipulations.

Recalling that the optimal protocol-based TKF derived in Theorem 5.2 has a random filtering gain K_k, the following theorem is dedicated to the derivation of a self-propagating upper bound on M_k via constructing a suboptimal protocol-based TKF with a deterministic filtering gain K_k^u.

Theorem 5.3 Let the initial condition $M_0^u > 0$ be given. Calculate the matrix sequence $\left\{M_{k+1}^u\right\}_{k\geq 0}$ according to the following difference equation:

$$
\begin{aligned}
M_{k+1}^u = {}& \mathcal{B}_k Q_k \mathcal{B}_k^T - \mathcal{A}_k M_k^u \left(\bar{\gamma}_k^u \sum_{s=1}^{p}\mathfrak{C}_{s,k}\right)^T \Bigg\{\bar{\gamma}_k^u \sum_{m=1}^{p}\sum_{n=1}^{p}\mathfrak{C}_{m,k} M_k^u \left(\bar{\gamma}_k^u \mathfrak{C}_{n,k}\right)^T \\
& + \bar{\gamma}_k^u \sum_{m=1}^{p}\sum_{n=1}^{p}\Gamma_{m,h_k}\tilde{\Lambda}_{m,k}C_{m,k}P_{\xi_k}\left(\bar{\gamma}_k^u \Gamma_{n,k}C_{n,k}\right)^T + \mathcal{R}_k\left[1-\bar{\varphi}_k^u\right]\Bigg\}^{-1} \\
& \times \bar{\gamma}_k^u \sum_{t=1}^{p}\mathfrak{C}_{t,k}M_k^u \mathcal{A}_k^T + \mathcal{A}_k M_k^u \mathcal{A}_k^T .
\end{aligned}
\tag{5.31}
$$

Then, the calculated matrix M_{k+1}^u satisfies

$$
M_{k+1} \leq M_{k+1}^u,
\tag{5.32}
$$

i.e. M_{k+1}^u is a self-propagating upper bound on M_{k+1}, where $\bar{\gamma}_k^u$ and $\bar{\varphi}_k^u$ are the suboptimal counterparts of $\bar{\gamma}_k$ and $\bar{\varphi}_k$, respectively.

Proof Now, let us concentrate on designing a suboptimal protocol-based TKF whose gain matrix K_k^u does not hinge on γ_k. Let $\bar{\gamma}_k^u$, $\tilde{\xi}_k^{u-}$, $\tilde{y}_k^{u-}$, $\bar{\varphi}_k^u$, $P_{\tilde{\xi}_k}^u$, $P_{\tilde{\xi}_k^-}^u$, K_k^u, $P_{\tilde{\xi}_k^- \tilde{y}_k}^u$, $P_{\tilde{y}_k}^u$ and M_k^u, respectively, be the suboptimal counterparts of $\bar{\gamma}_k$, $\tilde{\xi}_k^-$, $\tilde{y}_k^-$, $\bar{\varphi}_k$, $P_{\tilde{\xi}_k}$, $P_{\tilde{\xi}_k^-}$, K_k, $P_{\tilde{\xi}_k^- \tilde{y}_k}$, $P_{\tilde{y}_k}$ and M_k. In this respect, we have

$$
\begin{aligned}
P_{\tilde{\xi}_k}^u &= \mathbb{E}\left\{ \left(\tilde{\xi}_k^{u-} - K_k^u \tilde{y}_k^{u-} \right)\left(\tilde{\xi}_k^{u-} - K_k^u \tilde{y}_k^{u-} \right)^T \right\} \\
&= P_{\tilde{\xi}_k^-}^u - P_{\tilde{\xi}_k^- \tilde{y}_k^-}^u \left(K_k^u \right)^T - K_k^u \left(P_{\tilde{\xi}_k^- \tilde{y}_k^-}^u \right)^T + K_k^u P_{\tilde{y}_k^-}^u \left(K_k^u \right)^T ,
\end{aligned}
\tag{5.33}
$$

$$
\begin{aligned}
P_{\tilde{\xi}_{k+1}^-}^u &= \mathbb{E}\left\{ \left(\mathcal{A}_k \tilde{\xi}_k^{u-} + \mathcal{B}_k \omega_k - \mathcal{A}_k K_k^u \tilde{y}_k^{u-} \right)\left(\mathcal{A}_k \tilde{\xi}_k^{u-} + \mathcal{B}_k \omega_k - \mathcal{A}_k K_k^u \tilde{y}_k^{u-} \right)^T \right\} \\
&= \mathcal{A}_k P_{\tilde{\xi}_k^-}^u \mathcal{A}_k^T + \mathcal{B}_k Q_k \mathcal{B}_k^T - \mathcal{A}_k P_{\tilde{\xi}_k^- \tilde{y}_k^-}^u (\mathcal{A}_k K_k^u)^T \\
&\quad - \left[\mathcal{A}_k P_{\tilde{\xi}_k^- \tilde{y}_k^-}^u (\mathcal{A}_k K_k^u)^T \right]^T + \mathcal{A}_k K_k^u P_{\tilde{y}_k^-}^u (\mathcal{A}_k K_k^u)^T .
\end{aligned}
\tag{5.34}
$$

Taking expectation on both sides of (5.33) yields

$$
\begin{aligned}
\mathbb{E}\left\{ P_{\tilde{\xi}_k}^u \right\} &= \mathbb{E}\left\{ P_{\tilde{\xi}_k^-}^u \right\} - \mathbb{E}\left\{ P_{\tilde{\xi}_k^- \tilde{y}_k^-}^u \right\}\left(K_k^u \right)^T \\
&\quad - K_k^u \mathbb{E}\left\{ \left(P_{\tilde{\xi}_k^- \tilde{y}_k^-}^u \right)^T \right\} + K_k^u \mathbb{E}\left\{ P_{\tilde{y}_k^-}^u \right\}\left(K_k^u \right)^T .
\end{aligned}
\tag{5.35}
$$

In this regard, the suboptimal gain K_k^u can be determined by minimizing the trace of $\mathbb{E}\left\{ P_{\tilde{\xi}_k}^u \right\}$. Taking the matrix trace and derivative with respect to K_k^u on both sides of (5.35), we have

$$
\frac{\partial tr\left\{ \mathbb{E}\left\{ P_{\tilde{\xi}_k}^u \right\} \right\}}{\partial K_k^u} = -2tr\left\{ \mathbb{E}\left\{ P_{\tilde{\xi}_k^- \tilde{y}_k^-}^u \right\} \right\} + 2tr\left\{ K_k^u \mathbb{E}\left\{ P_{\tilde{y}_k^-}^u \right\} \right\}.
\tag{5.36}
$$

Letting (5.36) be equal to zero generates

$$
K_k^u = \mathbb{E}\left\{ P_{\tilde{\xi}_k^- \tilde{y}_k^-}^u \right\} \mathbb{E}^{-1}\left\{ P_{\tilde{y}_k^-}^u \right\}.
\tag{5.37}
$$

Taking expectation on both sides of s, we arrive at

$$
\begin{aligned}
M_{k+1}^u &= \mathcal{A}_k M_k^u \mathcal{A}_k^T + \mathcal{B}_k Q_k \mathcal{B}_k^T - \mathcal{A}_k \mathbb{E}\left\{ P_{\tilde{\xi}_k^- \tilde{y}_k^-}^u (K_k^u)^T \right\} \mathcal{A}_k^T \\
&\quad - \left[\mathcal{A}_k \mathbb{E}\left\{ P_{\tilde{\xi}_k^- \tilde{y}_k^-}^u (K_k^u)^T \right\} \mathcal{A}_k^T \right]^T + \mathcal{A}_k \mathbb{E}\left\{ K_k^u P_{\tilde{\xi}_k^- \tilde{y}_k^-}^u (K_k^u)^T \right\} \mathcal{A}_k^T .
\end{aligned}
\tag{5.38}
$$

Bearing in mind the non-randomness of K_k^u and substituting (5.37) into (5.38), we have

$$
\begin{aligned}
M_{k+1}^u &= \mathcal{A}_k M_k^u \mathcal{A}_k^T + \mathcal{B}_k Q_k \mathcal{B}_k^T - \mathcal{A}_k K_k^u \mathbb{E}^T\left\{ P_{\tilde{\xi}_k^- \tilde{y}_k^-}^u \right\} \mathcal{A}_k^T \\
&= -\mathcal{A}_k M_k^u \left(\overline{\gamma}_k^u \sum_{s=1}^{p} \mathfrak{C}_{s,k} \right)^T \left\{ \overline{\gamma}_k^u \sum_{m=1}^{p}\sum_{n=1}^{p} \mathfrak{C}_{m,k} M_k^u \left(\overline{\gamma}_k^u \mathfrak{C}_{n,k} \right)^T \right. \\
&\quad \left. + \overline{\gamma}_k^u \sum_{m=1}^{p}\sum_{n=1}^{p} \Gamma_{m,h_k} \tilde{\Lambda}_{m,k} \mathcal{C}_{m,k} P_{\xi_k} \left(\overline{\gamma}_k^u \Gamma_{n,h_k} \mathcal{C}_{n,k} \right)^T + \mathcal{R}_k \left[1 - \overline{\varphi}_k^u \right] \right\}^{-1} \\
&\quad \times \overline{\gamma}_k^u \sum_{t=1}^{p} \mathfrak{C}_{t,k} M_k^u \mathcal{A}_k^T + \mathcal{A}_k M_k^u \mathcal{A}_k^T + \mathcal{B}_k Q_k \mathcal{B}_k^T ,
\end{aligned}
$$

which is exactly the same as (5.31). Assuming that the initial condition of (5.31) is independent of the realization in regard to the censoring sequence $\gamma_{1:k}$, we set $M_0^u = M_0 > 0$. For $k > 0$, since the performance of the optimal protocol-based TKF must be no less than any of its suboptimal counterparts, it can be concluded that $M_k \le M_k^u$. As such, M_k^u is an upper bound on M_k for all $k \ge 0$. This completes the proof.

On account of the randomness of the gain matrix K_k in Theorem 5.1, a straightforward idea is to construct a suboptimal protocol-based TKF by setting the gain matrix to be deterministic. Consequently, Theorem 5.3 presents an upper bound M_k^u on M_k, and the suboptimal gain K_k^u is obtained via minimizing the trace of M_k^u. Provided the initial condition, such an upper bound holds for all $k \ge 0$ due to the fact that the filtering performance of the optimal protocol-based TKF should be no less than any of its suboptimal counterparts. It needs to be pointed out that the self-propagation of M_k^u holds regardless of the fact that M_k is not self-propagating. This provides a feasible way to the online recursive computation of M_k^u. In addition, it can be observed from (5.31) that the calculation of M_k^u has close relationships with the augmented coefficients $\mathcal{A}_k$ and $\mathfrak{C}_{m,h_k}$, updated coefficient $\Gamma_{m,k}$, failure variance $\tilde{\Lambda}_{m,k}$ and censoring probability $\bar{\gamma}_k^u$, which explicitly elucidates the impacts from the integral measurements, RRP, sensor failures and measurement censoring on the upper bound.

Theorem 5.4. Let the initial condition $M_0^l > 0$ be given. Calculate the matrix sequence $\left\{ M_{k+1}^l \right\}_{k \ge 0}$ according to the following difference equation:

$$M_{k+1}^l = (1 - \bar{\gamma}_k)\mathcal{A}_k M_k^l \mathcal{A}_k^T + \mathcal{B}_k Q_k \mathcal{B}_k^T, \tag{5.39}$$

Then, the calculated matrix M_{k+1}^l satisfies

$$M_{k+1}^l \le M_{k+1}, \tag{5.40}$$

i.e. M_{k+1}^l is a self-propagating lower bound on M_{k+1}.

Proof Denoting $\mathfrak{A}_k \triangleq \mathcal{A}_k + \gamma_k \mathfrak{K}_k \sum_{m=1}^{p} \mathfrak{C}_{m,k}$ and $\mathfrak{A}_k \triangleq -\mathcal{A}_k K_k$, we have

$$\mathfrak{A}_k P_{\xi_k^-} \left(\gamma_k \sum_{m=1}^{p} \mathfrak{C}_{m,k} \right)^T$$

$$+ \mathfrak{K}_k \left\{ \gamma_k \sum_{m=1}^{p} \sum_{n=1}^{p} \Gamma_{m,h_k} \tilde{\Lambda}_{m,k} \mathcal{C}_{m,k} P_{\xi_k} \left(\gamma_k \Gamma_{n,h_k} \mathcal{C}_{n,k} \right)^T + \mathcal{R}_k [1 - \bar{\varphi}_k] \right\}$$

$$= \left(\mathcal{A}_k + \gamma_k \mathfrak{K}_k \sum_{m=1}^{p} \mathfrak{C}_{m,k} \right) P_{\xi_k^-} \left(\gamma_k \sum_{n=1}^{p} \mathfrak{C}_{n,k} \right)^T \tag{5.41}$$

$$+ \mathfrak{K}_k \left\{ \gamma_k \sum_{m=1}^{p} \sum_{n=1}^{p} {}^{\prime\prime}{}_{m,h_k} \tilde{\Lambda}_{m,k} \mathcal{C}_{m,k} P_{\xi_k} \left(\gamma_k \Gamma_{n,h_k} \mathcal{C}_{n,k} \right)^T + \mathcal{R}_k [1 - \bar{\varphi}_k] \right\}$$

$$= \mathcal{A}_k P_{\xi_k^-} \left(\gamma_k \sum_{n=1}^{p} \mathfrak{C}_{n,k} \right)^T + \mathfrak{K}_k P_{\tilde{y}_k^-}$$

$$= 0.$$

It follows from (5.29) that

$$
\begin{aligned}
P_{\xi_{k+1}^-} &= (1-\gamma_k)\left(\mathcal{A}_k P_{\xi_k^-}\mathcal{A}_k^T + \mathcal{B}_k Q_k \mathcal{B}_k^T\right) + \gamma_k(\mathcal{A}_k P_{\xi_k^-}\mathcal{A}_k^T + \mathcal{B}_k Q_k \mathcal{B}_k^T) - \mathcal{A}_k P_{\xi_k^-} \\
&\quad \times \left(\gamma_k \sum_{m=1}^{p}\mathfrak{C}_{m,k}\right)^T P_{\tilde{y}_k^-}^{-1}\gamma_k \sum_{n=1}^{p}\mathfrak{C}_{n,k}P_{\xi_k^-}\mathcal{A}_k^T \\
&= (1-\gamma_k)\left(\mathcal{A}_k P_{\xi_k^-}\mathcal{A}_k^T + \mathcal{B}_k Q_k \mathcal{B}_k^T\right) + \gamma_k\left(\mathfrak{A}_k P_{\xi_k^-}\mathcal{A}_k^T + \mathcal{B}_k Q_k \mathcal{B}_k^T\right).
\end{aligned}
\tag{5.42}
$$

Inserting (5.41) into (5.42) generates

$$
\begin{aligned}
P_{\xi_{k+1}^-} &= (1-\gamma_k)\left(\mathcal{A}_k P_{\xi_k^-}\mathcal{A}_k^T + \mathcal{B}_k Q_k \mathcal{B}_k^T\right) \\
&\quad + \gamma_k\left(\mathfrak{A}_k P_{\xi_k^-}\mathcal{A}_k^T + \mathcal{B}_k Q_k \mathcal{B}_k^T\right) \\
&\quad + \gamma_k \mathfrak{A}_k P_{\xi_k^-}\left(\gamma_k \sum_{m=1}^{p}\mathfrak{C}_{m,k}\right)^T \mathfrak{K}_k^T \\
&\quad + \gamma_k \mathfrak{A}_k \left\{\gamma_k \sum_{m=1}^{p}\sum_{n=1}^{p}\Gamma_{m,h_k}\tilde{\Lambda}_{m,k}\mathcal{C}_{m,k}\right. \\
&\quad \left. \times P_{\xi_k}\left(\gamma_k \Gamma_{n,h_k}\mathcal{C}_{n,k}\right)^T + \mathcal{R}_k[1-\bar{\varphi}_k]\right\}\mathfrak{K}_k^T \\
&= (1-\gamma_k)\left(\mathcal{A}_k P_{\xi_k^-}\mathcal{A}_k^T + \mathcal{B}_k Q_k \mathcal{B}_k^T\right) \\
&\quad + \gamma_k(\mathfrak{A}_k P_{\xi_k^-}\mathfrak{A}_k^T + \mathcal{B}_k Q_k \mathcal{B}_k^T) \\
&\quad + \gamma_k \mathfrak{K}_k \left\{\gamma_k \sum_{m=1}^{p}\sum_{n=1}^{p}\Gamma_{m,h_k}\tilde{\Lambda}_{m,k}\mathcal{C}_{m,k}\right. \\
&\quad \left. \times P_{\xi_k}\left(\gamma_k \Gamma_{n,h_k}\mathcal{C}_{n,k}\right)^T + \mathcal{R}_k[1-\bar{\varphi}_k]\right\}\mathfrak{K}_k^T.
\end{aligned}
\tag{5.43}
$$

Inspired by (5.43), let us define

$$
M_{k+1}^l \triangleq (1-\bar{\gamma}_k)\mathcal{A}_k M_k^l \mathcal{A}_k^T + \bar{\gamma}_k \mathcal{B}_k Q_k \mathcal{B}_k^T,
\tag{5.44}
$$

which is initialized at $M_0^l = 0$. Now, let us prove $M_k^l \le M_k$ via mathematical induction. Noticeably, at the initial time k = 0, we have $M_0^l = 0 \le M_0$. Supposing that $M_k^l \le M_k$ holds at time k, we have

$$
\begin{aligned}
M_{k+1}^l &= (1-\bar{\gamma}_k)\mathcal{A}_k M_k^l \mathcal{A}_k^T + \bar{\gamma}_k \mathcal{B}_k Q_k \mathcal{B}_k^T \\
&\le (1-\bar{\gamma}_k)\mathcal{A}_k M_k \mathcal{A}_k^T + \bar{\gamma}_k \mathcal{B}_k Q_k \mathcal{B}_k^T \\
&\le M_{k+1}.
\end{aligned}
\tag{5.45}
$$

As a result, it can be concluded that $M_k^l \le M_k$ holds for all k ≥ 0, i.e. M_k^l is a lower bound on M_k. This completes the proof.

Making full use of some subtle matrix manipulations, a self-propagating lower bound M_k^l on M_k is acquired in Theorem 5.4 based on the semi-positive definiteness of $P_{\tilde{\xi}_k^-}$, $P_{\tilde{\xi}_k}$, $\tilde{\Lambda}_{m,k}$ and $\mathcal{R}_k$. *Casting insights into Theorem 5.4, a remarkable finding is* that the calculation of M_k^l tightly hinges on the augmented coefficients $\mathcal{A}_k$ and $\mathcal{B}_k$ and censoring probability $\bar{\gamma}_k$, which noticeably characterizes the effects of the RRP, integral measurements, sensor failures and measurement censoring.

Keeping in mind the relationship between $P_{\tilde{\xi}_k^-}$ and $P_{\tilde{x}_k^-}$ as shown in Theorem 5.2, we arrive at the following theorem.

Theorem 5.5 There exist an upper bound N_k^u and a lower bound N_k^l on the mean error covariance N_k such that

$$N_k^l \leq N_k \leq N_k^u, \tag{5.46}$$

holds for all $k \geq 0$, where $N_k^u = \Pi M_k^u \Pi^T$ and $N_k^l = \Pi M_k^l \Pi^T$.

Proof The result follows noticeably from Theorems 5.2–5.4.

Remark 5.5 Theorems 5.5 manifests the upper and lower bounds on the mean estimation error covariance N_k under the RRP and multiple measurement uncertainties (e.g. measurement censoring, integral measurements and sensor failures). The concurrence of the RRP and uncertainties provokes substantial difficulties in the assessment of the filtering performance, which can be encapsulated from the following two aspects. (1) The exact acknowledgement of the censoring phenomenon gives rise to a time-varying and stochastic protocol-based TKF as shown in Theorems 5.1–5.2, which prevents us from carrying out a rigorous convergence analysis on the proposed filter. As such, we turn to explore the self-propagating upper and lower bounds on the mean error covariance N_k. The upper bound is achieved by establishing a suboptimal protocol-based TKF whose gain matrix is independent of γ_k, and the lower bound is acquired by resorting to subtle matrix manipulations based on the semi-positive definiteness of matrices $P_{\tilde{\xi}_k^-}$, $P_{\tilde{\xi}_k}$, $\tilde{\Lambda}_{m,k}$ and $\mathcal{R}_k$. (2) The conjunction of the RRP and measurement uncertainties gives rise to a set of protocol-induced and uncertainty-induced terms, significantly sophisticating the design of the suboptimal filter and the implementation of the involved matrix manipulations. To sum up, the simultaneous presence of the RRP and measurement uncertainties ineluctably results in distinctive filter design techniques and performance analysis procedures, which are elegantly orchestrated through Theorems 5.1–5.5.

5.4   AN ILLUSTRATIVE EXAMPLE

In this section, we leverage an oscillator example to elucidate the applicability of the presented filter design strategy and performance analysis mechanism.

Denote the root mean-squared errors (RMSEs) of x_k^1 and x_k^2, respectively, as

$$\text{RMSE1} \triangleq \sqrt{\frac{1}{M}\sum_{i=1}^{M}\left(x_k^{1(i)} - \hat{x}_k^{1(i)}\right)^2} \ and \ \text{RMSE2} \triangleq \sqrt{\frac{1}{M}\sum_{i=1}^{M}\left(x_k^{2(i)} - \hat{x}_k^{2(i)}\right)^2}, \ and \ the$$

mean error covariance trace (MECT) as $\text{MECT} \triangleq (1/M)\sum_{i=1}^{M}\text{tr}\left(P_{\tilde{x}_k}^{(i)}\right)$ *where M is* the number of Monte Carlo trials.

Let the discrete time-varying system (5.1)–(5.6) have the following parameters:

$$A_k = \begin{bmatrix} \cos(w) & -\sin(w) \\ \sin(w) & \cos(w) \end{bmatrix}, w = 0.052\pi,$$

$$R_{1,k} = R_{2,k} = 1, \tau = 0, p = 2, \ell = 2,$$

$$C_{1,k} = \begin{bmatrix} 1 & 0 \end{bmatrix}, C_{2,k} = \begin{bmatrix} 0 & 1 \end{bmatrix}, \bar{x}_0 = \begin{bmatrix} 5 & 0 \end{bmatrix}^T,$$

$$Q_k = \mathrm{diag}\{0.0025, 0.0025\}, P_0 = I_2, M_0 = I_6,$$

where ℓ is the interval required for the measurement integral. The oscillator example concerns the estimation of ballistic roll rates in case of the noisy dynamic model and uncertain magnetometer data. Sensor failure rates $\Lambda_{m,k}$ ($m = 1, 2$) can be determined via statistical tests and are supposed to be regulated by the PDF $p(s) = 0.05\delta(s) + 0.10\delta(s - 0.5) + 0.85\delta(s - 1)$. Apparently, expectations and variances of $\Lambda_{m,k}$ are computed as $\bar{\Lambda}_{m,k} = 0.90$ and $\tilde{\Lambda}_{m,k} = 0.065$. Note that the dynamic model is corrupted by ambient disturbances entering the system via ω_k. The magnetometer sampling is subject to sensor failures and integral measurements, and the data transmission is scheduled by the RRP.

Figure 5.2 depicts the true state values and associate estimates generated by the protocol-based Tobit Kalman filter (which is named PBTKF and is capable of

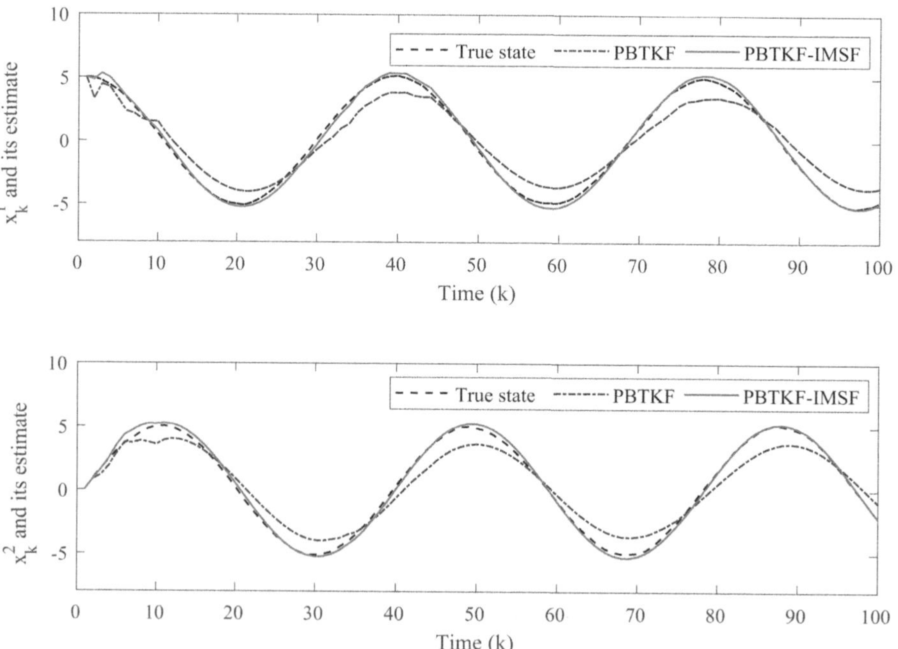

FIGURE 5.2 True values of the first and second dimensions of the state and their estimates.

tackling censored observations under the RRP) and the protocol-based Tobit Kalman filter with integral measurements and sensor failures (which is named PBTKF-IMSF and is capable of simultaneously tackling integral measurements, sensor failures and censored observations under the RRP). Figure 5.3 plots the comparison result in RMSE between the PBTKF and PBTKF-IMSF after 1000 independent Monte Carlo trials. It is witnessed from Figure 5.2 that the PBTKF-IMSF manages to track the true state values precisely, whilst the PBTKF appears to have considerable deviations from the true state values. Besides, it is sketched in Figure 5.3 that the RMSE curve of the PBTKF-IMSF resides lower than that of the PBTKF, indicating that issues of integral measurements and sensor failures are suitably addressed in the PBTKF-IMSF, whilst they are not settled in the PBTKF.

Next, letting $A_k = 0.9 \begin{bmatrix} \cos(w) & -\sin(w) \\ \sin(w) & \cos(w) \end{bmatrix}$ and $M = 1000$, the comparison result between the PBTKF and PBTKF-IMSF in MECT is manifested in Figure 5.4. Besides, relationships among the trace of the mean error covariance N_k, its upper bound N_k^u and lower bound N_k^l (calculated by our PBTKF-IMSF) are sketched in Figure 5.5. Due to the impossibility of analytically computing N_k, $\log_{10}\left(\text{tr}\{N_k\}\right)$ is approximated by $\log_{10}$ (MECT). It can be spotted from Figure 5.3 that the value of $\log_{10}\left(\text{tr}\{N_k\}\right)$ calculated by the PBTKF-IMSF is always smaller than that calculated by the PBTKF, certifying the superiority of our PBTKF-IMSF in simultaneously handling integral measurements and sensor failures over the PBTKF. Additionally, it is observed from Figure 5.4 that the curve of $\log_{10}\left(\text{tr}\{N_k\}\right)$ always resides between those of $\log_{10}\left(\text{tr}\left(N_k^u\right)\right)$ and $\log_{10}\left(\text{tr}\left(N_k^l\right)\right)$, which justifies the statement in Theorem 5.5 that N_k^u and N_k^l can be, respectively, treated as reasonable upper and lower bounds on N_k.

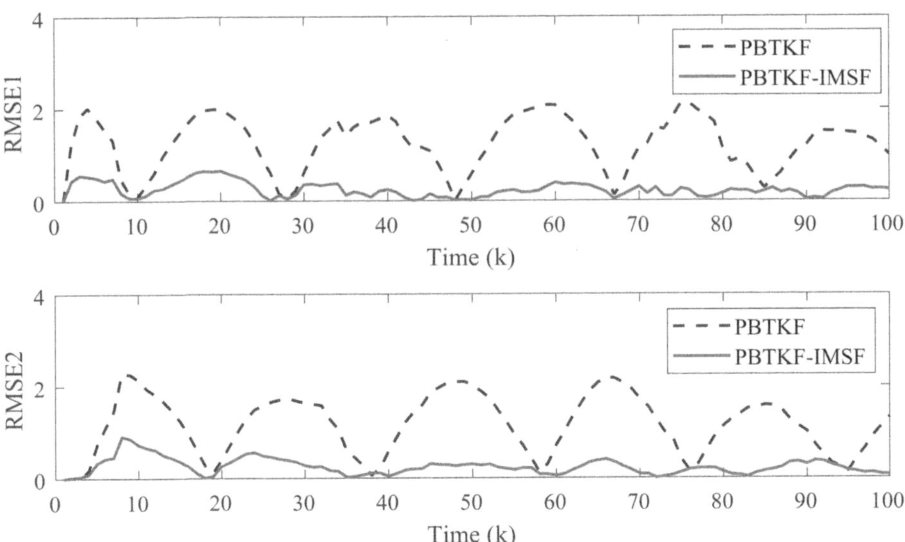

FIGURE 5.3 Performance comparison in RMSE1 and RMSE2.

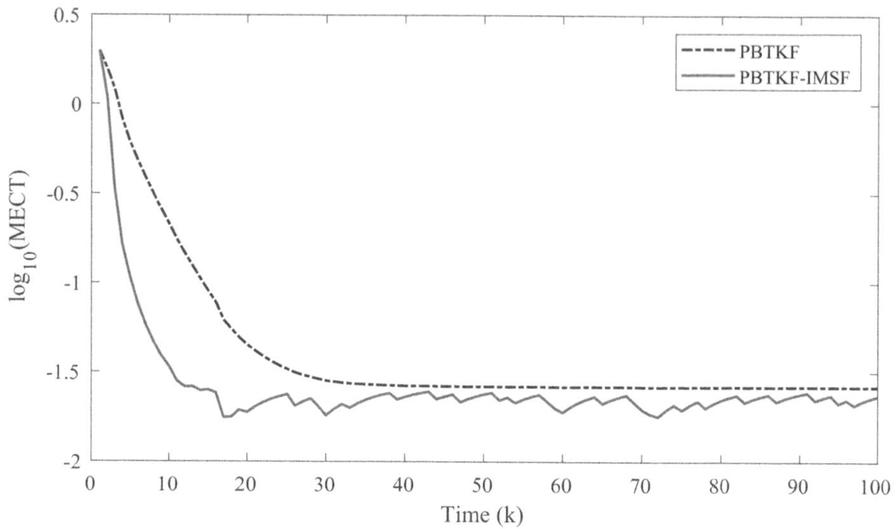

FIGURE 5.4   Performance comparison in $\log_{10}(\text{MECT})$.

In addition, to further demonstrate the filtering performance of our developed PBTKF-IMSF under different measurement censoring, integral measurements and sensor failures, simulation scenarios with different values of censoring thresholds τ, integral intervals ℓ and failure coefficients $\Lambda_{m,k}$ are tested. Let us consider two sets of censoring thresholds $\{-5,\ 0\}$, two sets of integral intervals $\{2,5\}$ and two sets of failure coefficients $\{\Lambda_{1,m,k}, \Lambda_{2,m,k}\}$ where $\Lambda_{1,m,k}$ and $\Lambda_{2,m,k}$ are, respectively, regulated by PDFs $p_1(s) = 0.35\delta(s) + 0.10\delta(s-0.5) + 0.55\delta(s-1)$ and $p_2(s) = 0.05\delta(s) + 0.10\delta(s-0.5) + 0.85\delta(s-1)$. Apparently, expectations and variances of $\Lambda_{1,m,k}$ and $\Lambda_{2,m,k}$ are computed as $\overline{\Lambda}_{1,m,k} = 0.6$, $\tilde{\Lambda}_{1,m,k} = 0.215$, $\overline{\Lambda}_{2,m,k} = 0.9$ and $\tilde{\Lambda}_{2,m,k} = 0.065$.

After 1000 independent Monte Carlo trials, RMSE results of our PBTKF-IMSF under different censoring thresholds, integral intervals and failure coefficients are, respectively, sketched in Figures 5.6–5.8. It is witnessed from Figure 5.6 that the RMSE curve generated in case of $\tau = -5$ always locates lower than that generated in case of $\tau = 0$. This is reasonable because a smaller censoring threshold τ indicates that less measurements are inclined to the censoring phenomenon and more measurement information can be utilized in state estimation. This undoubtedly leads to better filtering accuracy of the PBTKF-IMSF.

One observes from Figure 5.7 that the RMSE curves generated in cases of $\ell = 2$ and $\ell = 5$ are intertwined with each other, indicating that no deterministic relationship exists between the length of the integral interval and the filtering accuracy of our PBTKF-IMSF. As a matter of fact, in case of the integral measurement, the sensor observation is actually proportional to the integral of system states within a prescribed time interval ℓ, and hence the information from not only the current system

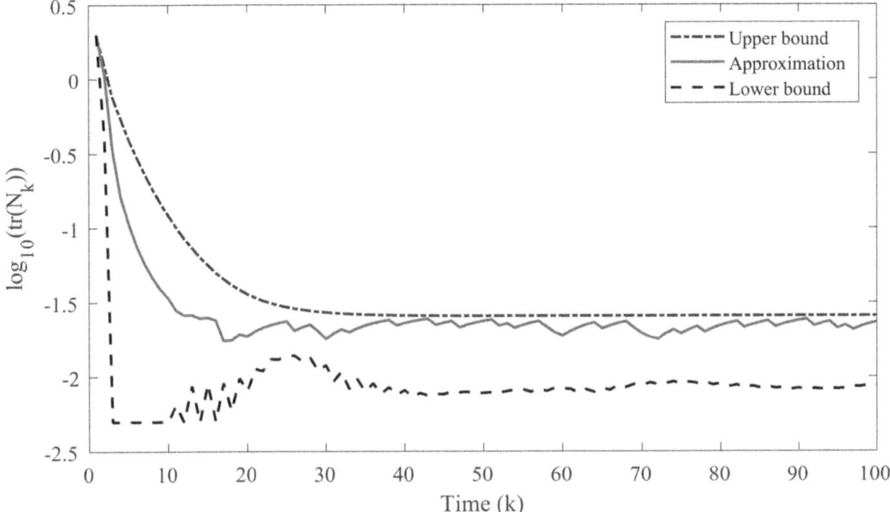

FIGURE 5.5 $\log_{10}\left(\mathrm{tr}\{N_k\}\right)$, $\log_{10}\left(\mathrm{tr}\{N_k^u\}\right)$ and $\log_{10}\left(\mathrm{tr}\{N_k^l\}\right)$ comparison.

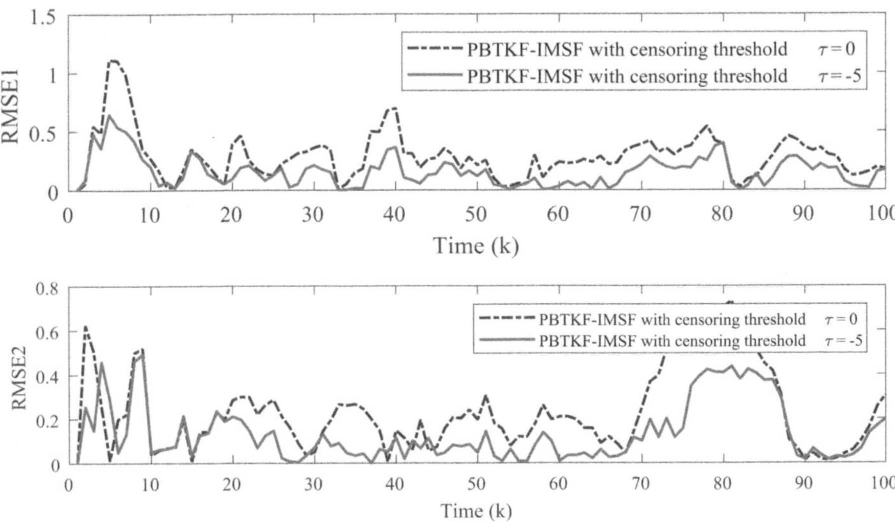

FIGURE 5.6   RMSE1 and RMSE2 comparison of the PBTKF-IMSFs with different censoring thresholds: $\tau = -5$ and $\tau = 0$.

state but also the past system states can be utilized for the estimation of the current system state. The variation of the integral interval ℓ (from $\ell = 2$ to $\ell = 5$) implies that more information on the past system states is introduced to the sensor observation. Nevertheless, the extra information about the past system states does not contain any knowledge about the current system state and makes no contribution to the estimation of the current system state. As a result, it can be concluded that the variation of the integral interval ℓ has no explicit influence on the filtering performance of our PBTKF-IMSF, which coincides with the observation made from Figure 5.7.

It can be spotted from Figure 5.8 that the RMSE curve generated by our PBTKF-IMSF with the failure coefficient $\Lambda_{1,m,k}$ always locates lower than that generated by

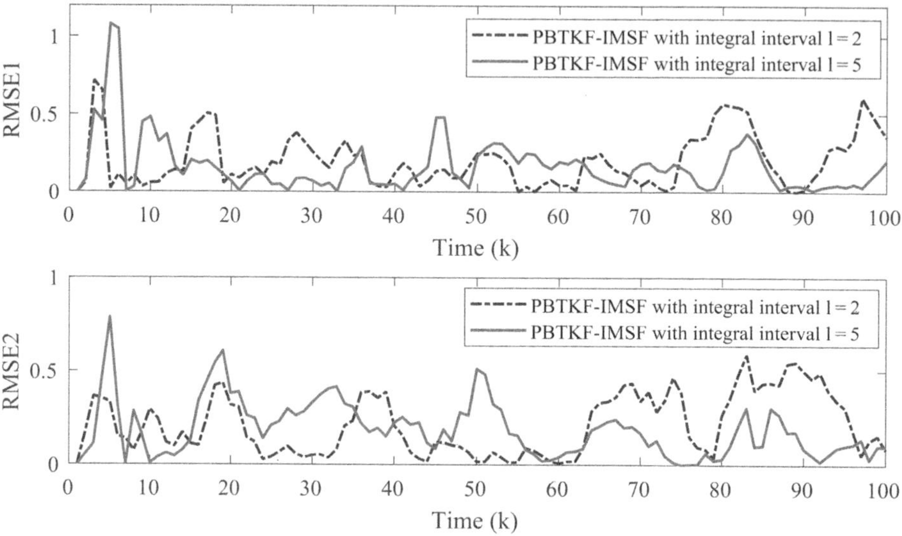

FIGURE 5.7   RMSE1 and RMSE2 comparison of the PBTKF-IMSFs with different integral intervals: $\ell = 2$ and $\ell = 5$.

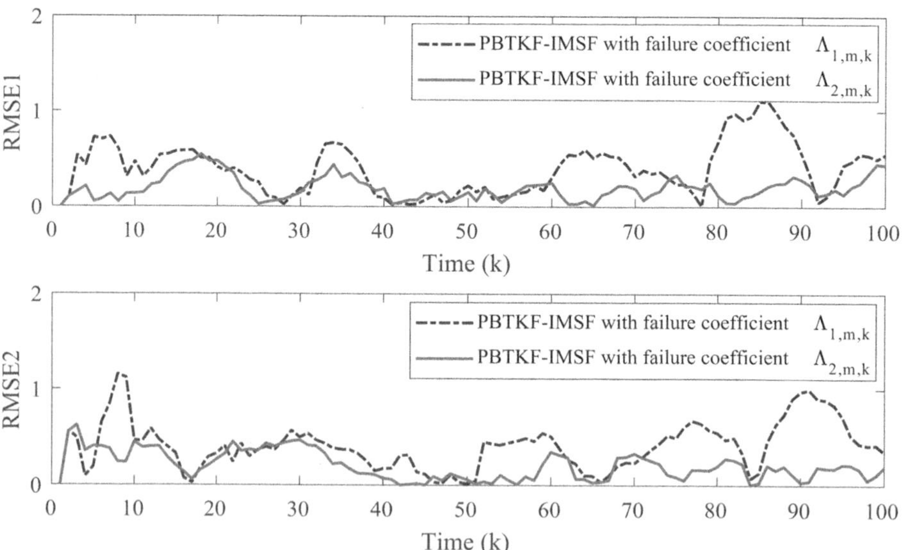

FIGURE 5.8   RMSE1 and RMSE2 comparison of the PBTKF-IMSFs with different failure coefficients: $\Lambda_{1,m,k}$ and $\Lambda_{2,m,k}$.

our PBTKF-IMSF with the failure coefficient $\Lambda_{2,m,k}$. This is reasonable because it can be observed from the expectations and variances of $\Lambda_{1,m,k}$ and $\Lambda_{2,m,k}$ that the sensor with the failure coefficient $\Lambda_{2,m,k}$ is more likely to have failures than the sensor with the failure coefficient $\Lambda_{1,m,k}$. This indicates that the sensor with the failure coefficient $\Lambda_{1,m,k}$ is capable of providing more state information to the filter than the sensor with the failure coefficient $\Lambda_{2,m,k}$. As a result, the PBTKF-IMSF with the failure coefficient $\Lambda_{1,m,k}$ outperforms the PBTKF-IMSF with the failure coefficient $\Lambda_{2,m,k}$ in filtering accuracy.

5.5 SUMMARY

In this chapter, we have settled the protocol-based Tobit Kalman filtering problem susceptible to phenomena of integral measurements and sensor failures. The integral measurements have been described as functions of states over a time period, the sensor failures have been characterized by random variables taking values on the interval [0,1] according to certain PDFs with known means and variances, and the data transmission in the network has been commanded by the RRP. These phenomena have been elaborately addressed via bringing on board a couple of new terms, which have provoked additional calculations in measurement predictions, gain matrices as well as error covariances. Fortunately, the increased calculations are recursive or can be conducted offline. Consequently, the devised filter is propitious for online scenarios. Further analysis has been performed to assess the filtering performance, and a sufficient condition has been pinned down to guarantee the existence of self-propagating upper and lower bounds on the mean estimation error covariance. Finally, an application case study has been exploited to verify the efficacy of the developed method.

6 Distributed Optimal Filtering Fusion over a Packet-Delaying Network Subject to Censored Data
A Probabilistic Perspective

It is well known that multiple sensor data fusion refers to the process of effectively combining data from multiple homogeneous/heterogeneous sensors so as to achieve a satisfactory performance. Over the past decades, the multiple sensor data fusion technique has been successfully applied in a wide range of areas including artificial intelligence, automatic vehicles, fault detection, health monitoring, image processing, navigation, robotics, sensor networks and target tracking. A remarkable mass of researchers have dedicated their time to the study on the multiple sensor data fusion problem for various complex systems and fruitful results have been produced in the literature which can be mainly divided into two categories, i.e. centralized data fusion and distributed data fusion.

In both scenarios of centralized and distributed data fusion, measurements have to be sent to the fusion center via a network which is typically subject to limited communication capacity, and this gives rise to the so-called networked-induced phenomena, for example, missing/fading measurements, signal quantization, sensor saturations, communication delays, randomly occurring incomplete information and so forth. The time delay, as one of the most frequently encountered networked-induced phenomena, has been recognized to be a major source for undesirable dynamic behaviors such as oscillation and instability. It should be noted that the majority of delay-related results focus only on the deterministic delays, whereas the networked-induced communication delay is often random yet time-varying by nature.

For networked control systems with randomly occurring time delays, the optimal LMV [172] and H_∞ [173] filters have been designed by modeling the delay phenomenon as a random Bernoulli process. By adopting the Gaussian sum to approximate the one-step posterior predictive probability density functions of the state and the delayed measurement, the filtering problem for nonlinear systems with randomly delayed measurements has been concerned in [174]. In addition, with the help of the covariance information, the prediction, filtering, and

DOI: 10.1201/9781003461623-6

smoothing problems for discrete-time systems with stochastic delays have been simultaneously solved in [175]. Nevertheless, the aforementioned results have adopted the expected value of the estimation error covariance matrix as the measure when analyzing the filter stability, which conceals the fact that events with any arbitrarily low probabilities may cause the divergence of the expected value. To deal with this issue, the modified Kalman filters for discrete-time systems with packet delays have been developed in [176], where the filter stability has been evaluated via a probabilistic way.

Summarizing these discussions, it can be concluded that (1) distributed filtering fusion, as an effective approach to multi-sensor fusion, has advantages of strong robustness, reliability and fault-tolerance, and the key issue for its implementation is the choice of the appropriate fusion rule; (2) the stability of the filtering algorithm (in terms of the mean value of the estimation error covariance) has been extensively studied for networked control systems with delayed measurements, but such a mean measure has the drawback of divergence for events with arbitrarily low probabilities; and (3) increasing research attention has been paid to the TKF because of its Kalman-type filter structure and its capability in handling censored measurements. Based on these conclusions, a seemingly natural research idea is to design a TKF to obtain the optimal state estimate for systems suffering from both censored and delayed measurements and then analyze the filter stability via an alternative approach, for example, the probabilistic approach. This appears to be a non-trivial job for the following three reasons: (1) it is unclear how the Tobit regression model in the standard TKF is modified/changed when presented with multiple sensors and packet delays; (2) it is quite hard to determine the distributed fusion rule which benefits the filter design and the subsequent performance analysis; and (3) it is theoretically difficult to quantify the impacts from multiple sensors and packet delays on the performance analysis of the filter. It is, therefore, the main motivation of this chapter to shorten the gap identified here.

In this chapter, we are devoted to the study of the distributed federated Tobit Kalman filtering problem for a class of discrete-time systems with censored measurements and packet delays. The packet delay phenomenon is characterized by a Poisson distribution, and the measurement censoring is described by the Tobit measurement model. Firstly, a modified Tobit regression model is derived by taking into account the packet delays, based on which a modified Tobit Kalman filter (MTKF) is developed. Under the assumption that each local estimator keeps a censoring detection device and a buffer at its input terminal, a local modified Tobit Kalman filter (LMTKF) is devised by running $s_i + 1$ MTKF algorithms where $s_i \in [0, \mathcal{Q}_i - 1]$ and $\mathcal{Q}_i$ is the buffer length of the ith LMTKF. Then, on the basis of the federated Kalman filtering fusion rule, a distributed federated modified Tobit Kalman filter (DFMTKF) is designed at the fusion center by fusing estimates from all the LMTKFs. Finally, given the buffer length, the probability distribution of the delay and the prescribed estimation error covariance bound, the filtering performance of the DFMTKF is evaluated in the pursuit of the upper and lower bounds of the probability, at which the estimation error covariance is within the prescribed bound. Simulation experiments are conducted to illustrate the effectiveness of the proposed filtering algorithm.

6.1 PROBLEM FORMULATION

Consider the distributed federated Tobit Kalman filtering problem over a packet-delaying network with l sensors, as shown in Figure 6.1. The process dynamics and sensor measurement equations are given as follows:

$$\mathbf{x}_{k+1} = \mathbf{A}\mathbf{x}_k + \mathbf{w}_k, \tag{6.1}$$

$$\mathbf{y}_{i,k}^* = \mathbf{C}_i\mathbf{x}_k + \mathbf{v}_{i,k}, i = 1,2,\ldots,l, \tag{6.2}$$

where $\mathbf{x}_k \in \mathbb{R}^{n_x}$ and $\mathbf{y}_{i,k}^* \in \mathbb{R}^{n_{yi}}$ are, respectively, the state and measurement vectors of the ith $(i = 1,2,\ldots,l)$ sensor, and l is the number of the sensors in the network. $\mathbf{A}$ and $\mathbf{C}_i$ are known matrices with appropriate dimensions. $\mathbf{w}_k \in \mathbb{R}^{n_\omega}$ and $\mathbf{v}_{i,k} \in \mathbb{R}^{n_{yi}}$ are zero-mean and Gaussian noises with covariance $\mathbf{Q}_k$ and $\mathcal{R}_i$, respectively.

Suppose that each sensor sends its time-stamped measurement $\mathbf{y}_{i,k}^*$ to its corresponding local estimator via a packet-delaying network with negligible quantization effects. Each sensor measurement $\mathbf{y}_{i,k}^*$ is delayed by $d_{i,k}$ times, where $d_{i,k}$ is a random variable described by the following probability mass function.

$$f(j) = \text{Prob}\{d_{i,k} = j\}, j = 0,1,\ldots. \tag{6.3}$$

Let a Bernoulli random variable $\alpha_i(k,k)$ be the indicator function for $\mathbf{y}_{i,k}^*$ at time k with probabilities $\text{Prob}\{\alpha_i(k,k) = 1\} = \bar{\alpha}_i(k,k)$ and $\text{Prob}\{\alpha_i(k,k) = 0\} = 1 - \bar{\alpha}_i(k,k)$. To be specific, if $\mathbf{y}_{i,k}^*$ arrives at the local estimator at time k, then $\alpha_i(k,k) = 1$, otherwise, $\alpha_i(k,k) = 0$. It follows easily from (6.3) that $\text{Prob}\{\alpha_i(k,k) = 1\} = \text{Prob}\{d_{i,k} = 0\} = f(0)$, which implies that $\bar{\alpha}_i(k,k) = f(0)$. In this case, the probability distribution function (pdf) of the measurement noise $\mathbf{v}_{i,k}$ is defined the following way.

$$p(\upsilon_{i,k} \mid \alpha_i(k,k)) = \begin{cases} \mathcal{N}(0,\mathcal{R}_i), & \alpha_i(k,k) = 1, \\ \mathcal{N}(0,\sigma^2\mathbf{I}), & \alpha_i(k,k) = 0, \end{cases} \tag{6.4}$$

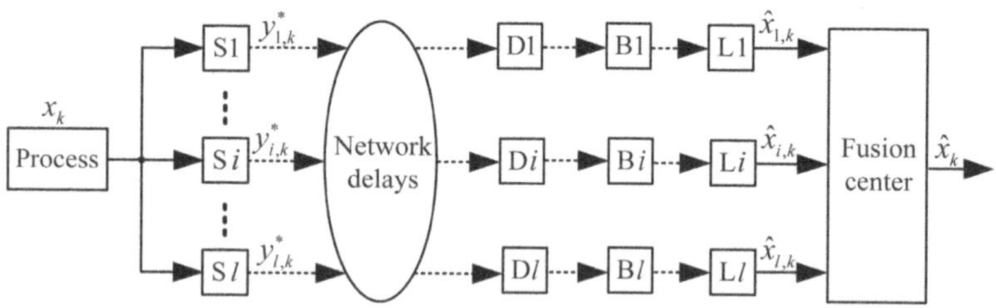

FIGURE 6.1 System block diagram of the distributed federated Tobit Kalman filtering problem with networked delays. The block Si $(i = 1,2,\ldots,l)$ denotes the ith sensor, and l is the number of sensors. The blocks Bi, Di and Li, respectively, stand for the ith buffer, censoring detection device and local estimator corresponding to the ith sensor.

where $\sigma \to \infty$ and $\mathcal{R}_i$ is the known variance. As a result, the variance of the measurement noise becomes time-varying and can be written as

$$\mathbf{R}_{i,k} = \alpha_i(k,k)\mathcal{R}_i + \left(1 - \alpha_i(k,k)\right)\sigma^2 \mathbf{I}, \qquad (6.5)$$

which indicates that the noise variance at time k is $\mathcal{R}_i$, if $\alpha_i(k,k) = 1$, and $\sigma^2 \mathbf{I}$ otherwise.

Suppose that each local estimator maintains an additional detection device at its input terminal to check whether a received measurement $\mathbf{y}_{i,k}^*$ needs to be censored or not. This leads to the following Tobit measurement model:

$$\mathbf{y}_{i,k} = \begin{cases} \mathbf{y}_{i,k}^*, & \mathbf{y}_{i,k}^* > \mathcal{I}_i, \\ \mathcal{I}_i, & \mathbf{y}_{i,k}^* \le \mathcal{I}_i, \end{cases} \qquad (6.6)$$

where $\mathbf{y}_{i,k}$ is the censored measurement, $\mathcal{I}_i = \begin{bmatrix} \mathcal{I}_i^1 & \mathcal{I}_i^2 & \cdots & \mathcal{I}_i^{n_{yi}} \end{bmatrix}^T \in \mathbb{R}^{n_{yi}}$ is the censoring threshold vector of $\mathbf{y}_{i,k}$ and $\mathcal{I}_i^{m_i}$ $\left(m_i = 1,2,\ldots,n_{y_i}\right)$ is the constant censoring threshold of $\mathbf{y}_{i,k}^{m_i}$. Depending on whether $\mathbf{y}_{i,k}$ is censored or not, it is assumed that (6.6) can be transformed into the following form:

$$\mathbf{y}_{i,k} = \Upsilon_{i,k}\mathbf{y}_{i,k}^* + \left(\mathbf{I} - \Upsilon_{i,k}\right)\mathcal{I}_i, \qquad (6.7)$$

where $\Upsilon_{i,k} = \mathrm{diag}\left\{\gamma_{i,k}^1, \gamma_{i,k}^2, \ldots, \gamma_{i,k}^{n_{yi}}\right\}$ and $\gamma_{i,k}^{m_i}$ $\left(m_i = 1,2,\ldots,n_{y_i}\right)$ are known Bernoulli random variables regulating the censoring phenomena of $\mathbf{y}_{i,k}^{m_i}$ with the following probability distribution:

$$\mathrm{Prob}\left\{\gamma_{i,k}^{m_i} = 1\right\} = \overline{\gamma}_i^{m_i}, \quad \mathrm{Prob}\left\{\gamma_{i,k}^{m_i} = 0\right\} = 1 - \overline{\gamma}_i^{m_i}, \qquad (6.8)$$

where $\overline{\gamma}_i^{m_i}$ are known non-negative constants. Similar to the definition of $\Upsilon_{i,k}$, we have $\overline{\Upsilon}_i = \mathrm{diag}\{\overline{\gamma}_i^1, \overline{\gamma}_i^2, \ldots, \overline{\gamma}_i^{n_{yi}}\}$. It is assumed that all the random variables $\gamma_{i,k}^{m_i}$ $\left(m_i = 1,2,\ldots,n_{y_i}\right)$ are mutually independent and are also uncorrelated with $d_{i,k}$ and other noise signals.

After censoring check, a buffer with length $\mathcal{D}_i$ $\left(\mathcal{D}_i \ge 2\right)$ is adopted to store the received measurements for each local estimator. The buffer is able to store all the received measurements from time $k - \mathcal{D}_i + 1$ to time k. The local estimator discards any measurement $\mathbf{y}_{i,k}$ that is delayed by $\mathcal{D}_i$ times or more, which means that if measurement $\mathbf{y}_{i,k-\mathcal{D}_i}$ does not arrive at the local estimator before or at time k, it will be discarded. Based on the stored measurements in the buffer, the local estimator calculates its local estimate $\hat{\mathbf{x}}_{i,k}$, which is further sent to the fusion center for the final fusion. We now have the following assumptions.

Assumption 6.1 The matrix $\overline{\Upsilon}_i \mathbf{C}_i$ is of full rank, and $\left(\overline{\Upsilon}_i \mathbf{C}_i\right)^{-1}$ exists.

Assumption 6.2 The pair $\left(\mathbf{A}, \overline{\Upsilon}_i \mathbf{C}_i\right)$ is observable, and the pair $\left(\mathbf{A}, \sqrt{\mathbf{Q}}\right)$ is controllable.

Assumption 6.3 $d_{i,k}$ and $d_{j,t}$ are independent of each other if $i \neq j$ or $k \neq t$, $\overline{\alpha}_i(k,k)$ is uncorrelated with $\overline{\alpha}_j(t,t)$ if $i \neq j$ or $k \neq t$, and $\mathbf{v}_{i,k}$ and $\mathbf{v}_{j,t}$ are mutually independent if $i \neq j$ or $k \neq t$.

Assumption 6.4 The initial state $\mathbf{x}_0$ has the mean $\overline{\mathbf{x}}_0$ and covariance $\mathbf{P}_0$, and $\mathbf{x}_0$, $\alpha_i(k,k)$, $d_{i,k}$, $\mathbf{w}_k$ and $\mathbf{v}_{i,k}$ are mutually independent.

Define $Y_{i,k}$ and Y_k, respectively, as all the measurements received by the ith local estimator and the fusion center up to time k. Note that the discarded measurements are not included in $Y_{i,k}$ and Y_k. Define $\hat{\mathbf{x}}_{i,k}^{-} \triangleq \mathbb{E}\{\mathbf{x}_k \mid Y_{i,k-1}, \alpha_i(k-1,k-1)\}$,

$$\hat{\mathbf{x}}_{i,k} \triangleq \mathbb{E}\{\mathbf{x}_k \mid Y_{i,k}, \alpha_i(k,k)\}, \quad \hat{\mathbf{x}}_k^{-} \triangleq \mathbb{E}\{\mathbf{x}_k \mid Y_{k-1}\}, \quad \hat{\mathbf{x}}_k \triangleq \mathbb{E}\{\mathbf{x}_k \mid Y_k\}, \quad \hat{\mathbf{y}}_k^{-} \triangleq \mathbb{E}\{\mathbf{y}_k \mid Y_{k-1}\},$$

$$\mathbf{e}_{i,k}^{-} \triangleq \mathbf{x}_k - \hat{\mathbf{x}}_{i,k}^{-} \quad \mathbf{e}_{i,k} \triangleq \mathbf{x}_k - \hat{\mathbf{x}}_{i,k}, \quad \mathbf{e}_k^{-} \triangleq \mathbf{x}_k - \hat{\mathbf{x}}_k^{-}, \quad \mathbf{e}_k \triangleq \mathbf{x}_k - \hat{\mathbf{x}}_k, \quad \mathbf{P}_{ij,k}^{-} \triangleq \mathbb{E}\left\{\mathbf{e}_{i,k}^{-}\left(\mathbf{e}_{j,k}^{-}\right)^T\right\},$$

$$\mathbf{P}_{ij,k} \triangleq \mathbb{E}\left\{\mathbf{e}_{i,k}^T, \mathbf{e}_{j,k}\right\}, \mathbf{P}_k^{-} \triangleq \mathbb{E}\left\{\mathbf{e}_k^{-}\left(\mathbf{e}_k^{-}\right)^T\right\} \text{ and } \mathbf{P}_k \triangleq \mathbb{E}\left\{\mathbf{e}_k^T \mathbf{e}_k\right\}.$$

Remark 6.1 Notice that $Y_{i,k}$ and Y_k are functions of $\mathscr{D}_i$ and $f(j)$, which indicates that for different buffer lengths and delay distributions, $Y_{i,k}$, Y_k as well as the received measurements that are not discarded, will be different. As a result, all the variables defined earlier are functions of $\mathscr{D}_i$ and $f(j)$ and should be denoted as

$$\hat{\mathbf{x}}_{i,k}^{-}\left(\mathscr{D}_i, f(j)\right), \hat{\mathbf{x}}_{i,k}\left(\mathscr{D}_i, f(j)\right), \hat{\mathbf{x}}_k^{-}\left(\mathscr{D}_i, f(j)\right), \hat{\mathbf{x}}_k\left(\mathscr{D}_i, f(j)\right),$$

$$\hat{\mathbf{y}}_k^{-}\left(\mathscr{D}_i, f(j)\right), \mathbf{e}_{i,k}^{-}\left(\mathscr{D}_i, f(j)\right), \mathbf{e}_{i,k}\left(\mathscr{D}_i, f(j)\right), \mathbf{e}_k^{-}\left(\mathscr{D}_i, f(j)\right),$$

$$\mathbf{P}_{ij,k}^{-}\left(\mathscr{D}_i, \mathscr{D}_j, f(j)\right), \mathbf{P}_{ij,k}\left(\mathscr{D}_i, \mathscr{D}_j, f(j)\right), \mathbf{P}_k^{-}\left(\mathscr{D}_i, f(j)\right),$$

and $\mathbf{P}_k\left(\mathscr{D}_i, f(j)\right)$. For simplicity, such variables are written as $\hat{\mathbf{x}}_i$, $\hat{\mathbf{x}}_{i,k}^{-}$, $\hat{\mathbf{x}}_{i,k}$, $\hat{\mathbf{x}}_k^{-}$, $\hat{\mathbf{x}}_k$, $\hat{\mathbf{y}}_k^{-}$, $\mathbf{e}_{i,k}^{-}$, $\mathbf{e}_{i,k}$, $\mathbf{e}_k^{-}$, $\mathbf{e}_k$, $\mathbf{P}_{ij,k}^{-}$, $\mathbf{P}_{ij,k}$, $\mathbf{P}_k^{-}$ and $\mathbf{P}_k$ in the following sections. It should be noted that $\mathbf{P}_k$ reflects how well the global state estimate $\hat{\mathbf{x}}_k$ is close to the true state x_k, and $\mathbf{P}_k$ is a random variable whose randomness comes from the randomness of $d_{i,k}$ $(i = 1, 2, \ldots, l, k = 0, 1, \ldots)$. As such, in this chapter, we only focus on the statistic property of $\mathbf{P}_k$.

Remark 6.2 The filtering problem with delayed measurements has been extensively investigated even before the emergence of the networked control systems, and a vast variety of Kalman-type filters have been developed. When concerning the stability analysis of these Kalman-type filters, the expectation of the estimation error covariance matrix is often utilized as an evaluation index. However, the adoption of such an evaluation index neglects the fact that events with any arbitrarily low probabilities may cause expectation divergence. On the other hand, it is quite common in filtering application areas that sensors are prone to a specific type of measurement nonlinearities called censoring or censored measurements because of saturated sensors, limited detection effects and image frame effects. In the context of filtering with censored measurements, there have been some initial results. It is worth pointing out that the delayed measurements and censored measurements have not been considered simultaneously in the framework of the KF in the multi-sensor case, not to mention the case where the filter performance is evaluated via a probabilistic approach. This motivates us to conduct the present research to shorten such a gap.

Remark 6.3 As described in (6.8), the random variables $\gamma_{i,k}^{m_i}$ $\left(i=1,2,\ldots,l, m_i=1,2,\ldots,n_{y_i}\right)$ are utilized to characterize the censoring phenomena of $\mathbf{y}_{i,k}^{m_i}$. According to (6.8), if no censoring occurs for $\mathbf{y}_{i,k}^{m_i}$, i.e. $\gamma_{i,k}^{m_i}=1$, the measurement output is $\mathbf{y}_{i,k}^{m_i}=\left(\mathbf{y}_{i,k}^{*}\right)^{m_i}$, which implies that the output measurement is the same as the latent measurement. If the censoring phenomena occur for $\mathbf{y}_{i,k}^{m_i}$, i.e. $\gamma_{i,k}^{m_i}=0$, the measurement output is $\mathbf{y}_{i,k}^{m_i}=\mathcal{I}_i^{m_i}$, which implies that the threshold is assigned to the output measurement. Here, the censoring probabilities $\overline{\gamma}_i^{m_i}$ are assumed to be known a priori via some statistical experiments. Alternatively, $\overline{\gamma}_i^{m_i}$ can also be approximated by

$$\overline{\gamma}_i^{m_i} \approx \Phi\left(\frac{\sum_{n=1}^{n_x} \mathbf{C}_i^{m_i n}\left(\hat{\mathbf{x}}_k^-\right)^n - \mathcal{I}_i^{m_i}}{\sqrt{\alpha_i(k,k)\mathcal{R}_i^{m_i m_i}}}\right), \tag{6.9}$$

where $C_i^{m_i n}$ $\left(n=1,2,\ldots,n_x\right)$ and $\mathcal{R}_i^{m_i m_i}$ are, respectively, the (m_i,n) th and (m_i,m_i) th elements of $\mathbf{C}_i$ and $\mathbf{R}_i$; $\hat{\mathbf{x}}_{k|k-1}^n$ is the nth element of $\hat{\mathbf{x}}_{k|k-1}$, which is the one-step prediction of $\mathbf{x}_{k-1}$; and $\Phi(\cdot)$ is the cumulative distribution function of the random variable "·" which obeys the standard normal distribution.

Denote $\mathscr{D}_i$ and $\mathbf{M}_i$, respectively, as the buffer length and the estimation error bound corresponding to the ith local estimator. Define $\mathscr{D} \triangleq \max\left\{\mathscr{D}_i, i=1,2,\ldots,l\right\}$ and $\mathbf{M} \triangleq \min\left\{\mathbf{M}_i, i=1,2,\ldots,l\left(M_i \geq 0\right)\right.$.

Let the system (6.1)–(6.2), the network delay model (6.3) and the Tobit measurement model (6.6) be given. Our objective in this chapter is twofold: (1) for the given measurement collection Y_k, find the globally optimal estimate $\hat{\mathbf{x}}_k$ and its covariance $\mathbf{P}_k$; and (2) for the given delay distribution function f_j $\left(j=0,1,\ldots\right)$, the designated buffer length $\mathscr{D}$, the prescribed estimation error bound $\mathbf{M}$, and any $\theta_1,\theta_2 \in [0,1]$, find the lower and upper bounds for the probability of $\mathbf{P}_k$ such that

$$1-\theta_1 \leq \mathrm{Prob}\left\{\mathbf{P}_k \leq \mathbf{M}\right\} \leq 1-\theta_2 \tag{6.10}$$

6.2 DISTRIBUTED FEDERATED TOBIT KALMAN FILTER WITH PACKET DELAYS

In this section, we aim to establish a unified framework to address the presented distributed federated Tobit Kalman filtering problem over a packet-delaying network. Three types of filters are involved in such a framework, i.e. the MTKF, the LMTKF and the DFMTKF. The derivation of the MTKF has two main differences: (i) a modified Tobit regression model resulting from the packet delays is introduced, and (ii) additional computations of the measurement noise covariance, the gain matrix and the error covariance matrix are included. The LMTKF is devised through running s_i+1 MTKFs where s_i is the number of the measurements stored in the ith buffer. Besides, the DFMTKF is designed by fusing estimates from all the LMTKFs, taking advantage of the well-known federated Kalman fusion rule.

Let $\chi_{i,k}^{m_i} \triangleq \sum_{b=1}^{n_x} C_i^{m_i b} x_k^b$ and $\mathscr{R}_{i,k}^{m_i m_i} \triangleq \alpha_i(k,k)\mathcal{R}_i^{m_i m_i}$. Depending on whether $\mathbf{y}_{i,k}$ is received or not, i.e. $\alpha_i(k,k)=1$ or $\alpha_i(k,k)=0$, the following lemma is presented to modify the standard Tobit regression model to incorporate the packet delay phenomenon.

Lemma 6.1 The mathematical expectation and variance of $\mathbf{y}_{i,k}^{m_i}$ $\left(i=1,2,\ldots,l,\ m_i=1,2,\ldots,n_{y_i}\right)$ are derived as:

$$\mathbb{E}\left\{\mathbf{y}_{i,k}^{m_i}\mid\mathbf{x}_k,\mathbf{R}_{i,k}^{m_i m_i}\right\}=\Phi\left(\frac{\chi_{i,k}^{m_i}-\mathcal{I}_i^{m_i}}{\sqrt{\mathscr{R}_{i,k}^{m_i m_i}}}\right)\mathcal{I}\left[\chi_{i,k}^{m_i}+\sqrt{\mathscr{R}_{i,k}^{m_i m_i}}\,\lambda\left(\frac{\mathcal{I}_i^{m_i}-\chi_{i,k}^{m_i}}{\sqrt{\mathscr{R}_{i,k}^{m_i m_i}}}\right)\right]$$
$$+\Phi\left(\frac{\mathcal{I}_i^{m_i}-\chi_{i,k}^{m_i}}{\sqrt{\mathscr{R}_{i,k}^{m_i m_i}}}\right)\mathcal{I}_i^{m_i}, \tag{6.11}$$

$$Var\left\{\mathbf{y}_{i,k}^{m_i}\mid\mathbf{x}_k,\mathbf{R}_{i,k}^{m_i m_i}\right\}=\mathscr{R}_{i,k}^{m_i m_i}\left[1-\varphi\left(\frac{\mathcal{I}_i^{m_i}-\chi_{i,k}^{m_i}}{\sqrt{\mathscr{R}_{i,k}^{m_i m_i}}}\right)\right], \tag{6.12}$$

where

$$\varphi\left(\frac{\mathcal{I}_i^{m_i}-\chi_{i,k}^{m_i}}{\sqrt{\mathscr{R}_{i,k}^{m_i m_i}}}\right)=\lambda\left(\frac{\mathcal{I}_i^{m_i}-\chi_{i,k}^{m_i}}{\sqrt{\mathscr{R}_{i,k}^{m_i m_i}}}\right)$$
$$\times\left[\lambda\left(\frac{\mathcal{I}_i^{m_i}-\chi_{i,k}^{m_i}}{\sqrt{\mathscr{R}_{i,k}^{m_i m_i}}}\right)-\frac{\mathcal{I}_i^{m_i}-\chi_{i,k}^{m_i}}{\sqrt{\mathscr{R}_{i,k}^{m_i m_i}}}\right], \tag{6.13}$$

$$\lambda\left(\frac{\mathcal{I}_i^{m_i}-\chi_{i,k}^{m_i}}{\sqrt{\mathscr{R}_{i,k}^{m_i m_i}}}\right)=\frac{\phi\left(\dfrac{\mathcal{I}_i^{m_i}-\chi_{i,k}^{m_i}}{\sqrt{\mathscr{R}_{i,k}^{m_i m_i}}}\right)}{1-\Phi\left(\dfrac{\mathcal{I}_i^{m_i}-\chi_{i,k}^{m_i}}{\sqrt{\mathscr{R}_{i,k}^{m_i m_i}}}\right)}, \tag{6.14}$$

with the probability density function (pdf) and the cumulative distribution function (CDF) of the Gaussian random variable $\mathbf{y}_{i,k}^{m_i}$ given by the following equations:

$$\phi\left(\frac{\mathbf{y}_{i,k}^{m_i}-\chi_{i,k}^{m_i}}{\sqrt{\mathscr{R}_{i,k}^{m_i m_i}}}\right)=\frac{1}{\sqrt{2\pi}}e^{-\frac{\left(\eta_{i,k}^{m_i}-\chi_{i,k}^{m_i}\right)^2}{2\mathscr{R}_{i,k}^{m_i m_i}}}, \tag{6.15}$$

$$\Phi\left(\frac{\mathbf{y}_{i,k}^{m_i}-\chi_{i,k}^{m_i}}{\sqrt{\mathscr{R}_{i,k}^{m_i m_i}}}\right)=\int_{-\infty}^{\mathbf{y}_{i,k}^{m_i}}\frac{1}{\sqrt{2\pi\mathscr{R}_{i,k}^{m_i m_i}}}e^{-\frac{\left(\eta_{i,k}^{m_i}-\chi_{i,k}^{m_i}\right)^2}{2\mathscr{R}_{i,k}^{m_i m_i}}}\,d_{\eta_{i,k}^{m_i}}. \tag{6.16}$$

Proof According to (6.7), the pdf of $\mathbf{y}_{i,k}^{m_i}$ with a normally distributed noise $\mathbf{v}_{i,k}^{m_i}$ is given by

$$p\left(\mathbf{y}_{i,k}^{m_i} \mid \mathbf{x}_k, \alpha_{i,(k,k)}\right) = \frac{1}{\sqrt{\mathbf{R}_{i,k}^{m_i m_i}}} \phi\left(\frac{\mathbf{y}_{i,k}^{m_i} - \chi_{i,k}^{m_i}}{\sqrt{\mathbf{R}_{i,k}^{m_i m_i}}}\right) u\left(\mathbf{y}_{i,k}^{m_i} - \mathcal{I}_i^{m_i}\right)$$

$$+ \delta\left(\mathcal{I}_i^{m_i} - \mathbf{y}_{i,k}^{m_i}\right) \Phi\left(\frac{\mathcal{I}_i^{m_i} - \chi_{i,k}^{m_u}}{\sqrt{\mathbf{R}_{i,k}^{m_i m_i}}}\right), \tag{6.17}$$

where $u\left(\mathbf{y}_{i,k}^{m_i} - \mathcal{I}_i^{m_i}\right)$ is the unit step function and $\delta\left(\mathcal{I}_i^{m_i} - \mathbf{y}_{i,k}^{m_i}\right)$ is the Dirac delta function. For (6.17), taking the limit as $\sigma \to \infty$, we have

$$p\left(\mathbf{y}_{i,k}^{m_i} \mid \mathbf{x}_k, \alpha_{i,(k,k)}\right) = \frac{1}{\sqrt{\mathscr{R}_{i,k}^{m_i m_i}}} \phi\left(\frac{\mathbf{y}_{i,k}^{m_i} - \chi_{i,k}^{m_i}}{\sqrt{\mathscr{R}_{i,k}^{m_i m_i}}}\right) u\left(\mathbf{y}_{i,k}^{m_i} - \mathcal{I}_i^{m_i}\right)$$

$$+ \delta\left(\mathcal{I}_i^{m_i} - \mathbf{y}_{i,k}^{m_i}\right) \Phi\left(\frac{\mathcal{I}_i^{m_i} - \chi_{i,k}^{m_i}}{\sqrt{\mathscr{R}_{i,k}^{m_i m_i}}}\right), \tag{6.18}$$

where $\phi\left(\dfrac{\mathbf{y}_{i,k}^{m_i} - \chi_{i,k}^{m_i}}{\sqrt{\mathscr{R}_{i,k}^{m_i m_i}}}\right)$ and $\Phi\left(\dfrac{\mathcal{I}_i^{m_i} - \chi_{i,k}^{m_i}}{\sqrt{\mathscr{R}_{i,k}^{m_i m_i}}}\right)$ are, respectively, given by (6.15) and (6.16).

On the basis of (6.18), we have the following mathematical expectation and variance of $\mathbf{y}_{i,k}^{m_i}$:

$$\mathbb{E}\left\{y_{i,k}^{m_i} \mid x_k, R_{i,k}^{m_i, m_i}\right\} = Prob\left\{\mathbf{y}_{i,k}^{m_i} > \mathcal{I}_i^{m_i} \mid \mathbf{x}_k, \mathbf{R}_{i,k}^{m_i m_i}\right\}$$

$$\mathbb{E}\left\{\mathbf{y}_{i,k}^{m_i} \mid \mathbf{y}_{i,k}^{m_i} > \mathcal{I}_i^{m_i}, \mathbf{x}_k, \mathbf{R}_{i,k}^{m_i m_i}\right\}$$

$$+ Prob\left\{\mathbf{y}_{i,k}^{m_i} = \mathcal{I}_i^{m_i} \mid \mathbf{x}_k, \mathbf{R}_{i,k}^{m_i m_i}\right\} \tag{6.19}$$

$$\mathbb{E}\left\{\mathbf{y}_{i,k}^{m_i} \mid \mathbf{y}_{i,k}^{m_i} = \mathcal{I}_i^{m_i}, \mathbf{x}_k, \mathbf{R}_{i,k}^{m_i m_i}\right\},$$

It can be observed from (6.19) that the computation of $\mathbb{E}\left\{\mathbf{y}_{i,k}^{m_i} \mid \mathbf{x}_k, \mathbf{R}_{i,k}^{m_i m_i}\right\}$ requires the computation of the probabilities and expected values in (6.19).

$$Prob\left\{\mathbf{y}_{i,k}^{m_i} > \mathcal{I}_i^{m_i} \mid \mathbf{x}_k, \mathbf{R}_{i,k}^{m_i m_i}\right\} = Prob\left\{\left(\mathbf{y}_{i,k}^{*}\right)^{m_i} > \mathcal{I}_i^{m_i} \mid \mathbf{x}_k, \mathbf{R}_{i,k}^{m_i m_i}\right\}$$

$$= Prob\left\{\mathbf{v}_{i,k}^{m_i} > \mathcal{I}_i^{m_i} - \chi_{i,k}^{m_i} \mid \mathbf{x}_k, \mathbf{R}_{i,k}^{m_i m_i}\right\} \tag{6.20}$$

$$= \Phi\left(\frac{\chi_{i,k}^{m_i} - \mathcal{I}_i^{m_i}}{\sqrt{\mathscr{R}_{i,k}^{m_i m_i}}}\right),$$

As a result, we arrive at

$$\mathbb{E}\left\{y_{i,k}^{m_i} \mid y_{i,k}^{m_i} > \mathcal{I}_i^{m_i}, x_k, \mathbf{R}_{i,k}^{m_i m_i}\right\}$$

$$= \frac{1}{\sqrt{\mathscr{R}_{i,k}^{m_i m_i}}} \int_{\mathcal{I}_i^{m_i}}^{+\infty} z_{i,k}^{m_i} \frac{\Phi\left(\dfrac{z_{i,k}^{m_i} - \chi_{i,k}^{m_i}}{\sqrt{\mathscr{R}_{i,k}^{m_i m_i}}}\right)}{1 - \Phi\left(\dfrac{\mathcal{I}_i^{m_i} - \chi_{i,k}^{m_i}}{\sqrt{\mathscr{R}_{i,k}^{m_i m_i}}}\right)} d_{z_{i,k}^{m_i}} \qquad (6.21)$$

$$= \chi_{i,k}^{m_i} + \sqrt{\mathscr{R}_{i,k}^{m_i m_i}} \, \lambda\left(\frac{\mathcal{I}_i^{m_i} - \chi_{i,k}^{m_i}}{\sqrt{\mathscr{R}_{i,k}^{m_i m_i}}}\right),$$

where $\lambda\left(\dfrac{\mathcal{I}_i^{m_i} - \chi_{i,k}^{m_i}}{\sqrt{\mathscr{R}_{i,k}^{m_i m_i}}}\right)$ is given by (6.14). Similar to (6.20)–(6.21), we get

$$Prob\left\{\mathbf{y}_{i,k}^{m_i} = \mathcal{I}_i^{m_i} \mid \mathbf{x}_k, \mathbf{R}_{i,k}^{m_i m_i}\right\} = \Phi\left(\frac{\mathcal{I}_i^{m_i} - \chi_{i,k}^{m_i}}{\sqrt{\mathscr{R}_{i,k}^{m_i m_i}}}\right), \qquad (6.22)$$

$$\mathbb{E}\left\{\mathbf{y}_{i,k}^{m_i} \mid \mathbf{y}_{i,k}^{m_i} = \mathcal{I}_i^{m_i}, \mathbf{x}_k, \mathbf{R}_{i,k}^{m_i m_i}\right\} = \mathcal{I}_i^{m_i} \qquad (6.23)$$

Substituting (6.20)–(6.23) into (6.19) leads to (6.11). It follows easily from (6.18) that $Var\left\{\mathbf{y}_{i,k}^{m_i} \mid \mathbf{x}_k, \mathbf{R}_{i,k}^{m_i m_i}\right\} = 0$, and thus we have

$$Var\left\{\mathbf{y}_{i,k}^{m_i} \mid \mathbf{x}_k, \mathbf{y}_{i,k}^{m_i} = \mathcal{I}_i^{m_i}, \mathbf{R}_{i,k}^{m_i m_i}\right\} = Var\left\{\mathbf{y}_{i,k}^{m_i} \mid \mathbf{x}_k, \mathbf{y}_{i,k}^{m_i} > \mathcal{I}_i^{m_i}, \mathbf{R}_{i,k}^{m_i m_i}\right\}$$

$$= \mathbb{E}\left\{\left(\mathbf{y}_{i,k}^{m_i}\right)^2 \mid \mathbf{y}_{i,k}^{m_i} > \mathcal{I}_i^{m_i}, \mathbf{x}_k, \mathbf{R}_{i,k}^{m_i m_i}\right\}$$

$$- \left(\mathbb{E}\left\{\mathbf{y}_{i,k}^{m_i} \mid \mathbf{y}_{i,k}^{m_i} > \mathcal{I}_i^{m_i}, \mathbf{x}_k, \mathbf{R}_{i,k}^{m_i m_i}\right\}\right)^2 \qquad (6.24)$$

$$= \mathscr{R}_{i,k}^{m_i m_i}\left[1 - \varphi\left(\frac{\mathcal{I}_i^{m_i} - \chi_{i,k}^{m_i}}{\sqrt{\mathscr{R}_{i,k}^{m_i m_i}}}\right)\right].$$

which is exactly the same as (6.12), where $\varphi\left(\dfrac{\mathcal{I}_i^{m_i} - \chi_{i,k}^{m_i}}{\sqrt{\mathscr{R}_{i,k}^{m_i m_i}}}\right)$ is given by (6.13). This completes the proof.

Remark 6.4 Equations (6.11)–(6.12) in Lemma 6.1 demonstrate the expressions of the mathematic expectation and variance of the delayed and censored measurement $\mathbf{y}_{i,k}^{m_i}$. Compared with the case where only the single sensor case is considered, one observes that a set of $\mathbb{E}\left\{\mathbf{y}_{i,k}^{m_i} \mid \mathbf{x}_k, \mathbf{R}_{i,k}^{m_i m_i}\right\}$ and $Var\left\{\mathbf{y}_{i,k}^{m_i} \mid \mathbf{x}_k, \mathbf{R}_{i,k}^{m_i m_i}\right\}$ $(i = 1, 2, \dots, l)$, which represents the information from multiple sensors, needs to be calculated in this chapter. Moreover, due to the introduction of the delayed measurements, the known measurement noise variance R_k^{mm} is replaced by the unknown measurement noise variance $\alpha_i(k,k)\mathcal{R}_i^{m_i m_i}$, which characterizes the packet delay effect. If the multi-sensor and packet-delaying phenomena are not taken into account, (6.2) reduces to $\mathbf{y}_k^* = \mathbf{C}\mathbf{x}_k + \mathbf{v}_k$, and equations (6.3)–(6.5) disappear. Then, (6.11) and (6.12) in Lemma 6.1 will, respectively, degrade to (8) and (11) in the standard TKF paper.

Remark 6.5 Taking advantage of (6.11)–(6.16), the Tobit model with only censored measurements in the single-sensor case has been extended to the one subjecting to both delayed and censored measurements in the multi-sensor case. It should be noted that the extra multiple sensors and delayed measurements lead to (1) the computation of a set of expected values $\mathbb{E}\left\{\mathbf{y}_{i,k}^{m_i}|\mathbf{x}_k,\mathbf{R}_{i,k}^{m_i m_i}\right\}$ and variances $Var\left\{\mathbf{y}_{i,k}^{m_i}\mid\mathbf{x}_k,\mathbf{R}_{i,k}^{m_i m_i}\right\}$

$(i=1,2,\ldots,l)$; and (2) the replacement of $\mathbf{R}_k^{mm}$ by $\alpha_i(k,k)\mathcal{R}_i^{m_i m_i}$ in all the equations in relationship to measurement $\mathbf{y}_{i,k}^{m_i}$.

Before proceeding further, we first introduce the following federated Kalman information distribution and fusion principle.

Lemma 6.2 [177]: Suppose that $\hat{\mathbf{x}}_{i,k}$ $(i=1,2,\ldots,l)$ and $\mathbf{P}_{i,k}$ are, respectively, the estimate and covariance of a stochastic n_x-dimension vector $\mathbf{x}_k$, obtained by the ith local estimator, $\hat{\mathbf{x}}_k^f$ and $\mathbf{P}_k^f$ are, respectively, the global optimal estimate and covariance obtained by the fusion estimator under the federated Kalman fusion rule, and $\mathbf{Q}_k$ is the covariance of the process noise. Then, the information-sharing process among the local estimators and the fusion estimator is as follows:

$$\begin{cases} \mathbf{Q}_{i,k-1} = \beta_i^{-1}\mathbf{Q}_{k-1}, \\ \mathbf{P}_{i,k-1} = \beta_i^{-1}\mathbf{P}_{k-1}^f, \\ \hat{\mathbf{x}}_{i,k-1} = \hat{\mathbf{x}}_{k-1}^f, \end{cases} \tag{6.25}$$

where β_i $(i=1,2,\ldots,l)$ are the information-sharing coefficients satisfying $\sum_{i=1}^l \beta_i = 1$. Moreover, the fused estimate and covariance of x_k are given as

$$\begin{cases} \mathbf{P}_k^f = \left(\sum_{i=1}^l \mathbf{P}_{i,k}^{-1}\right)^{-1}, \\ \hat{\mathbf{x}}_k^f = \mathbf{P}_k^f\left(\sum_{i=1}^l \mathbf{P}_{i,k}^{-1}\hat{\mathbf{x}}_{i,k}\right). \end{cases} \tag{6.26}$$

6.2.1 Local Tobit Kalman Filter with Packet Delays

In this subsection, we first calculate the local state estimate $\hat{\mathbf{x}}_{i,k}$ depending on whether $\mathbf{y}_{i,k}$ is received by the local estimator or not at time k. Then, by utilizing the past measurements $\mathbf{y}_{i,k-s_i}$ $(s_i \in [0,\mathcal{D}_i-1])$ stored in the i th buffer, we can recalculate the current estimate $\hat{\mathbf{x}}_{i,k}$

Let

$$\hat{\chi}_{i,k}^- \triangleq \mathbf{C}_i\hat{\mathbf{x}}_{i,k}^-,$$

$$\left(\hat{\chi}_{i,k}^-\right)^{m_i} \triangleq \sum_{b=1}^{n_x}\mathbf{C}_i^{m,b}\left(\hat{\mathbf{x}}_{i,k}^-\right)^b,$$

and $\mathfrak{R}_i \triangleq col\left\{\sqrt{\mathcal{R}_i^{11}}, \sqrt{\mathcal{R}_i^{22}}, \ldots, \sqrt{\mathcal{R}_i^{n_{y_i} n_{y_i}}}\right\}$ col where $\mathcal{R}_i^{m_i m_i}$ $\left(m_i = 1, 2, \ldots, n_{y_i}\right)$ is the $\left(m_i, m_i\right)$ th element of $\mathcal{R}_i$. Suppose that the fused estimate and covariance given by the DFMTKF at time $k{-}1$ are, respectively, $\hat{x}_{k-1}$ and P_{k-1}. According to the information-sharing rule in (6.25), the local estimate $\hat{\mathbf{x}}_{i,k-1}$, covariance $\mathbf{P}_{i,k-1}$ and noise covariance $\mathbf{Q}_{i,k-1}$ initializing the LMTKF are assigned to be $\hat{\mathbf{x}}_{i,k-1} \triangleq \hat{\mathbf{x}}_{k-1}$, $\mathbf{P}_{i,k-1} \triangleq \beta_i^{-1} \mathbf{P}_k$ and $\mathbf{Q}_{i,k-1} \triangleq \beta_i^{-1} \mathbf{Q}_p$, where $\beta_{i,k-1} = \mathbf{P}_{k-1}^{-1} \mathbf{P}_{i,k-1}$. On the basis of Lemma 6.1, the optimal MTKF with packet delays can be derived as follows.

Theorem 6.1 Depending on whether $\mathbf{y}_{i,k}$ is received or not, i.e. $\alpha_i(k,k) = 1$ or $\alpha_i(k,k) = 0$, the values of $\hat{\mathbf{x}}_{i,k}$ and $\mathbf{P}_{i,k}$ for system (6.1)–(6.8) can be computed by

$$\left(\hat{\mathbf{x}}_{i,k}, \mathbf{P}_{i,k}\right) = MTKF\left(\hat{\mathbf{x}}_{i,k-1}, \mathbf{P}_{i,k-1}, \Upsilon_{i,k}, \alpha_i(k,k), \mathbf{y}_{i,k}\right) \tag{6.27}$$

which represents the following set of equations:

$$\hat{\mathbf{x}}_{i,k}^- = \mathbf{A}\hat{\mathbf{x}}_{i,k-1}, \tag{6.28}$$

$$\mathbf{P}_{i,k}^- = \mathbf{A}\mathbf{P}_{i,k-1}^T \mathbf{A}^T + \mathbf{Q}_{i,k-1}, \tag{6.29}$$

$$\mathbf{K}_{i,k} = \Sigma_{\tilde{x}_{i,k}^- \tilde{y}_{i,k}^-} \Sigma_{\tilde{y}_{i,k}^- \tilde{y}_{i,k}^-}^{-1}, \tag{6.30}$$

$$\hat{\mathbf{x}}_{i,k} = \hat{\mathbf{x}}_{i,k}^- + \alpha_i(k,k)\mathbf{K}_{i,k}\left(\mathbf{y}_{i,k} - \hat{\mathbf{y}}_{i,k}^-\right), \tag{6.31}$$

$$\mathbf{P}_{i,k} = \left(I - \alpha_i(k,k)\mathbf{K}_{i,k}\overline{\Upsilon}_{i,k}\mathbf{C}_i\right)\mathbf{P}_{i,k}^-, \tag{6.32}$$

where

$$\Sigma_{\tilde{x}_{i,k}^- \tilde{y}_{i,k}^-} = \mathbf{P}_{i,k}^-\left(\overline{\Upsilon}_{i,k}\mathbf{C}_i\right)^T \tag{6.33}$$

$$\Sigma_{\tilde{y}_{i,k}^- \tilde{y}_{i,k}^-} = \overline{\Upsilon}_{i,k}\mathbf{C}_i\mathbf{P}_{i,k}^-\left(\overline{\Upsilon}_{i,k}\mathbf{C}_i\right)^T$$
$$+ \overline{\alpha}_i(k,k)\mathcal{R}_i\left[I - \varphi\left(\frac{\mathcal{I}_i - \hat{\chi}_{i,k}^-}{\sqrt{\overline{\alpha}_i(k,k)\mathcal{R}_i}}\right)\right]^{-1}, \tag{6.34}$$

$$\varphi\left(\frac{\mathcal{I}_i - \hat{\chi}_{i,k}^-}{\sqrt{\overline{\alpha}_i(k,k)\mathcal{R}_i}}\right) = diag\left\{\varphi\left(\frac{\mathcal{I}^{m_i} - \left(\hat{\chi}_{i,k}^-\right)^{m_i}}{\sqrt{\overline{\alpha}_i(k,k)\mathcal{R}_i^{m_i m_i}}}\right)\right\}. \tag{6.35}$$

The one-step prediction of $\mathbf{y}_{i,k}$ in (6.31) has the following form:

$$\overline{\mathbf{y}}_{i,k}^- = \overline{\Upsilon}_{i,k}\left[\hat{\chi}_{i,k}^- + \sqrt{\overline{\alpha}_i(k,k)\mathcal{R}_i}\,\lambda\left(\frac{\mathcal{I}_i - \hat{\chi}_{i,k}^-}{\sqrt{\overline{\alpha}_i(k,k)\mathcal{R}_i}}\right)\mathfrak{R}_i\right] + \left(\mathbf{I} - \overline{\Upsilon}_{i,k}\right)\mathcal{I}_i, \tag{6.36}$$

where

$$\lambda\left(\frac{\mathcal{I}_i - \hat{\chi}_{i,k}^-}{\sqrt{\bar{\alpha}_i(k,k)\mathcal{R}_i}}\right) = diag\left\{\lambda\left(\frac{\mathcal{I}_i^{m_i} - \left(\hat{\chi}_{i,k}^-\right)^{m_i}}{\sqrt{\bar{\alpha}_i(k,k)\mathcal{R}_i^{m_i m_i}}}\right)\right\}. \tag{6.37}$$

Proof Let $\bar{\mathscr{R}}_{i,k} \triangleq \alpha_i(k,k)\mathcal{R}_i$. Applying the orthogonality projection principle to system (6.1)–(6.8), taking the limit as $\sigma \to \infty$, and noting Assumptions 6.3–6.4, we can directly get (6.28)–(6.32). Referencing (6.11) and noting that $\mathbf{y}_{i,k} = col\left\{\mathbf{y}_{i,k}^1, \mathbf{y}_{i,k}^2, \ldots, \mathbf{y}_{i,k}^{n_{y_i}}\right\}$, we have

$$\hat{\mathbf{y}}_{i,k}^- = \bar{\mathbf{\Upsilon}}_{i,k}\left[\hat{\chi}_{i,k}^- + \sqrt{\bar{\mathscr{R}}_{i,k}}\,\lambda\left(\frac{\mathcal{I}_i - \hat{\chi}_{i,k}^-}{\sqrt{\bar{\mathscr{R}}_{i,k}}}\right)\mathfrak{R}_i\right] + \left(\mathbf{I} - \bar{\mathbf{\Upsilon}}_{i,k}\right)\mathcal{I}_i,$$

which is exactly the same as (6.36) where $\lambda\left(\dfrac{\mathcal{I}_i - \hat{\chi}_{i,k}^-}{\sqrt{\bar{\mathscr{R}}_{i,k}}}\right)$ is given by (6.37). Subtracting (6.36) from (6.7), we have

$$\tilde{\mathbf{y}}_{i,k}^- = \mathbf{\Upsilon}_{i,k}\mathbf{C}_{i,k}\tilde{\mathbf{x}}_{i,k}^- + \mathbf{\Upsilon}_k\left(\upsilon_k - \sqrt{\bar{\mathscr{R}}_{i,k}}\,\lambda\left(\frac{\mathcal{I}_i - \hat{\chi}_{i,k}^-}{\sqrt{\bar{\mathscr{R}}_{i,k}}}\right)\mathfrak{R}_i\right). \tag{6.38}$$

Substituting (6.38) into the definitions of $\Sigma_{\tilde{x}_{i,k}^-\tilde{y}_{i,k}^-}$ and $\Sigma_{\tilde{y}_{i,k}^-\tilde{y}_{i,k}^-}$, we have

$$\Sigma_{\tilde{x}_{i,k}^-\tilde{y}_{i,k}^-} = \mathbb{E}\left\{\tilde{\mathbf{x}}_{i,k}^-\left(\tilde{\mathbf{y}}_{i,k}^-\right)^T\right\}$$

$$= \mathbb{E}\left\{\tilde{\mathbf{x}}_{i,k}^-\left[\mathbf{\Upsilon}_{i,k}\mathbf{C}_{i,k}\tilde{\mathbf{x}}_{i,k}^- + \mathbf{\Upsilon}_k\left(\mathbf{v}_k - \sqrt{\bar{\mathscr{R}}_{i,k}}\,\lambda\left(\frac{\mathcal{I}_i - \hat{\chi}_{i,k}^-}{\sqrt{\bar{\mathscr{R}}_{i,k}}}\right)\mathfrak{R}_i\right)\right]^T\right\}$$

$$= \mathbf{P}_{i,k}^-\left(\bar{\mathbf{\Upsilon}}_{i,k}\mathbf{C}_{i,k}\right)^T,$$

which is exactly the same as (6.33). Meanwhile,

$$\Sigma_{\tilde{y}_{i,k}^-\tilde{y}_{i,k}^-} = \mathbb{E}\left\{\tilde{\mathbf{y}}_{k|k-1}\tilde{\mathbf{y}}_{k|k-1}^T\right\}$$

$$= \bar{\mathbf{\Upsilon}}_{i,k}\mathbf{C}_{i,k}\mathbf{P}_{i,k}^-\left(\bar{\mathbf{\Upsilon}}_{i,k}\mathbf{C}_{i,k}\right)^T$$

$$+ \mathbb{E}\left\{\mathbf{\Upsilon}_k\left(\mathbf{v}_k - \sqrt{\bar{\mathscr{R}}_{i,k}}\,\lambda\left(\frac{\mathcal{I}_i - \hat{\chi}_{i,k}^-}{\sqrt{\mathcal{I}_{i,k}}}\right)\mathfrak{R}_i\right) \right. \tag{6.39}$$

$$\left. \times\left(\mathbf{v}_k - \sqrt{\bar{\mathscr{R}}_{i,k}}\,\lambda\left(\frac{\mathcal{I}_i - \hat{\chi}_{i,k}^-}{\sqrt{\bar{\mathscr{R}}_{i,k}}}\right)\mathfrak{R}_i\right)^T \mathbf{\Upsilon}_k^T\right\}.$$

For simplicity of the deduction, we assume that $cov\{\mathbf{y}_{i,k}^m, \mathbf{y}_{i,k}^t\} = 0$ for $m \neq t$, $m, t = 1, 2, \ldots, n_{y_i}$. The generalization of the result to the case where $cov\{\mathbf{y}_{i,k}^m, \mathbf{y}_{i,k}^t\} \neq 0$ for $m \neq t$ is straightforward but will bring notational burdens. Recalling the definition of the variance of $\mathbf{y}_k$ and noting (6.12), we have

$$
\mathbb{E}\left\{ \boldsymbol{\Upsilon}_k \left(\mathbf{v}_k - \sqrt{\mathscr{R}_{i,k}}\, \lambda \left(\frac{\mathcal{I}_i - \hat{\mathcal{X}}_{i,k}^-}{\sqrt{\mathscr{R}_{i,k}}} \right) \mathfrak{R}_i \right) \right.
$$

$$
\left. \times \left(\mathbf{v}_k - \sqrt{\mathscr{R}_{i,k}}\, \lambda \left(\frac{\mathcal{I}_i - \hat{\mathcal{X}}_{i,k}^-}{\sqrt{\mathscr{R}_{i,k}}} \right) \mathfrak{R}_i \right)^T \boldsymbol{\Upsilon}_k^T \right\}
$$

$$
= Var\{\mathbf{y}_{i,k} \mid \mathbf{x}_k, \mathbf{R}_{i,k}\}
$$

$$
= diag\left\{ Var\{\mathbf{y}_{i,k}^1 \mid \mathbf{x}_k, \mathbf{R}_{i,k}^{11}\}, \ldots, Var\{\mathbf{y}_{i,k}^{n_{y_i}} \mid \mathbf{x}_k, \mathbf{R}_{i,k}^{n_{y_i} n_{y_i}}\} \right\}
$$

$$
= \bar{\alpha}_i(k,k)\, \mathcal{R}_i \left[I - \varphi\left(\frac{\mathcal{I}_i - \hat{\mathcal{X}}_{i,k}^-}{\sqrt{\mathscr{R}_{i,k}}} \right) \right],
$$

$$ \tag{6.40} $$

where the second equality holds from the fact that $cov\{\mathbf{y}_{i,k}^m, \mathbf{y}_{i,k}^t\} \neq 0$ for $m \neq t$, and $\varphi\left(\frac{\mathcal{I}_i - \hat{\mathcal{X}}_{i,k}^-}{\mathscr{R}_{i,k}} \right)$ is given by (6.35). Substituting (6.40) into (6.39) yields (6.34). This completes the proof.

Remark 6.6 Two main differences can be found when comparing the standard TKF with the established MTKF in Theorem 6.1. One is the replacement of the gain matrix $\mathbf{K}_{i,k}$ by $\alpha_i(k,k)\mathbf{K}_{i,k}$ in the calculations of the state estimate in (6.32) as well as its covariance in (6.33). The other is the replacement of $\mathbf{R}_i$ by $\alpha_i(k,k)\mathbf{R}_i$ in all the equations with respect to the calculations of the one-step measurement prediction as well as its noise covariance. Both differences clearly characterize the effect of the packet-delaying phenomenon on the design of the filtering algorithm.

Since each local estimator maintains a buffer with length $\mathscr{D}_i$ to store measurements (delayed no more than $\mathscr{D}_i$ times) to tackle the network delays, there is a possibility that measurement $\mathbf{y}_{i,k-s_i}$ $(s_i \in [0, \mathscr{D}_i - 1])$ arrives at the local estimator at time k. As a result, the local estimation accuracy can be improved by recalculating $\hat{\mathbf{x}}_{i,k-s_i}$ with the newly arriving measurement $\mathbf{y}_{i,k-s_i}$. Once $\hat{\mathbf{x}}_{i,k-s_i}$ is updated, $\hat{\mathbf{x}}_{i,k-s_i+1}, \hat{\mathbf{x}}_{i,k-s_i+2}, \ldots, \hat{\mathbf{x}}_{i,k}$ can be updated in a similar way.

Similar to the definition of $\alpha_i(k,k)$, we let $\alpha_i(k,t)$ be the indicator function for $\mathbf{y}_{i,t}$ at time k, $t \leq k$. If $\mathbf{y}_{i,t}$ arrives at the local estimator at time k, $\alpha_i(k,t) = 1$, otherwise, $\alpha_i(k,t) = 0$. Further define $\alpha_{i,k-s_i} \triangleq \sum_{\kappa_i=0}^{s_i} \alpha_i(k - \kappa_i, k - s_i)$. Then, $\alpha_{i,k-s_i} = 1$ indicates that $\mathbf{y}_{i,k-s_i}$ arrives at the local estimator at or before time k, while $\alpha_{i,k-s_i} = 0$ indicates that $\mathbf{y}_{i,k-s_i}$ does not arrive at the local estimator at or before time k and will be

TABLE 6.1

The Pseudocode of the Distributed Federated Modified TKF

Algorithm: Distributed federated modified TKF

Input: $\overline{\mathbf{x}}_0$, $\mathbf{P}_0$, Y_k

Output: $\hat{\mathbf{x}}_k$, $\mathbf{P}_k$

1: let $\hat{\mathbf{x}}_{i,0} = \overline{\mathbf{x}}_0$, $\mathbf{P}_{i,0} = \mathbf{P}_0\beta_i^{-1}$ and $\mathbf{Q}_{i,0} = \mathbf{Q}_0\beta_i^{-}$.

2: **for** $k = 1 : N$ **do**

3: **If** $k \leq \mathscr{D}_i$

4: $s_i = k - 1$;

5: **else** $s_i = \mathscr{D}_i - 1$;

6: **else if**

7: **for** $j = s_i : -1 : 0$ **do**

8: calculate $\hat{\mathbf{x}}_{i,k-j}$ and $\mathbf{P}_{i,k-j}$ by Eqs. (6.28)–(6.37);

9: **end for**

10: calculate $\hat{\mathbf{x}}_k$ and $\mathbf{P}_k$ by Eq. (6.42);

11: calculate $\mathbf{Q}_{i,k}$, $\mathbf{P}_{i,k}$ and $\hat{\mathbf{x}}_{i,k}$ by Eq. (6.25);

12: **end for**

---

discarded by the local estimator even if it arrives after time k. The following theorem summarizes the recalculation process using measurement $\mathbf{y}_{i,k-s_i}$ $\left(s_i \in [0, \mathscr{D}_i - 1]\right)$.

Theorem 6.2 Let $\mathbf{y}_{i,k-s_i}$ $\left(s_i \in [0, \mathscr{D}_i - 1]\right)$ be the oldest measurement arriving at the ith local estimator at time k. Then, $\hat{\mathbf{x}}_{i,k}$ and $\mathbf{P}_{i,k}$ can be recalculated utilizing the following LMTKF which consists of $s_i + 1$ MTKFs:

$$\begin{cases} \left(\hat{\mathbf{x}}_{i,k-s_i}, \mathbf{P}_{i,k-s_i}\right) \\ = MTKF\left(\hat{\mathbf{x}}_{i,k-s_i-1}, \mathbf{P}_{i,k-s_i-1}, \Upsilon_{i,k-s_i}, 1, \mathbf{y}_{i,k-s_i}\right) \\ \left(\hat{\mathbf{x}}_{i,k-s_i+1}, \mathbf{P}_{i,k-s_i+1}\right) \\ = MTKF\left(\hat{\mathbf{x}}_{i,k-s_i}, \mathbf{P}_{i,k,-s_i}, \Upsilon_{i,k-s_i+1}, \alpha_{i,k-s_i}, \mathbf{y}_{i,k-s_i+1}\right), \\ \vdots \\ \left(\hat{\mathbf{x}}_{i,k}, \mathbf{P}_{i,k}\right) \\ = MTKF\left(\hat{\mathbf{x}}_{i,k-1}, \mathbf{P}_{i,k-1}, \Upsilon_{i,k}, \alpha_{i,k}, \mathbf{y}_{i,k}\right). \end{cases} \tag{6.41}$$

Proof Since $\alpha_{i,k-s_i} = \sum_{\kappa_i=0}^{s_i} \alpha_i\left(k-\kappa_i, k-s_i\right)$, we know that $\alpha_i(k,k) = \alpha_{i,k}$. Paying attention to the MTKF in Theorem 6.1, we find that the computation of the state estimate $\hat{\mathbf{x}}_{i,k}$ needs the predicted value $\hat{\mathbf{x}}_{i,k}^-$, the measurement $\mathbf{y}_{i,k}$, the censoring probability $\overline{\Upsilon}_{i,k}$ together with $\alpha_{i,k}$ at time k. Similarly, the computation of the state estimate $\hat{\mathbf{x}}_{i,k-1}$ needs the predicted value $\hat{\mathbf{x}}_{i,k-1}^{-}$, the measurement $\mathbf{y}_{i,k-1}$, the censoring probability $\overline{\Upsilon}_{i,k-1}$ together with $\alpha_{i,k-1}$ at time $k-1$. Along this recursive process, we finally find that the computation of the state estimate $\hat{\mathbf{x}}_{i,k-s_i}$ needs the predicted value $\hat{\mathbf{x}}_{i,k-s_i}^{-}$, the measurement $\mathbf{y}_{i,k-s_i}$, the censoring probability $\overline{\Upsilon}_{i,k-s_i}$, together with $\alpha_{i,k-s_i}$ at time $k-s_i$. Such a recursive process from time k to time $k-s_i$ with s_i+1 steps is equivalent to s_i+1 LMTKFs, which are presented in (6.41). This completes the proof.

Remark 6.7 By defining a new Bernoulli random variable $\alpha_{i,k-s_i}$, the relationship between the current estimate $\hat{\mathbf{x}}_{i,k}$ and the past measurement $\mathbf{y}_{i,k-s_i}$ is established as shown in the presented LMTKF. It is worth mentioning that though measurement $\mathbf{y}_{i,k-s_i}$ cannot be used in calculating $\hat{\mathbf{x}}_{i,k-s_i}$ at time $k-s_i$ if $\alpha_i\left(k-s_i,k-s_i\right)=0$, $\mathbf{y}_{i,k-s_i}$ may arrive at the local estimator in the period from time $k-s_i$ to k. Once $\mathbf{y}_{i,k-s_i}$ arrives before or at time k, it will be stored in the buffer and used to recalculate $\hat{\mathbf{x}}_{i,k-s_i},\hat{\mathbf{x}}_{i,k-s_i+1},\ldots,\hat{\mathbf{x}}_{i,k}$ recursively using the s_i+1 MTKFs in Theorem 6.2, which undoubtedly improves the estimation accuracy of the local estimator.

6.2.2   Distributed Tobit Kalman Filter with Packet Delays

Based on the developed LMTKF, in this subsection, we further study the distributed modified Tobit Kalman filter fusion, taking advantage of the federated Kalman fusion criterion presented in Lemma 6.2.

With the help of Lemma 6.2 local estimates $\hat{\mathbf{x}}_{i,k}$ $(i=1,2,\ldots,l)$ from LMTKFs are further collected at the fusion center, and a DFMTKF is established to generate the fused estimate $\hat{\mathbf{x}}_k$ and covariance $\mathbf{P}_k$ as shown in the following theorem.

Theorem 6.3 Given system (6.1)–(6.2), network delay model (6.3), and Tobit measurement model (6.6), its optimal DFMTKF is given by

$$\begin{cases} \mathbf{P}_k = \left(\displaystyle\sum_{i=1}^{l} \mathbf{P}_{i,k}^{-1} \right)^{-1}, \\[2ex] \hat{\mathbf{x}}_k = \mathbf{P}_k \left(\displaystyle\sum_{i=1}^{l} \mathbf{P}_{i,k}^{-1}\hat{\mathbf{x}}_{i,k} \right). \end{cases} \tag{6.42}$$

Proof Theorem 6.3 follows easily from Theorem 6.2 and Lemma 6.2.

Lemma 6.1 and Theorems 6.1–6.3 constitute the design of the DFMTKF with multi-sensor structure, measurement censoring and packet delays, and Table 6.1 demonstrates its Pseudocode.

Remark 6.8 According to Lemma 6.1 and Theorems 6.1–6.3, a distributed recursive filtering scheme is established to solve the novel filtering problem for the discrete time-varying systems subject simultaneously to measurement censoring and packet-delaying phenomena. System (6.1)–(6.8) under consideration is comprehensive that covers two important measurement uncertainties, i.e. censored measurements and delayed measurements, which are often encountered in engineering applications including networked control, remote sensing and vehicle positioning. The two measurement uncertainties are dealt with in a unified yet effective framework. It should be pointed out that the developed DFMTKF in Theorem 6.3 is an extension of the existing filters. In the case that the measurement is not censored, i.e. $y_{i,k}^* = y_{i,k}$ and $\Upsilon_{i,k} = I$, the LMTKF in Theorem 6.2 reduces to the filter in [176], while the DFMTKF in Theorem 6.3 reduces to the distributed counterpart of the filter in [176]. In the case that the delay phenomenon does not exist, i.e. $\alpha_i(k,k) = 1$ for $k = 0,1,\ldots$, the filters in Theorem 6.1 and Theorem 6.2 become the same and are reduced to the standard TKF. Accordingly, the DFMTKF in Theorem 6.3 reduces to the distributed counterpart of the TKF.

6.3 A PROBABILISTIC PERSPECTIVE

In this section, we try to give a complete characterization of the filtering performance of the proposed DFMTKF via a probabilistic approach under the assumption that the probability distribution $f(j)$ $(j = 0,1,\ldots)$ of the delay and the buffer length $\mathcal{D}_i$ $(i = 1,2,\ldots,l)$ are known. Given the prescribed bound $\mathbf{M}$ for $\mathbf{P}_k$, we present the lower and upper bounds for the probability at which the filtering error covariance of the DFMTKF is within a prescribed bound, i.e. $\mathrm{Prob}\{\mathbf{P}_k \leq \mathbf{M}\}$.

For convenience of the subsequent derivation, let $\bar{\alpha}_i$, φ_i, $\mathbf{Q}_i$ and $\mathcal{R}_i$, respectively, be the time-invariant counterparts of $\bar{\alpha}_i(k,k)$, $\varphi\left(\dfrac{\mathcal{I}_i - C_i \hat{x}_{i,k}^-}{\sqrt{\bar{\alpha}_i \mathcal{R}_i}}\right)$, $\mathbf{Q}_{i,k}$ and $\mathcal{R}_{i,k}$. Denote $X_i \in \mathbb{S}_+^{n_x}$ $(i = 1,2,\ldots,l)$ where $\mathbb{S}_+^{n_x}$ is the set of n_x by n_x positive semi-definite matrices. Let $h(X_i)$, $g(X_i)$ and $\tilde{g}(X_i)$ be the following matrix functions:

$$h(\mathbf{X}_i) = \mathbf{A}\mathbf{X}_i\mathbf{A}^T + \mathbf{Q}_i,$$

$$g(\mathbf{X}_i) = h(\mathbf{X}_i) - \mathbf{A}\mathbf{X}_i(\bar{\Upsilon}_i C_i)^T \bar{\Upsilon}_i^T\left[(\bar{\Upsilon}_i C_i)\mathbf{X}_i(\bar{\Upsilon}_i C_i)^T\right.$$
$$\left. + \mathcal{R}_i(\mathbf{I} - \varphi_i)\right]^{-1}(\bar{\Upsilon}_i C_i)\mathbf{X}_i\mathbf{A}^T,$$

$$\tilde{g}(\mathbf{X}_i) = \mathbf{X}_i - \mathbf{X}_i(\bar{\Upsilon}_i C_i)^T\left[(\bar{\Upsilon}_i C_i)\mathbf{X}_i(\bar{\Upsilon}_i C_i)^T\right.$$
$$\left. + \mathcal{R}_i(\mathbf{I} - \varphi_i)\right]^{-1}(\bar{\Upsilon}_i C_i)\mathbf{X}_i,$$

$$h \circ g(\mathbf{X}_i) = h(g(\mathbf{X}_i)),$$

$$h^t(\mathbf{X}_i) = \underbrace{h \circ \cdots \circ h(\mathbf{X}_i)}_{t \text{ times}}.$$

If we assume $\sum_{i=1}^{l} \alpha_i(k,k) = l$, then the LMTKF in Theorem 6.1 will reduce to the standard TKF. In this case, $\mathbf{P}_{i,k}^-$ in (6.30) and $\mathbf{P}_{i,k}$ in (6.33) will, respectively, satisfy $\mathbf{P}_{i,k}^- = g(\mathbf{P}_{i,k-1}^-)$ and $\mathbf{P}_{i,k} = \tilde{g} \circ h(\mathbf{P}_{i,k-1})$. Since $(\mathbf{A}, \bar{\Upsilon}_i C_i)$ is observable and $(\mathbf{A}, \sqrt{\mathbf{Q}})$ is controllable, from standard Tobit Kalman filtering analysis, we can draw the conclusion that there exists a unique solution $\mathbf{P}_i^*$ to $g(\mathbf{X}_i) = \mathbf{X}_i$. Let $\mathbf{P}_i^*$ be the unique solution to $g(\mathbf{X}_i) = \mathbf{X}_i$, i.e. $\mathbf{P}_i^* = g(\mathbf{P}_i^*)$. Defining $\bar{\mathbf{P}}_i \triangleq \tilde{g}(\mathbf{P}_i^*)$, we have

$$\tilde{g} \circ h\left(\bar{\mathbf{P}}_i\right) = \tilde{g} \circ h \circ \tilde{g}\left(\mathbf{P}_i^*\right) = \tilde{g} \circ g\left(\mathbf{P}_i^*\right) - \tilde{g}\left(\mathbf{P}_l^*\right) = \bar{\mathbf{P}}_i, \qquad (6.43)$$

where the second equality holds from the fact that $h \circ \tilde{g} = g$. (6.43) implies that $\mathbf{P}_i^* = \lim_{k \to \infty} \mathbf{P}_{i,k}^-$ and $\bar{P}_i = \lim_{k \to \infty} \mathbf{P}_{i,k}$. Defining $\bar{\mathbf{P}} \triangleq \lim_{k \to \infty}$, it follows easily from (6.43) that

$$\bar{\mathbf{P}} = \left(\sum_{i=1}^{l} \bar{\mathbf{P}}_i^{-1} \right)^{-1}. \qquad (6.44)$$

Let $F(j) \triangleq \sum_{\lambda=0}^{j} f(\lambda)$ be the CDF of $d_{i,k}$,

$$\mathrm{Prob}\left\{\alpha_{i,k-j} = 1\right\} \triangleq \begin{cases} F(j), & \text{if } j \leq \mathscr{D}_i, \\ F\left(\mathscr{D}_i - 1\right), & \text{if } j > \mathscr{D}_i, \end{cases}$$

$$\bar{\mathbf{M}}_i \triangleq \left(\bar{\Upsilon}_i \mathbf{C}_i\right)^{-1} \mathscr{R}_i \left(\mathbf{I} - \varphi_i\right)\left[\left(\bar{\Upsilon}_i \mathbf{C}_i\right)^{-1}\right]^T,$$

$$\bar{\mathbf{M}} \triangleq \min\left\{\bar{\mathbf{M}}_1, \bar{\mathbf{M}}_2, \ldots, \bar{\mathbf{M}}_l\right\},$$

For $\mathbf{M}_i \geq \bar{\mathbf{M}}_i \ (i = 1, 2, \ldots, l)$, define

$$k_1(\mathbf{M}) \triangleq \min\left\{t \geq 1 : h^t\left(\bar{\mathbf{M}}\right) > \mathbf{M}\right\},$$

$$k_2(\mathbf{M}) \triangleq \min\left\{t \geq 1 : h^t\left(\bar{\mathbf{P}}\right) > \mathbf{M}\right\}.$$

We then have the following lemma describing the relationships between $\mathbf{P}_k$ and $\bar{\mathbf{M}}$, $\bar{P}$ and $\bar{\mathbf{M}}$, $k_1(\mathbf{M})$ and $k_2(\mathbf{M})$ and $\mathbf{P}_k$ and $\bar{\mathbf{P}}$.

Lemma 6.3 (1) For any $k \geq 1$ and $i = 1, 2, \ldots, l$, if $\alpha_{i,k} = 1$, then $\mathbf{P}_k \leq \bar{\mathbf{M}}$; (2) $\bar{\mathbf{P}} \leq \bar{\mathbf{M}}$; (3) if $k_1(\mathbf{M}) \leq \infty$ then $k_1(\mathbf{M}) \leq k_2(\mathbf{M})$; (4) for $i = 1, 2, \ldots, l$, if $\mathbf{P}_{i,0} \geq \bar{\mathbf{P}}_i$, then $\mathbf{P}_k \leq \bar{\mathbf{P}}$ for all $k \geq 0$.

Proof (1) If $\alpha_{i,k} = 1$, by noticing Assumption 6.1, we obtain that $\mathbf{P}_{i,k} \leq \bar{\mathbf{M}}_i$ similar to the proof of Lemma 3.4 in [176]. It follows directly from (6.42) that $\mathbf{P}_k \leq \mathbf{P}_{i,k}$, and hence, we have $\mathbf{P}_k \leq \mathbf{P}_{i,k} \leq \bar{\mathbf{M}}_i$ for all $i = 0, 1, \ldots, l$. Since $\bar{\mathbf{M}} = \min\left\{\bar{\mathbf{M}}_1, \bar{\mathbf{M}}_2, \ldots, \bar{\mathbf{M}}_p\right\}$, we have $\mathbf{P}_k \leq \bar{\mathbf{M}}$ from the fact that $\mathbf{P}_k \leq \bar{\mathbf{M}}_i$ for all $i = 0, 1, \ldots, l$.

(2) According to Lemma 3.6 in [176], we have $\bar{\mathbf{P}}_i \leq \bar{\mathbf{M}}_i$ for all $i = 0, 1, \ldots, l$. Recalling that $\bar{P}_i$ and $\bar{P}$ are, respectively, the steady estimation error covariance matrices for the ith LMTKF in Theorem 6.2 and the DMTKF in Theorem 6.3, we arrive at $\bar{\mathbf{P}} \leq \mathbf{P}_k \leq \bar{\mathbf{P}}_i \leq \mathbf{P}_{i,k}$, i.e. $\bar{\mathbf{P}} \leq \bar{\mathbf{P}}_i$. As a result, we have $\bar{\mathbf{P}} \leq \bar{\mathbf{P}}_i \leq \bar{M}_i$ for all $i = 0, 1, \ldots, l$. Since $\bar{\mathbf{M}} = \min\left\{\bar{\mathbf{M}}_1, \bar{\mathbf{M}}_2, \ldots, \bar{\mathbf{M}}_p\right\}$, we have $\bar{\mathbf{P}} \leq \bar{\mathbf{P}}_i \leq \bar{\mathbf{M}} \leq \bar{\mathbf{M}}_i$ for all $i = 0, 1, \ldots, l$. Therefore, we have $\bar{\mathbf{P}} \leq \bar{\mathbf{M}}$.

(3) From Lemmas A.1–A.2 in [176], we know that function $h(\cdot)$ is nondecreasing and $\bar{\mathbf{P}}_i \leq h\left(\bar{\mathbf{P}}_i\right)$. As $\bar{\mathbf{P}}$ is the steady estimation error covariance matrix for the DMTKF, $h\left(\bar{\mathbf{P}}\right) = \mathbf{A}\bar{\mathbf{P}}\mathbf{A}^T + \mathbf{Q}$ becomes the steady prediction error covariance matrix for the LMTKF as well as the DMTKF, according to the information-sharing theory. Due to the fact that the steady estimation error covariance should always be smaller than the

steady prediction error covariance, we have $\overline{\mathbf{P}} \leq h(\overline{\mathbf{P}})$. Without loss of generality, we assume that $k_2(\mathbf{M}) < \infty$. If $k_1(\mathbf{M}) > k_2(\mathbf{M})$, we have the following equation by making reference to the nondecreasing property of function $h(\cdot)$ and the fact that $\overline{\mathbf{P}} \leq h(\overline{\mathbf{P}})$.

$$\mathbf{M} > h^{k_1(\mathbf{M})-1}\left(\overline{\mathbf{M}}\right) \geq h^{k_1(\mathbf{M})-1}\left(\overline{\mathbf{P}}\right) \geq h^{k_2(\mathbf{M})}\left(\overline{\mathbf{P}}\right),$$

which goes against the definition of $k_2(\mathbf{M})$. As such, we have $k_1(\mathbf{M}) \leq k_2(\mathbf{M})$.

(4) According to Lemma 3.7 in [176], we know that if $\mathbf{P}_{i,0} \geq \overline{\mathbf{P}}_i$, then $\mathbf{P}_{i,k} \geq \overline{\mathbf{P}}_i$ for all $k \geq 0$. Making reference to Theorem 6.3 and (6.44), we have

$$\mathbf{P}_k = \left(\sum_{i=1}^{l} \mathbf{P}_{i,k}^{-1}\right)^{-1} \geq \left(\sum_{i=1}^{l} \overline{\mathbf{P}}_i^{-1}\right)^{-1} = \overline{\mathbf{P}},$$

for all $k \geq 0$. This completes the proof.

Let N_k be the number of consecutive packets that are not received by any LMTKF at time k, i.e. $N_k \triangleq \min\left\{t \geq 0 : \sum_{i=1}^{l} \alpha_{i,k-t} = 1\right\}$. Then, N_k is the minimum of a sequence of independent Bernoulli random variables. Thus, if we define $\theta\left(k_p(\mathbf{M}), \mathscr{D}\right) \triangleq \prod_{i=1}^{l} \prod_{j=0}^{k_p(\mathbf{M})-1}\left(1 - \mathrm{Prob}\{\alpha_{i,k-j} = 1\}\right)$, $p = 1, 2$, we can easily obtain $\mathrm{Prob}\{N_k \geq k_p(\mathbf{M})\} = \theta\left(k_p(\mathbf{M}), \mathscr{D}\right)$.

With the definitions of $k_1(\mathbf{M}), k_2(\mathbf{M})$ and N_k, we are now in the position to present the lower and upper bounds for $\mathrm{Prob}\{\mathbf{P}_k \leq \mathbf{M}\}$.

Theorem 6.4 Assume that Lemma 6.3 holds and $\overline{\mathbf{P}} \leq \mathbf{P}_0 \leq \overline{\mathbf{M}}$, and given the prescribed bound $\mathbf{M}$, the buffer length $\mathscr{D}$, and the delay distribution function $f(j)$ in (6.3). Then, for any $\mathbf{M} \geq \overline{\mathbf{M}}$, we have

$$1 - \theta\left(k_1(\mathbf{M}), \mathscr{D}\right) \leq Prob\{\mathbf{P}_k \leq \mathbf{M}\} \leq 1 - \theta\left(k_2(\mathbf{M}), \mathscr{D}\right). \tag{6.45}$$

Proof Let us first prove $1 - \theta\left(k_1(\mathbf{M}), \mathscr{D}\right) \leq Prob\{\mathbf{P}_k \leq \mathbf{M}\}$, which is equivalent to $1 - Prob\{N_k \geq k_1(\mathbf{M})\} \leq Prob\{\mathbf{P}_k \leq \mathbf{M}\}$. Since $\alpha_{i,k} = 0$ or 1 for $i = 1, 2, \ldots, l$, there are totally $\left(2^k\right)^l$ possible combinations of $\alpha_{i,1}$ to $\alpha_{i,k}$. Define a matrix $\Gamma \in \mathbb{R}^{l \times k}$ as follows to represent the packet arrival sequences with respect to $\alpha_{i,1}, \alpha_{i,2}, \ldots, \alpha_{i,k}$.

$$\Gamma = \begin{bmatrix} \alpha_1(k,1) & \alpha_1(k,2) & \cdots & \alpha_1(k,k) \\ \alpha_2(k,1) & \alpha_2(k,2) & \cdots & \alpha_2(k,k) \\ \vdots & \vdots & \ddots & \vdots \\ \alpha_l(k,1) & \alpha_l(k,2) & \cdots & \alpha_l(k,k) \end{bmatrix}.$$

Let Σ_1 and Σ_2 be subsets of Γ where Σ_1 and Σ_2, respectively, stand for the packet arrival sequences such that $N_k \geq k_1(\mathrm{M})$ and $N_k < k_1(\mathrm{M})$. Let $\sigma_p \in \Sigma_p$ $(p = 1, 2)$ be the packet arrival sequence with respect to $\alpha_{i,1}, \alpha_{i,2}, \ldots, \alpha_{i,k}$ and $\mathbf{P}_k\left(\sigma_p\right)$ be the corresponding estimation error covariance matrix at time k. Consider a particular packet arrival sequence $\sigma_2 \in \Sigma_2$ where $\sum_{i=1}^{l} \alpha_{i,k-k_1(\mathrm{M})+1} = l$, i.e. $\alpha_{i,k-k_1(\mathrm{M})+1} = 1$. For $\sigma_2 \in \Sigma_2$, we know that $N_k \le k_1(\mathrm{M}) - 1$, and thus we have

$$\mathbf{P}_k\left(\sigma_2\right) = h^{N_k}\left(\mathbf{P}_{k-N_k}\right) \le h^{k_1(\mathrm{M})-1}\left(\mathbf{P}_{k-k_1(\mathrm{M})+1}\right), \tag{6.46}$$

where the equality holds due to the fact that no measurements arrive from time $k - k_1(\mathbf{M}) + 1$ to time k at any LMTKFs, and the inequality holds from the fact that function $h(\cdot)$ is nondecreasing. As $\alpha_{i,k-k_1(\mathbf{M})+1} = 1$, we have the following equation from Assumption 6.1 and Lemma 6.3.

$$P_{k-k_1(\mathbf{M})+1} \le \bar{\mathbf{M}}. \tag{6.47}$$

Combining (6.46)–(6.47), we certainly have

$$P_k\left(\sigma_2\right) \le h^{k_1(\mathbf{M})-1}\left(P_{k-k_1(\mathbf{M})+1}\right) \le h^{k_1(\mathbf{M})-1}\left(\bar{\mathbf{M}}\right) \le \mathbf{M}, \tag{6.48}$$

where the last inequality holds from the definition of $k_1(\mathbf{M})$. (6.48) implies that $Prob\{P_k \le \mathbf{M} \mid \sigma_2\} = 1$. As a result, we have

$$\begin{aligned}
Prob\{P_k \le \mathbf{M}\} &= \sum_{\sigma_1 \in \Sigma_1} Prob\{\mathbf{P}_k \le \mathbf{M} \mid \sigma_1\} Prob\{\sigma_1\} \\
&\quad + \sum_{\sigma_2 \in \Sigma_2} Prob\{\mathbf{P}_k \le \mathbf{M} \mid \sigma_2\} Prob\{\sigma_2\} \\
&\ge \sum_{\sigma_2 \in \Sigma_2} Prob\{\mathbf{P}_k \le \mathbf{M} \mid \sigma_2\} Prob\{\sigma_2\} \\
&= \sum_{\sigma_2 \in \Sigma_2} Prob\{\sigma_2\} \\
&= Prob\{\Sigma_2\} \\
&= 1 - Prob\{\Sigma_1\} \\
&= 1 - Prob\{N_k \ge k_1(\mathbf{M})\}.
\end{aligned}$$

where the first equality holds from the total probability theorem and the fact that σ_1 and σ_2 are disjoint, and the inequality holds from the fact that $\sum_{\sigma_1 \in \Sigma_1} Prob\{P_k \le \mathbf{M} \mid \sigma_1\} Prob\{\sigma_1\}$ is non-negative. This completes the proof of $1 - \theta\left(k_1(\mathbf{M}), \mathscr{D}\right) \le Prob\{P_k \le \mathbf{M}\}$. Similarly, we can prove $Prob\{P_k \le \mathbf{M}\} \le 1 - \theta\left(k_2(\mathbf{M}), \mathscr{D}\right)$. This completes the proof.

Remark 6.9 We assume in Theorem 3.16 that $\bar{\mathbf{P}} \le \mathbf{P}_0 \le \bar{\mathbf{M}}$. This is reasonable from the fact that if $\alpha_{i,k} = 1$ $(i = 1,2,\ldots,l)$, then the DFMTKF in Theorem 6.4 is equivalent to the distributed TKF. As a result, the steady estimation error covariance matrix $\bar{P}$ should be no bigger than $\mathbf{P}_0$ (the initial error covariance matrix) since the

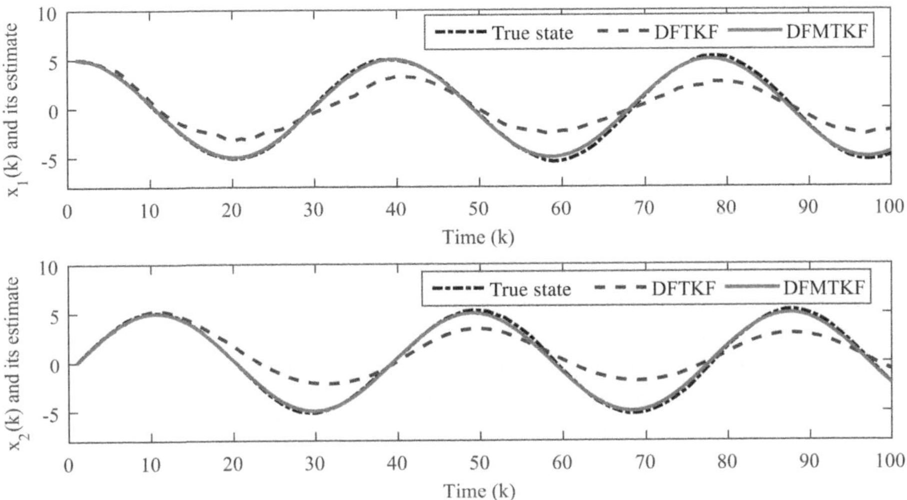

FIGURE 6.2   True values of states and their estimates under fault-free case.

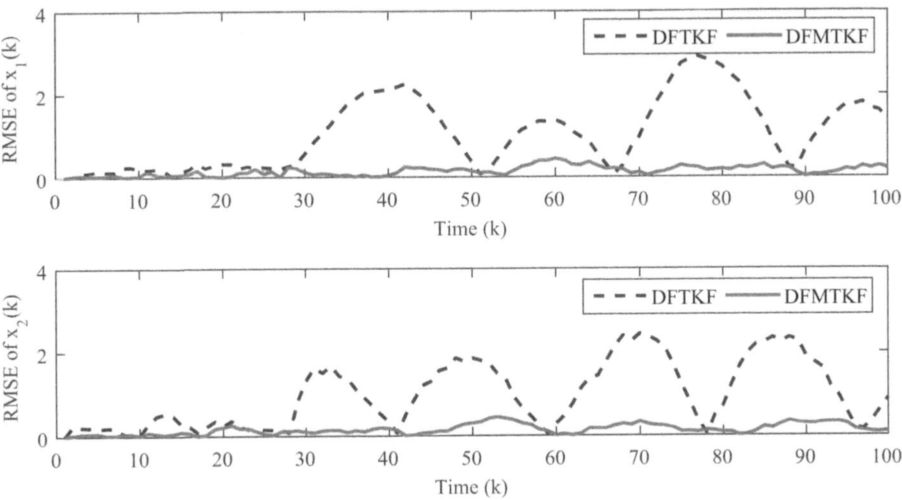

FIGURE 6.3   Comparison of DFTKF and DFMTKF in RMSE under fault-free case.

distributed TKF is optimal in the LMV sense among the set of all linear and distributed filters with censored measurements. In addition, it follows from Lemma 6.3 that $\mathbf{P}_k \leq \overline{\mathbf{M}}$ for all $k = 0, 1, \ldots$, which indicates that $\mathbf{P}_0 \leq \overline{\mathbf{M}}$.

Remark 6.10 Because of the complex relationship between $\mathbf{P}_{i,k}$ and $\mathbf{P}_k$ in (6.42), it is quite difficult to find the upper and lower bounds for $Prob\{\mathbf{P}_k \leq \mathbf{M}\}$ directly from the upper and lower bounds for $Prob\{\mathbf{P}_{i,k} \leq \mathbf{M}_i\}$. However, by defining some new matrices and variables such as $\mathscr{D}$, $\bar{\mathbf{M}}$, $\mathbf{M}$, $k_p(\mathbf{M})$, N_k and $\theta(k_p(\mathbf{M}),\mathscr{D})$ $(p=1,2)$, we can directly analyze the filtering performance of the DFMTKF and explore the bounds for $Prob\{\mathbf{P}_k \leq \mathbf{M}\}$. It should be noted that the newly obtained bounds heavily depend on the censoring probability and the multi-sensor feature. To be specific, to calculate $\theta(k_p(\mathbf{M}),\mathscr{D})$ in (6.45), we need to calculate $k_p(\mathbf{M})$, which requires the calculation of $\mathbf{M}$. Meanwhile, the determination of $\mathbf{M}$ should satisfy $\mathbf{M} \geq \bar{\mathbf{M}}$ where $\bar{M} = min\{\bar{\mathbf{M}}_1,\bar{\mathbf{M}}_2,\ldots,\bar{\mathbf{M}}_l\}$, $\bar{\mathbf{M}}_i = (\bar{\Upsilon}_i\mathbf{C}_i)^{-1}\mathscr{R}_i(\mathbf{I}-\varphi_i)\left[(\bar{\Upsilon}_i\mathbf{C}_i)^{-1}\right]^T$, and $\theta(k_p(\mathbf{M}),\mathscr{D}) = \prod_{i=1}^{l}\prod_{j=0}^{k_p(\mathbf{M})-1}(1-Prob\{\alpha_{i,k-j}=1\})$. All these terms imply the effects of the censoring probability as well as the multi-sensor nature of the determination of the bounds for $Prob\{\mathbf{P}_k \leq \mathbf{M}\}$.

Remark 6.11 It should be noted that although Assumption 6.1 is given to ensure the boundedness of the probability of the error covariance as shown in Theorem 6.4, Theorem 6.4 still holds in the case that $(\bar{\Upsilon}_i\mathbf{C}_i)^{-1}$ does not exist as long as $(\mathbf{A},\bar{\Upsilon}_i\mathbf{C}_i)$ is observable. Assume that $\bar{\Upsilon}_i\mathbf{C}_i$ is not invertible. Since $(\mathbf{A},\bar{\Upsilon}_i\mathbf{C}_i)$ is observable, there exists $r\,(2 \leq r \leq n_x)$ such that $\left[(\bar{\Upsilon}_i\mathbf{C}_i)^T \quad (\bar{\Upsilon}_i\mathbf{C}_i\mathbf{A})^T \quad \cdots \quad (\bar{\Upsilon}_i\mathbf{C}_i\mathbf{A}^{r-1})^T\right]^T$ is of full rank. In the sequel, let us assume $r=2$ and the obtained result can be easily generalized to cases of $(r=3,4,\ldots,n)$.

Assume that sensor i is equipped with a buffer that stores the previous measurement $\mathbf{y}_{i,k-1}$, and hence $\mathbf{y}_{i,k-1}$ can be sent along with $\mathbf{y}_{i,k}$ via the network at time k. Then, if $\alpha_{i,k}=1$, both $\mathbf{y}_{i,k-1}$ and $\mathbf{y}_{i,k}$ are delivered successfully through the network, giving rise to the following estimator, which is constructed in parallel to the Tobit Kalman filter.

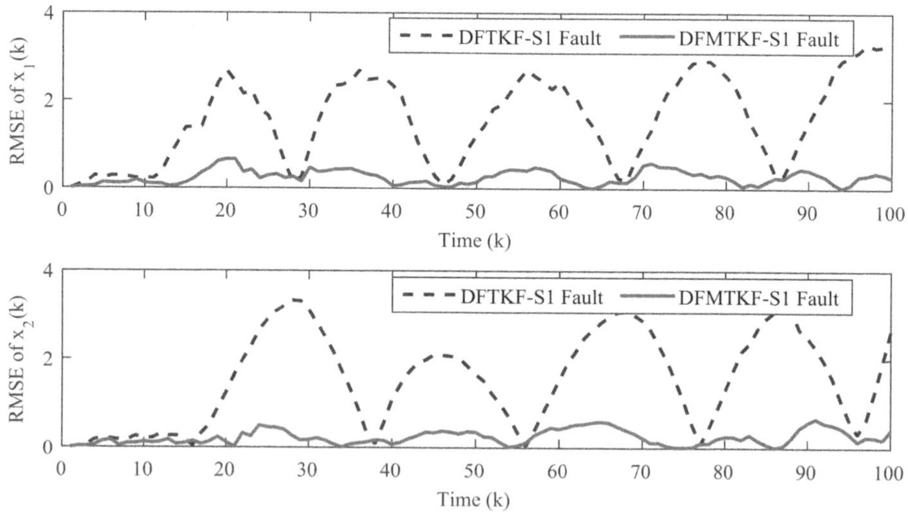

FIGURE 6.4 Comparison of DFTKF and DFMTKF in RMSE with S1 fault.

$$\tilde{\mathbf{x}}_{i,k} = \mathbf{A} \begin{bmatrix} \bar{\mathbf{Y}}_i \mathbf{C}_i \mathbf{A} \\ \bar{\bar{\mathbf{Y}}}_i \mathbf{C}_i \end{bmatrix}^{-1} \begin{bmatrix} \mathbf{y}_{i,k} \\ \mathbf{y}_{i,k-1} \end{bmatrix},$$

with its error covariance being

$$\tilde{\mathbf{P}}_{i,k} = \bar{\mathbf{M}}_i \triangleq \mathbf{A}\mathbf{M}_{1,i}\mathbf{A}^T + \mathbf{Q} - \mathbf{A} \begin{bmatrix} \bar{\mathbf{Y}}_i \mathbf{C}_i \mathbf{A} \\ \bar{\bar{\mathbf{Y}}}_i \mathbf{C}_i \end{bmatrix}^{-1} \begin{bmatrix} \bar{\mathbf{Y}}_i \mathbf{C}_i \mathbf{Q} \\ 0 \end{bmatrix},$$

$$- \begin{bmatrix} \bar{\mathbf{Y}}_i \mathbf{C}_i \mathbf{Q} \\ 0 \end{bmatrix}^T \left(\begin{bmatrix} \bar{\mathbf{Y}}_i \mathbf{C}_i \mathbf{A} \\ \bar{\bar{\mathbf{Y}}}_i \mathbf{C}_i \end{bmatrix}^{-1} \right)^T \mathbf{A}^T,$$

where

$$M_{1,i} = \begin{bmatrix} \bar{\mathbf{Y}}_i \mathbf{C}_i \mathbf{A} \\ \bar{\bar{\mathbf{Y}}}_i \mathbf{C}_i \end{bmatrix}^{-1} \begin{bmatrix} \tilde{\mathbf{C}}_i & 0 \\ 0 & \mathscr{R}_i(\mathbf{I}-\varphi_i) \end{bmatrix} \left(\begin{bmatrix} \bar{\mathbf{Y}}_i \mathbf{C}_i \mathbf{A} \\ \bar{\bar{\mathbf{Y}}}_i \mathbf{C}_i \end{bmatrix}^{-1} \right)^T,$$

$$\tilde{\mathbf{C}}_i = \bar{\mathbf{Y}}_i \mathbf{C}_i \mathbf{Q} \left(\bar{\mathbf{Y}}_i \mathbf{C}_i \right)^T + \mathscr{R}_i(\mathbf{I}-\varphi_i).$$

Due to the fact that the Tobit Kalman filter is optimal among the set of linear estimators with censored measurements, we get $\mathbf{P}_{i,k} \leq \tilde{\mathbf{P}}_{i,k}$ and hence $\mathbf{P}_{i,k} \leq \bar{\mathbf{M}}_i$. The inequality $\mathbf{P}_{i,k} \leq \bar{M}$ can be acquired similar to the derivation in Lemma 6.4. Hence, the same result as that in Theorem 3.16 can still be obtained.

6.4 AN ILLUSTRATIVE EXAMPLE

In this section, the oscillator example is adopted to demonstrate the feasibility of the proposed filter design approach and the filtering performance. Let RMSE1 denote the root mean-squared error (RMSE) of $\mathbf{x}_k^1$ (the first dimension of $\mathbf{x}_k$), i.e. $\text{RMSE1} = (1/M) \sum_{j=1}^{M} \left(\mathbf{x}_j^1 - \hat{\mathbf{x}}_j^2 \right)^2$, where M is the number of simulation tests. Similarly, RMSE2 is the RMSE for the estimation of $\mathbf{x}_k^2$ (the second dimension of $\mathbf{x}_k$), i.e. $\text{RMSE2} = (1/M) \sum_{j=1}^{M} \left(\mathbf{x}_j^2 - \hat{\mathbf{x}}_j^2 \right)^2$. Consider the following class of discrete-time system given by (6.1)–(6.8) with two sensors (Sensor 1 and Sensor 2) and the following parameters:

$$\mathbf{A} = \begin{bmatrix} \cos(\omega) & -\sin(\omega) \\ \sin(\omega) & \cos(\omega) \end{bmatrix}, \mathbf{C}_1 = \begin{bmatrix} 1 & 0 \end{bmatrix}, \mathbf{C}_2 = \begin{bmatrix} 0 & 1 \end{bmatrix},$$

$$\mathbf{Q}_k = \text{diag}\{0.0025, 0.0025\}, d_{1,k} = 3, d_{2,k} = 2,$$

$$\mathcal{I}_1 = \mathcal{I}_2 = 0, \mathbf{P}_0 = I_2, \bar{\mathbf{x}}_0 = \begin{bmatrix} 5 & 0 \end{bmatrix}^T, \beta_1 = \beta_2 = 0.5,$$

$$\mathcal{R}_1 = \mathcal{R}_2 = 1, \mathscr{D}_1 = 5, \mathscr{D}_2 = 3, \omega = 0.052\pi.$$

Similar to [176], the packet delay in (6.3) is modeled as a Poisson distribution with mean $d_{i,k}$, i.e.

$$f(j) = \mathrm{Prob}\{d_{i,k} = j\} = \frac{\left(d_{i,k}\right)^j e^{-d_{i,k}}}{j!}, \quad j = 0, 1, \ldots.$$

In the first part of this section, we compare the distributed federated Tobit Kalman filter (DFTKF) (which only handles the multi-sensor and measurement censoring phenomena) with the proposed distributed federated modified Tobit Kalman filter (DFMTKF) (which simultaneously handles the multi-sensor, measurement censoring and packet delay phenomena).

One thousand times of Monte Carlo simulations are conducted, and the results of the RMSEs are shown in Figures 6.2–6.3. Figure 6.2 sketches the true values of the states and the estimates given by the two methods, while Figure 6.3 illustrates the comparison of the RMSE curves between the two methods. It can be found from Figure 6.2 that our proposed DFMTKF is able to track the true values of the system states while the TKF deviates from the true state values. Figure 6.3 shows that the RMSE of our DFMTKF is always smaller than that of the TKF. This is because in our DFMTKF, the multi-sensor and packet delay phenomena are tackled by establishing a distributed and modified federated Kalman filtering fusion scheme, while they are not addressed in the TKF.

In addition, to further demonstrate the uncertainty tolerance and robustness property of our DFMTKF, we first assume that the considered system is subject to uncertainties caused by the fault $\mathbf{f}_{1,k}$ that occurred in Sensor 1 (S1). In this case, the measurement equation for S1 becomes $\mathbf{y}_{1,k}^* = \mathbf{C}_1 x_k + \mathbf{v}_{1,k} + \mathbf{f}_{1,k}$, where $\mathbf{f}_{1,k}$ is a random fault with zero mean and unit covariance and is independent of ω_k, $\tilde{\mathbf{v}}_{1,k}$ and $\tilde{\mathbf{v}}_{2,k}$. Then, we have $\mathbf{y}_{1,k}^* = \mathbf{C}_1 x_k + \tilde{\mathbf{v}}_{1,k}$, where $\tilde{\mathbf{v}}_{1,k} = \mathbf{v}_{1,k} + \mathbf{f}_{1,k}$ can be regarded as the equivalent S1 noise with zero mean and covariance $\tilde{\mathcal{R}}_1 = 2$. Similarly, we can also assume that the considered system suffers from uncertainties resulting from fault $f_{2,k}$ occurred in Sensor 2 (S2). In this regard, the measurement equation for S2 is $\mathbf{y}_{2,k}^* = \mathbf{C}_2 x_k + \mathbf{v}_{2,k} + \mathbf{f}_{2,k}$, where $\mathbf{f}_{2,k} = 0.5 + 0.5\mathbf{f}_{1,k}$. Then, we have $\mathbf{y}_{2,k}^* = \mathbf{C}_2 x_k + \tilde{\mathbf{v}}_{2,k}$, where $\tilde{\mathbf{v}}_{2,k} = \mathbf{v}_{2,k} + \mathbf{f}_{2,k}$ is regarded as the equivalent S2 noise with mean 0.5 and covariance $\tilde{\mathcal{R}}_2 = 1.25$.

From this description, it is known that the observation noise covariances used to generate measurements are the prescribed $\mathcal{R}_1$ and $\mathcal{R}_2$, while the ones adopted in the design of the filter are the uncertainty corrupted $\tilde{\mathcal{R}}_1$ and $\tilde{\mathcal{R}}_2$. After 1000 Monte Carlo simulations, Figures 6.4–6.5 describe the RMSE results of the DFTKF and DFMTKF, respectively, for cases of faulty S1 and S2. It can be observed from Figures 6.4–6.5 that regardless of S1 fault or S2 fault, both DFTKF and DFMTKF remain appropriate estimation performances compared with the fault-free case presented by Figure 6.3. This is due to the fact that both DFTKF and DFMTKF are distributed filtering approaches implemented in a parallel structure and hence have strong uncertainty tolerance and robustness. Nonetheless, it should be pointed out that the RMSE curves of our DFMTKF are always lower than those of DFTKF. The

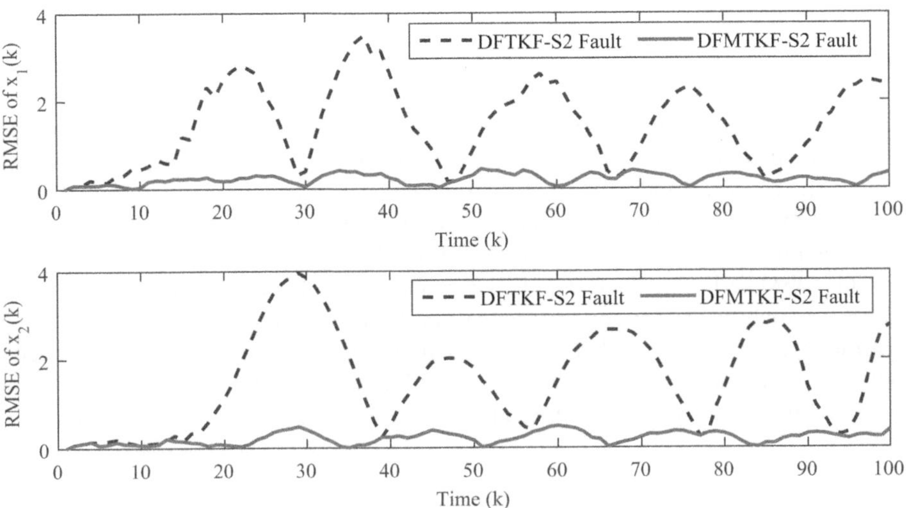

FIGURE 6.5   Comparison of DFTKF and DFMTKF in RMSE with S2 fault.

reason is that packet delays are handled properly in our DFMTKF, while they are not considered in the DFTKF.

Remark 6.12 Note that the DFMTKF fusion method presented in this chapter is based on approaches of federated Kalman filtering fusion, modified Kalman filtering with packet delays and Tobit Kalman filtering. These three approaches, respectively, have a time complexity of $\mathcal{O}\left(ln_x^3\right)$, $\mathcal{O}\left(s_i + 1n_x^3\right)$, and $\mathcal{O}\left(n_x^3\right)$, where n_x is the dimension of system state, l is the total number of sensors in the system, $si \in \left[0, \mathscr{D}_i - 1\right]$ and $\mathscr{D}_i$ is the buffer length with respect to the ith sensor. As a result, the computational complexity of the developed DFMTKF algorithm can be easily obtained via the time complexities of the aforementioned three algorithms. For example, let us look at the modified Tobit Kalman filter as described in Theorem 6.1; it has a computational complexity of $\mathcal{O}\left(n_x^3\right)$. Then, let us move forward to the LMTKF given in Theorem 6.2 which is carried out by running $s_i + 1$ modified Tobit Kalman filters, and hence it has a computational complexity of $\mathcal{O}\left((s_i + 1)n_x^3\right)$. Finally, it follows from Theorem 6.3 that the developed DFMTFK needs to incorporate estimates and covariances from all the l LMTKFs at the fusion center. Therefore, its complexity can be represented as $\mathcal{O}\left(\sum_{i=1}^{l}(s_i + 1)n_x^3\right)$. Similarly, the complexity of the DFTKF is obtained as $\mathcal{O}\left(ln_x^3\right)$. Obviously, the computational complexity of our DFMTKF algorithm depends linearly on $\sum_{i=1}^{l}(s_i + 1)$ and polynomially on n_x, while that of the DFTKF relies linearly on l and polynomially on n_x. Generally speaking, when l is quite large, for instance, in large-scale systems, we have $\sum_{i=1}^{l}(s_i + 1) \geq l$, and thus the DFTKF has much less computation burden than our DFMTKF. However, our DFMTKF has better computational accuracies and filtering performances than the DFTKF, though at the cost of extra calculations.

6.5 SUMMARY

In this chapter, we have investigated the distributed recursive filtering problem for a class of discrete-time systems in the simultaneous presence of multiple sensors, measurement censoring and packet delay. The delay of the measurement obeys the Poisson distribution and has been addressed using buffers of finite lengths. For the ith sensor, the censored and delayed measurement has been handled in a unified framework via the obtained LMTKF. Then, by resorting to the federated Kalman filtering fusion approach, local estimates from different LMTKFs have been fused to establish the DFMTKF. The upper and lower bounds have also been given for the probability at which the estimation error covariance matrix of the proposed DFMTKF is within a prescribed bound. The obtained bounds depend not only on the packet delay but also on the multiple-sensor nature as well as the censoring probability. Finally, the feasibility of the proposed DFMTKF has been proven by a numerical example.

7 Federated Tobit Kalman Filtering Fusion with Dead-Zone-Like Censoring and Dynamical Bias under the Round-Robin Protocol

It has been broadly acknowledged that the MSFF requires not only sensor observations but also mathematical models describing the dynamics of system behavior along with observation models relating sensor outputs to system states. Roughly speaking, the dynamic model is an approximation of the evolution process in relation to the system state that usually cannot be observed directly. In other words, internal disturbance signals and model perturbations are not uncommon in the MSFF, and the resultant dynamical uncertainties would severely restrict the application scope of the MSFF. To be more specific, the inherent characteristics of dynamical uncertainties, if inadequately handled, could seriously degrade the system performance, and this has then triggered lasting research enthusiasm towards the analysis and synthesis of uncertainty-corrupted dynamic state estimation. Note that most of the forgoing studies have been dedicated to the worst-case scenarios through using the upper bound of the uncertainty, which is in addition to an already crude approximation of uncertainty bounds resulting from the often poor prior knowledge. As a result, the final state estimate might be excessively conservative, thereby limiting its application scope to a large extent.

Apart from the worst-case investigation, another way of handling the dynamical uncertainty is to carry out the simultaneous state and uncertainty estimation with hope to jointly estimate the state together with the dynamical uncertainty. Stochastic bias, as a specific class of dynamical uncertainty, is often induced by unmodeled dynamics, neglected model nonlinearities, parameter variations and cyber-attacks. Stochastic bias is typically described as a dynamical stochastic process (driven by certain white noises) and can be mathematically dealt with by the state augmentation and the dual Kalman filter (KF) approaches [178].

DOI: 10.1201/9781003461623-7

To handle the stochastic bias, the augmentation approach lifts the system state to contain the bias and applies a KF to the resultant system for later state estimation, whilst the dual KF approach employs two KFs to estimate the state and bias simultaneously and update each other's estimates mutually. Very recently, the joint estimation problem of state and dynamical bias has been investigated in [179] for two-dimensional systems with shift-varying parameters, where the augmentation technique has been adopted to tackle the stochastic bias and the prescribed filtering performance has been ensured by minimizing the upper bound of the acquired error covariance in the sense of matrix-trace.

Following the discussions made thus far, we conclude that there is a lack of systematic investigation on bias-corrupted multi-sensor Tobit Kalman filtering fusion problems subject to dead-zone-like censoring under the RRP. As such, the primary objective of this chapter is to fill in such a gap by designing a protocol-based federated Tobit Kalman filter (FTKF) that is insensitive to dynamical biases and censored observations.

7.1 PROBLEM FORMULATION

Consider the Tobit Kalman filtering problem for a networked system, as shown in Figure 7.1. In this framework, the sensor is susceptible to dynamical biases, the signal transmission between the sensor and the local filter is implemented through a communication network under the RRP, and the measurement arriving at the filter is inclined to dead-zone-like censoring. State estimates collected from local filters are later sent to the fusion center for further process so as to achieve the integrated filtering result. In what follows, let us introduce the plant, dynamical bias, communication network and dead-zone-like censoring in a mathematical way.

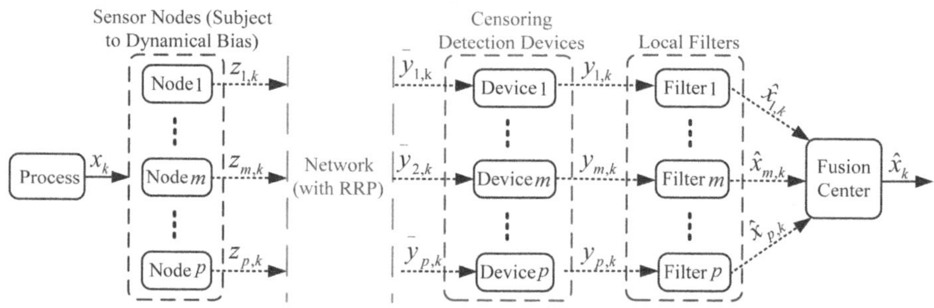

FIGURE 7.1 Schematic diagram for the concerned Tobit Kalman filtering problem.

Consider a multi-sensor system corrupted by dynamical biases:

$$x_{k+1} = A_k x_k + B_k b_k + \omega_k, \tag{7.1}$$

$$z_{m,k} = C_{m,k} x_k + \upsilon_{m,k}, \, m = 1, 2, \ldots, p, \tag{7.2}$$

where $x_k \in \mathbb{R}^{n_x}$ is the state vector, $z_{m,k} \in \mathbb{R}$ is the uncensored observation of the mth $(m = 1, 2, \ldots, p)$ sensor, and p is the total number of sensors. A_k, B_k and $C_{m,k}$ are known matrices with eligible dimensions. $\omega_k \in \mathbb{R}^{n_x}$ and $\upsilon_{m,k} \in \mathbb{R}$ are zero-mean white Gaussian noises with covariances Q_k and $R_{m,k}$, respectively. Here, $b_k \in \mathbb{R}^{n_b}$ is the stochastic bias driven by the following dynamic equation:

$$b_k = H_{k-1} b_{k-1} + \varepsilon_{k-1}, \tag{7.3}$$

where H_k is the known bias transition matrix, and ε_k is the white Gaussian noise with zero mean and covariance Ξ_k.

In maneuvering target tracking, the target manoeuver is usually modeled as an unknown acceleration bias to the constant velocity model as shown in (7.1), where x_k consists of the target position and velocity and b_k is the acceleration bias. By closely observing the target manoeuver, one is capable of deciding whether the dynamics of the unknown acceleration bias given by (7.3) should be characterized by a constant acceleration model or a variable acceleration model. In the event that (7.3) is a constant acceleration model, one certainly has $H_k = I$ for all $k \geq 1$. This example explicitly tells us that the bias transition matrix H_k can be determined beforehand according to the concerned engineering scenario. As such, H_k is assumed to be known a priori in this chapter.

Note that the observations $z_{m,k} \in \mathbb{R}$ $(m = 1, 2, \ldots, p)$ are transmitted to the remote estimator via a shared communication network. Due to limited communication bandwidth, the communication protocol is often deployed, with which only one single sensor is granted to propagate its output, at each communication time, through the shared network. As discussed in the introduction, in this chapter, the RRP is employed to orchestrate the transmission order of the sensors for the purpose of circumventing data collisions.

Denote mod($k-m$, p) as the unique non-negative remainder on division of $k-m$ by p, and $\hbar_k \in \{1, 2, \ldots, p\}$ as the sensor that has access to the network at time k. Let

$$\Gamma_{\hbar_k} \triangleq \mathrm{diag}\{\Gamma_{m,\hbar_k}\}, \quad m = 1, 2, \ldots, p$$

be the observation update coefficient where

$$\Gamma_{m,\hbar_k} \triangleq \delta(\hbar_k - m).$$

Abiding by the RRP and the zero-order holder strategy, for the mth sensor, the actual measurement $\bar{y}_{m,k}$ arriving at the remote estimator after network transmission is updated as follows:

$$\bar{y}_{m,k} = \begin{cases} z_{m,k}, & \text{if } \mathrm{mod}(k-m, p) = 0, \\ \bar{y}_{m,k-1}, & \text{otherwise.} \end{cases} \tag{7.4}$$

Taking advantage of the update coefficient Γ_{h_k}, (7.4) can be rearranged into

$$\overline{y}_{m,k} = \sum_{l=0}^{p-1} \Gamma_{m,h_{k-l}} z_{m,k-l},$$
(7.5)

where $\hbar_{k-l} \triangleq l$ and $z_{m,k-l} \triangleq z_{m,0}$ for $k-l \leq 0$.

Let the input terminal of the remote estimator be equipped with an additional detection device whose function is to check whether $\overline{y}_{m,k}$ is censored or not. In this sense, the Tobit observation model with dead-zone-like censoring is given as follows:

$$y_{m,k} = \begin{cases} \tau_m^l, & \overline{y}_{m,k} \leq \tau_m^l, \\ \overline{y}_{m,k}, & \tau_m^l < \overline{y}_{m,k} < \tau_m^r, \\ \tau_m^r, & \overline{y}_{m,k} \geq \tau_m^r, \end{cases}$$
(7.6)

where $y_{m,k}$ is the censored observation, and τ_m^l and τ_m^r are, respectively, the left- and right-censoring thresholds.

According to whether $\overline{y}_{m,k}$ is left censored, right censored or uncensored, the observation model (7.6) is rewritten as

$$y_{m,k} = \left(1 - \alpha_{m,k} - \beta_{m,k}\right) \overline{y}_{m,k} + \alpha_{m,k} \tau_m^l + \beta_{m,k} \tau_m^r,$$
(7.7)

where $\alpha_{m,k}$ and $\beta_{m,k}$ are, respectively, known Bernoulli random variables governing the left- and right-censoring phenomena of $\overline{y}_{m,k}$ with the following probability distributions:

$$\begin{cases} \text{Prob}\{\alpha_{m,k} = 1\} = \overline{\alpha}_{m,k}, \\ \text{Prob}\{\beta_{m,k} = 1\} = \overline{\beta}_{m,k}, \\ \text{Prob}\{\alpha_{m,k} = 0\} = 1 - \overline{\alpha}_{m,k}, \\ \text{Prob}\{\beta_{m,k} = 0\} = 1 - \overline{\beta}_{m,k}. \end{cases}$$
(7.8)

Here, $\overline{\alpha}_{m,k}$ and $\overline{\beta}_{m,k}$ are non-negative constants that are known *a priori*, and $\alpha_{m,k}$ and $\beta_{m,k}$ are uncorrelated with ω_k and ε_k.

Let

$$y_{m,1:k} \triangleq \{y_{m,0}, y_{m,1}, \ldots, y_{m,k}\}, \hat{x}_{m,k}^- \triangleq \mathbb{E}\{x_k \mid y_{m,1:k-1}\},$$

$$\hat{x}_{m,k} \triangleq \mathbb{E}\{x_k \mid y_{m,1:k}\}, \tilde{x}_{m,k}^- \triangleq x_k - \hat{x}_{m,k}^-, \tilde{x}_{m,k} \triangleq x_k - \hat{x}_{m,k},$$

$$P_{\tilde{x}_{m,k}} \triangleq \mathbb{E}\{\tilde{x}_{m,k} \tilde{x}_{m,k}^T\}, P_{\tilde{x}_{m,k}^-} \triangleq \mathbb{E}\{\tilde{x}_{m,k}^- \left(\tilde{x}_{m,k}^-\right)^T\}, \hat{y}_{m,k}^- \triangleq \mathbb{E}\{y_{m,k} \mid y_{m,1:k-1}\}.$$

Assumption 7.1 (1) The initial state x_0 and the bias b_0 have means $\bar{x}_0$ and $\bar{b}_0$, and covariances $P_{\bar{x}_0}$ and $P_{\bar{b}_0}$, respectively. (2) The random variables x_0, b_0, ω_k, ε_k and $\upsilon_{m,k}$ are mutually independent.

Remark 7.1 In system modeling, ambient disturbances provoked by environmental variations are often regarded as process noises taking the form of Gaussian white sequences. Such a treatment ignores the case where the disturbance might exhibit itself as a dynamically varying process, namely, the stochastic bias. As a matter of fact, stochastic biases (stemming from random frictions, wind resistance and/or electromagnetic interferences) might behave according to a dynamical manner similar to the evolution of the target system. Spurred by this fact, the random variable b_k is introduced in this chapter to characterize the stochastic bias, with its evolution kinetics being governed by (7.3).

Remark 7.2 Under the scheduling of the RRP, equal priority is assigned to each sensor, and the observation from each individual sensor is admitted to enter the network in a fixed circular manner. In comparison with other communication protocols (e.g. the try-once-discard protocol and random access protocol), the RRP predefines a periodic transmission rule, which makes the scheduling easy to implement. Therefore, the RRP is adopted in this chapter as the desired communication protocol for data transmission.

In the case that sensor m has no access to the network, two strategies (i.e. the zero-input and zero-order holder strategies) are often leveraged to establish the actual observation $\bar{y}_{m,k}$ that is received by the remote estimator. As suggested by their names, the null and previous measurements are, respectively, employed as compensatory inputs to the remote estimator in the two strategies in case of transmission failures. The selection between the two strategies rides on the actual network condition and the performance requirements. Here, the zero-order holder strategy is utilized to generate $\bar{y}_{m,k}$ for the purpose of offsetting $\bar{y}_{m,k}$. Accordingly, at time $k-l$, only the observation $\bar{y}_{h_{k-l},k-l}$ is adopted for later filter update, whilst the rest of the sensor observations $\bar{y}_{m,k-l}$ $(m = 1,2,...,p,\ m \neq h_{k-l})$ remain the same as their counterparts in $\bar{y}_{k-l-1}$. Obeying the *fixed circular* order for information propagation, $\bar{y}_{m,k}$ $(m = 1,2,...,p)$ can be represented by the sum of $\Gamma_{m,h_{k-l}} z_{m,k-l}$ $(l = 0,1,...,p-1)$, as shown in (7.5).

In practice, there exist two typical communication channels in multi-sensor fusion application scenarios, i.e. the sensor-to-estimator and estimator-to-fusion-center communication channels. Frankly speaking, the sensor-to-estimator channel is long, as the estimator is normally located far away from the sensor in order to obtain a reliable environment for the implementation of the estimation algorithm, whilst the estimator-to-fusion-center channel is short, as local estimators are placed close to the fusion center to guarantee that local estimates are perfectly transmitted for final fusion. Accordingly, the sensor-to-estimator channel is easily susceptible to noises tond disturbances, while the estimator-to-fusion-center channel normally appears to be noise free and perfect. In this regard, comparing with the estimator-to-fusion-center channel, the sensor-to-estimator channel is more likely to be subject to bandwidth limitations. As such, the Round-Robin protocol is implemented in the sensor-to-estimator channel with a view point to regulating the data transmission and mitigating the data collision.

Remark 7.3 The dead-zone-like censoring modeled by (7.6) is widely encountered in systems equipped with low-cost commercial off-the-shelf sensors. The distinctive feature of such censoring lies in its censored Gaussian (rather than pure Gaussian) noise distribution at/near the censored region, and this circumvents the direct employment of the conventional KF. The necessity of formulating a censoring-oriented KF gives birth to the celebrated TKF. By bringing in Bernoulli random variables to regulate the censoring phenomenon, the TKF perfectly handles the non-Gaussianity of the observation noise via developing a so-called Tobit regression model, based at which an explicit paradigm of the observation expectation and variance is provided. When the target plant is prone to dynamical biases, the standard TKF fails to take effect, accounting for the bias-induced variations that enter into the system dynamics, the regression model as well as the filter equations. Moreover, the adoption of the RRP would ineluctably pose significant impacts on the structure of the regression model, censoring probability together with other observation-related terms. As such, a holistic multi-sensor Tobit Kalman filtering fusion framework is constructed in this article to solve the aforementioned problems.

It is observed from (7.7) that random variables $\alpha_{m,k}$ and $\beta_{m,k}$ are exploited to describe the censoring phenomena of $\bar{y}_{m,k}$. In accordance with (7.7), if no censoring occurs for $\bar{y}_{m,k}$, i.e. $\alpha_{m,k} = 0$ and $\beta_{m,k} = 0$, the measurement becomes $y_{m,k} = \bar{y}_{m,k}$, which means that the observation is equivalent to the latent one. If the left censoring occurs for $\bar{y}_{m,k}$, i.e. $\alpha_{m,k} = 1$ and $\beta_{m,k} = 0$, the measurement becomes $y_{m,k} = \tau_m^l$, which means that the left-censoring threshold is allocated to the observation. If the right censoring occurs for $\bar{y}_{m,k}$, i.e. $\alpha_{m,k} = 0$ and $\beta_{m,k} = 1$, the measurement becomes $y_{m,k} = \tau_m^r$, which means that the right-censoring threshold is allocated to the observation. Here, we suppose that censoring probabilities $\bar{\alpha}_{m,k}$ and $\bar{\beta}_{m,k}$ are known *a priori* via some statistical experiments. Alternatively, $\bar{\alpha}_{m,k}$ and $\bar{\beta}_{m,k}$ can also be approximated by

$$\begin{cases} \bar{\alpha}_{m,k} \approx \Phi\left(\dfrac{\tau_m^l - \hat{\zeta}_{\bar{m},k}}{\sqrt{\mathcal{R}_{m,k}}}\right), \\[4mm] \bar{\beta}_{m,k} \approx \Phi\left(\dfrac{\hat{\zeta}_{\bar{m},k} - \tau_m^r}{\sqrt{\mathcal{R}_{m,k}}}\right), \end{cases} \tag{7.9}$$

where

$$\mathcal{R}_{m,k} \triangleq \sum_{l=0}^{p-1} \Gamma_{m,h_{k-l}}^2 R_{m,k-l}, \quad \hat{\zeta}_{\bar{m},k} \triangleq \Gamma_{m,h_k} C_{m,k} \hat{x}_{\bar{k}} + \sum_{l=1}^{p-1} \Gamma_{m,h_{k-l}} C_{m,k-l} \hat{x}_{k-l},$$

and $\Phi(\cdot)$ is the cumulative distribution function (CDF) of the random variable "$\bullet$" that obeys the standard normal distribution.

The objective of this chapter is to design a protocol-based FTKF for system (7.1)–(7.8) subject to dead-zone-like censoring and dynamical biases under the RRP.

7.2 MAIN RESULTS

In this section, we are devoted to formalizing a federated Tobit Kalman filtering fusion framework to overcome the identified multi-faceted challenges caused by the concurrence of dead-zone-like censoring, dynamical bias and RRP. Such a framework displays its distinctive characters from two viewpoints: (1) an elaborately designed local Tobit Kalman filter (LTKF) built on an enhanced regression model that gives holistic consideration of the impacts incurred by the dead-zone-like censoring, dynamical bias and RRP, where additional endeavor is demanded to compute the observation prediction, filter gain as well as associate error covariances; and (2) elegantly opted fusion rule which productively integrates available estimates from LTKFs, where the global optimality of the fused estimate is ensured.

Denoting

$$\xi_k \triangleq \begin{bmatrix} x_k^T & b_k^T \end{bmatrix}^T, \, w_k \triangleq \begin{bmatrix} \omega_k^T & \varepsilon_k^T \end{bmatrix}^T,$$

system (7.1)–(7.8) is augmented as

$$\xi_{k+1} = \mathcal{A}_k \xi_k + \mathcal{B}_k w_k, \tag{7.10}$$

$$z_{m,k} = \mathcal{C}_{m,k} \xi_k + \upsilon_{m,k}, \tag{7.11}$$

where

$$\mathcal{A}_k = \begin{bmatrix} A_k & B_k \\ 0 & H_k \end{bmatrix}, \mathcal{B}_k = \begin{bmatrix} I & 0 \\ 0 & I \end{bmatrix}, \mathcal{C}_{m,k} = \begin{bmatrix} C_{m,k} & 0 \end{bmatrix}.$$

Let

$$\bar{y}_{m,k} \triangleq \zeta_{m,k} + V_{m,k}, \zeta_{m,k} \triangleq \sum_{l=0}^{p-1} \Gamma_{m,\hbar_{k-l}} C_{m,k-l} \xi_{k-l},$$

$$V_{m,k} \triangleq \sum_{l=0}^{p-1} \Gamma_{m,\hbar_{k-l}} \upsilon_{m,k-l}, \vartheta_{m,k}^l \triangleq \frac{\tau_m^l - \zeta_{m,k}}{\mathcal{R}_{m,k}}, \vartheta_{m,k}^r \triangleq \frac{\tau_m^r - \zeta_{m,k}}{\mathcal{R}_{m,k}}.$$

On the basis of the augmented model (7.10)–(7.11), an enhanced Tobit regression is formulated as follows to accommodate the dead-zone-like censoring, dynamical bias and RRP influences.

Lemma 7.1 The expectation and variance of $y_{m,k}$ $(m=1,2,\ldots,p)$ are, respectively,

$$\mathbb{E}\{y_{m,k}\mid x_k,\mathcal{R}_{m,k}\} = \Phi\left(\vartheta_{m,k}^l\right)\tau_m^l + \left(1-\Phi\left(\vartheta_{m,k}^r\right)\right)\tau_m^r$$
$$+\left[\Phi\left(\vartheta_{m,k}^r\right)-\Phi\left(\vartheta_{m,k}^l\right)\right]\left[\zeta_{m,k}-\sqrt{\mathcal{R}_{m,k}}\,\lambda\left(\vartheta_{m,k}^r,\vartheta_{m,k}^l\right)\right], \tag{7.12}$$

$$var\{y_{m,k}\mid x_k,\mathcal{R}_{m,k}\} = \mathcal{R}_{m,k}\left[1+\varphi\left(\vartheta_{m,k}^r,\vartheta_{m,k}^l\right)\right], \tag{7.13}$$

where

$$\lambda\left(\vartheta_{m,k}^r,\vartheta_{m,k}^l\right) = \frac{\phi\left(\vartheta_{m,k}^r\right)-\phi\left(\vartheta_{m,k}^l\right)}{\Phi\left(\vartheta_{m,k}^r\right)-\Phi\left(\vartheta_{m,k}^l\right)}, \tag{7.14}$$

$$\varphi\left(\vartheta_{m,k}^r,\vartheta_{m,k}^l\right) = \frac{\vartheta_{m,k}^l\phi\left(\vartheta_{m,k}^l\right)-\vartheta_{m,k}^r\phi\left(\vartheta_{m,k}^r\right)}{\Phi\left(\vartheta_{m,k}^r\right)-\Phi\left(\vartheta_{m,k}^l\right)} - \lambda^2\left(\vartheta_{m,k}^r,\vartheta_{m,k}^l\right). \tag{7.15}$$

Here, $\phi(\cdot)$ and $\Phi(\cdot)$ are, respectively, the probability density function (PDF) and CDF of the Gaussian random variable $\vartheta_{m,k}\in\{\vartheta_{m,k}^r,\vartheta_{m,k}^l\}$ with the following forms:

$$\phi\left(\vartheta_{m,k}\right) = \frac{1}{\sqrt{2\pi}}\,e^{-\frac{(\tau_m-\zeta_{m,k})^2}{2\mathcal{R}_{m,k}}}, \tag{7.16}$$

$$\Phi\left(\vartheta_{m,k}\right) = \int_{-\infty}^{\tau_m}\frac{1}{\sqrt{2\pi\mathcal{R}_{m,k}}}\,e^{-\frac{(y_{m,k}-\zeta_{m,k})^2}{2\mathcal{R}_{m,k}}}\,d_{y_{m,k}}, \tag{7.17}$$

where $\tau_m\in\left\{\tau_m^r,\tau_m^l\right\}$.

Proof It is follows evidently from definitions

$$\bar{y}_{m,k} \triangleq \zeta_{m,k}+v_{m,k},\quad \zeta_{m,k}\triangleq\sum_{l=0}^{p-1}\Gamma_{m,h_{k-l}}C_{m,k-l}\xi_{k-l},\quad v_{m,k}\triangleq\sum_{l=0}^{p-1}\Gamma_{m,h_{k-l}}\upsilon_{m,k-l}$$

that $v_{m,k}$ is a zero-mean Gaussian noise with covariance

$$\mathcal{R}_{m,k} \triangleq \sum_{l=0}^{p-1}\Gamma_{m,h_{k-l}}^2 R_{m,k-l}.$$

Accordingly, the PDF of $y_{m,k}$ has the following expression:

$$f\left(y_{m,k}\mid x_k,\mathcal{R}_{m,k}\right) = \frac{1}{\sqrt{\mathcal{R}_{m,k}}}\phi\left(\frac{y_{m,k}-\xi_{m,k}}{\sqrt{\mathcal{R}_{m,k}}}\right)u\left(y_{m,k}-\tau_m^l\right)u\left(\tau_m^r-y_{m,k}\right)$$
$$+\delta\left(y_{m,k}-\tau_m^l\right)\Phi\left(\vartheta_{m,k}^l\right)+\delta\left(y_{m,k}-\tau_m^r\right)\left(1-\Phi\left(\vartheta_{m,k}^r\right)\right), \tag{7.18}$$

where $u\left(y_{m,k}-\tau_m^l\right)$ and $u\left(\tau_m^r-y_{m,k}\right)$ are unit step functions, and $\phi\left(\dfrac{y_{m,k}-\xi_{m,k}}{\sqrt{\mathcal{R}_{m,k}}}\right)$ and $\Phi\left(\vartheta_{m,k}\right)$ are calculated by (7.16)–(7.17).

In the light of (7.18), the mean of $y_{m,k}$ is

$$
\begin{aligned}
\mathbb{E}\left\{y_{m,k} \mid x_{k,\mathcal{R}_{m,k}}\right\} = &\; Prob\left\{\tau_m^l < y_{m,k} < \tau_m^r \mid x_k\right\}\mathbb{E}\left\{y_{m,k} \mid \tau_m^l < y_{m,k} < \tau_m^r, x_k\right\} \\
&+ Prob\left\{y_{m,k}=\tau_m^l \mid x_k\right\}\mathbb{E}\left\{y_{m,k} \mid y_{m,k}=\tau_m^l, x_k\right\} \\
&+ Prob\left\{y_{m,k}=\tau_m^r \mid x_k\right\}\mathbb{E}\left\{y_{m,k} \mid y_{m,k}=\tau_m^r, x_k\right\}
\end{aligned}
\tag{7.19}
$$

To compute $\mathbb{E}\left\{y_{m,k} \mid x_k, \mathcal{R}_{m,k}\right\}$, the probabilities and expectations on the right-hand side of (7.9) must be provided.

$$
\begin{aligned}
Prob\left\{\tau_m^l < y_{m,k} < \tau_m^r \mid x_k\right\} &= Prob\left\{\tau_m^l < \bar{y}_{m,k} < \tau_m^r \mid x_k\right\} \\
&= Prob\left\{\tau_m^l - \zeta_{m,k} < v_{m,k} < \tau_m^r - \zeta_{m,k} \mid x_k\right\} \\
&= \Phi\left(\vartheta_{m,k}^r\right) - \Phi\left(\vartheta_{m,k}^l\right).
\end{aligned}
\tag{7.20}
$$

In line with (7.18), we have

$$
\begin{aligned}
\mathbb{E}&\left\{y_{m,k} \mid \tau_m^l < y_{m,k} < \tau_m^r, x_k\right\} \\
&= \frac{1}{\sqrt{\mathcal{R}_{m,k}}}\int_{\tau_m^l}^{\tau_m^r} y_{m,k}\frac{\phi\left(\dfrac{y_{m,k}-\xi_{m,k}}{\sqrt{\mathcal{R}_{m,k}}}\right)}{\Phi\left(\vartheta_{m,k}^r\right)-\Phi\left(\vartheta_{m,k}^l\right)}d_{y_{m,k}} \\
&= \xi_{m,k} - \sqrt{\mathcal{R}_{m,k}}\,\lambda\left(\vartheta_{m,k}^r, \vartheta_{m,k}^l\right),
\end{aligned}
\tag{7.21}
$$

where $\lambda\left(\vartheta_{m,k}^r, \vartheta_{m,k}^l\right)$ is computed by (7.14).

Parallel to (7.20)–(7.21), we attain

$$
Prob\left\{y_{m,k}=\tau_m^l \mid x_k\right\} = \Phi\left(\vartheta_{m,k}^l\right),
\tag{7.22}
$$

$$
Prob\left\{y_{m,k}=\tau_m^r \mid x_k\right\} = 1 - \Phi\left(\vartheta_{m,k}^r\right),
\tag{7.23}
$$

$$
\mathbb{E}\left\{y_{m,k} \mid y_{m,k}=\tau_m^l, x_k\right\} = \tau_m^{\,l},
\tag{7.24}
$$

$$
\mathbb{E}\left\{y_{m,k} \mid y_{m,k}=\tau_m^r, x_k\right\} = \tau_m^{\,r}.
\tag{7.25}
$$

Inserting (7.20)–(7.25) into (7.19) yields (7.12). Referring to (7.18), (7.21), (7.24) and (7.25) produces

$$var\left\{y_{m,k} \mid y_{m,k} = \tau_m^l, x_k\right\} = var\left\{y_{m,k} \mid y_{m,k} = \tau_m^r, x_k\right\} = 0,$$

and therefore

$$\begin{aligned}
var\left\{y_{m,k} \mid x_k\right\} &= var\left\{y_{m,k} \mid \tau_m^l < y_{m,k} < \tau_m^r, x_k\right\} \\
&= \mathbb{E}\left\{y_{m,k}^2 \mid \tau_m^l < y_{m,k} < \tau_m^r, x_k\right\} \\
&\quad - \left(\mathbb{E}\left\{y_{m,k}^2 \mid \tau_m^l < y_{m,k} < \tau_m^r, x_k\right\}\right)^2 .
\end{aligned} \tag{7.26}$$

The notice of (7.18) along with (7.21) tells

$$\begin{aligned}
&\mathbb{E}\left\{y_{m,k}^2 \mid \tau_m^l < y_{m,k} < \tau_m^r, x_k\right\} \\
&= \frac{1}{\sqrt{\mathcal{R}_{m,k}}} \int_{\tau_m^l}^{\tau_m^r} y_{m,k}^2 \frac{\phi\left(\dfrac{y_{m,k} - \xi_{m,k}}{\sqrt{\mathcal{R}_{m,k}}}\right)}{\Phi\left(\vartheta_{m,k}^r\right) - \Phi\left(\vartheta_{m,k}^l\right)} d_{y_{m,k}} \\
&= \xi_{m,k}^2 + \mathcal{R}_{m,k} - \sqrt{\mathcal{R}_{m,k}}\, \xi_{m,k} \lambda\left(\vartheta_{m,k}^r, \vartheta_{m,k}^l\right) \\
&\quad + \sqrt{\mathcal{R}_{m,k}} \frac{\left(\tau_m^l \phi\left(\vartheta_{m,k}^l\right) - \tau_m^r \phi\left(\vartheta_{m,k}^r\right)\right)}{\Phi\left(\vartheta_{m,k}^r\right) - \Phi\left(\vartheta_{m,k}^l\right)} .
\end{aligned} \tag{7.27}$$

Inserting (7.21) and (7.27) into (7.26) gives

$$var\left\{y_{m,k} \mid x_k, \mathcal{R}_{m,k}\right\} = \mathcal{R}_{m,k}\left[1 + \varphi\left(\vartheta_{m,k}^r, \vartheta_{m,k}^l\right)\right],$$

which is precisely the same as (7.13), where $\varphi\left(\vartheta_{m,k}^r, \vartheta_{m,k}^l\right)$ is provided by (7.15).

Let

$$\hat{\xi}_{m,k}^- \triangleq \mathbb{E}\left\{x_k \mid y_{m,1:k-1}\right\}, \hat{\xi}_{m,k} \triangleq \mathbb{E}\left\{\xi_k \mid y_{m,1:k}\right\},$$

$$\tilde{\xi}_{\bar{m},k} \triangleq \xi_k - \hat{\xi}_{\bar{m},k}, \hat{y}_{\bar{m},k}^- \triangleq \mathbb{E}\left\{y_{m,k} \mid y_{m,1:k-1}\right\},$$

$$\bar{\vartheta}_{m,k}^l \triangleq \frac{\tau_m^l - \hat{\xi}_{\bar{m},k}}{\mathcal{R}_{m,k}}, \bar{\vartheta}_{m,k}^r \triangleq \frac{\tau_m^r - \hat{\xi}_{\bar{m},k}}{\mathcal{R}_{m,k}},$$

$$\hat{\zeta}_{m,k}^- \triangleq \sum_{l=1}^{p-1} \Gamma_{m,\hbar_{k-l}} C_{m,k-l} \hat{\xi}_{k-l} + \Gamma_{m,\hbar_k} C_{m,k} \hat{\xi}_k^-, \tilde{y}_{m,k}^- \triangleq y_{m,k} - \hat{y}_{m,k}^-,$$

$$\tilde{\xi}_{m,k} \triangleq \xi_k - \hat{\xi}_{m,k}, P_{\tilde{\xi}_{m,k}^-} \triangleq \mathbb{E}\left\{\tilde{\xi}_{m,k}^- \left(\tilde{\xi}_{m,k}^-\right)^T\right\},$$

$$P_{\tilde{\xi}_{m,k}} \triangleq \mathbb{E}\left\{\tilde{\xi}_{m,k} \tilde{\xi}_{m,k}^T\right\}, P_{\tilde{\xi}_{m,k}^- \tilde{y}_{m,k}^-} \triangleq \mathbb{E}\left\{\tilde{\xi}_{m,k}^- \left(\tilde{y}_{m,k}^-\right)^T\right\}.$$

Before proceeding further, we first introduce the following federated information distribution and fusion principle.

Lemma 7.2 [33] Suppose that $\hat{x}_{m,k}$ $(m = 1,2,\ldots,p)$ and $P_{m,k}$ are, respectively, the estimate and covariance of a stochastic n_x-dimension vector x_k obtained by the mth local estimator, $\hat{x}_k$ and P_k are, respectively, the global optimal estimate and covariance obtained by the fusion estimator under the federated Kalman fusion rule, and Q_k is the covariance of the process noise. Then, the information-sharing process among the local estimators and the fusion estimator is as follows:

$$\begin{cases} Q_{i,k-1} = \epsilon_m^{-1} Q_{k-1}, \\ P_{m,k-1} = \epsilon_m^{-1} P_{k-1}^f, \\ \hat{x}_{m,k-1} = \hat{x}_{k-1}, \end{cases}$$

where ϵ_m are the information sharing coefficients satisfying $\sum_{m=1}^{p} \epsilon_m = 1$. Moreover, the fused estimate and covariance of x_k are given as

$$\begin{cases} P_k = \left(\sum_{m=1}^{p} P_{m,k}^{-1} \right)^{-1}, \\ \hat{x}_k = P_k \left(\sum_{m=1}^{p} P_{m,k}^{-1} \hat{x}_{m,k} \right). \end{cases}$$

In line with Lemmas 7.1–7.2, the customized LTKF (susceptible to dead-zone-like censoring, dynamical bias and RRP) is presented as follows.

Theorem 7.1 The LTKF for system (7.10)–(7.11) is of the following structure:

$$\hat{\xi}_{m,k}^{-} = \mathcal{A}_{k-1} \hat{\xi}_{m,k-1}, \tag{7.28}$$

$$P_{\bar{\xi}_{m,k}^{-}} = \mathcal{A}_{k-1} P_{\bar{\xi}_{m,k-1}}^{T} \mathcal{A}_{k-1}^{T} + \mathcal{B}_{k-1} Q_{m,k-1} \mathcal{B}_{k-1}^{T}, \tag{7.29}$$

$$\hat{\xi}_{m,k} = \hat{\xi}_{m,k}^{-} + K_{m,k} \left(y_{m,k} - \hat{y}_{m,k}^{-} \right), \tag{7.30}$$

$$P_{\bar{\xi}_{m,k}} = P_{\bar{\xi}_{m,k}^{-}} - K_{m,k} P_{\bar{\xi}_{m,k} \tilde{y}_{m,k}}^{T}. \tag{7.31}$$

The local filtering gain matrix $K_{m,k}$ and one-step measurement prediction are

$$K_{m,k} = P_{\bar{\xi}_{m,k}^{-} \tilde{y}_{m,k}} P_{\tilde{y}_{m,k}}^{-1}, \tag{7.32}$$

$$\hat{y}_{m,k}^{-} = \bar{\alpha}_{m,k} \tau_m^l + \bar{\beta}_{m,k} \tau_m^r + \left(1 - \bar{\alpha}_{m,k} - \bar{\beta}_{m,k} \right)$$
$$\times \left[\hat{\xi}_{m,k}^{-} + \sqrt{\mathcal{R}_{m,k}} \lambda \left(\bar{\vartheta}_{m,k}^{r} - \bar{\vartheta}_{m,k}^{l} \right) \right], \tag{7.33}$$

where

$$P_{\bar{\xi}_{m,k}^{-} \tilde{y}_{m,k}} = \left(1 - \bar{\alpha}_{m,k} - \bar{\beta}_{m,k} \right) P_{\bar{\xi}_k^{-}} \left(\Gamma_{m,h_k} C_{m,k} \right)^{T}, \tag{7.34}$$

$$P_{\tilde{y}_{m,k}^-} = \left(1 - \bar{\alpha}_{m,k} - \bar{\beta}_{m,k}\right)^2 \Gamma_{m,h_k} C_{m,k} P_{\xi_k^-} C_{m,k}^T \Gamma_{m,h_k}$$

$$+ \left(1 - \bar{\alpha}_{m,k} - \bar{\beta}_{m,k}\right)^2 \sum_{l=1}^{p-1} \Gamma_{m,h_{k-l}} C_{m,k-l} P_{\xi_{k-l}^-} C_{m,k-l}^T \Gamma_{m,h_{k-l}} \qquad (7.35)$$

$$+ \mathcal{R}_{m,k} \left[1 + \varphi\left(\bar{\vartheta}_{m,k}^r, \bar{\vartheta}_{m,k}^l\right)\right].$$

Proof For the sake of derivation brevity, denote

$$P_{\xi_{m,k}\xi_{m,k-l}} \triangleq \mathbb{E}\left\{\xi_{m,k}\xi_{m,k-l}^T\right\}, m = 0,1\ldots,p-1,$$

and suppose

$$P_{\xi_k^- \xi_t^-} = 0 \ for \ k \neq t, k,t = 1,2,\ldots,$$

$$cov\left\{y_{m,k}, y_{s,k}\right\} = 0 \ for \ m \neq s, m,s = 1,2,\ldots,p.$$

The extension to the setting where

$$P_{\xi_k^- \xi_t^-} \neq 0 \ for \ k \neq t, k,t = 1,2,\ldots,$$

$$cov\left\{y_{m,k}, y_{s,k}\right\} \neq 0 \ for \ m \neq s, m,s = 1,2,\ldots,p.$$

is straightforward but rotationally cumbersome.

The orthogonality projection principle tells us that for system (7.10), its linear minimum variance estimate (based on the measurements from sensor m) is given by

$$\hat{\xi}_{m,k} = \mathbb{E}\left\{\xi_k \mid y_{m,1:k}\right\}$$

$$= \mathbb{E}\left\{\xi_k \mid y_{m,1:k-1}\right\} + P_{\xi_{m,k}^- \tilde{y}_{m,k}^-} P_{\tilde{y}_{m,k}^-}^{-1} \left(y_{m,k} - \mathbb{E}\left\{y_{m,k} \mid y_{m,1:k-1}\right\}\right)$$

$$= \hat{\xi}_{m,k}^- + K_{m,k}\left(y_{m,k} - \hat{y}_{m,k}^-\right),$$

which is exactly the same as (7.30), where $K_{m,k}$ has the structure of (7.32). Meanwhile, we have

$$\hat{\xi}_{m,k}^- = \mathbb{E}\left\{\xi_k \mid y_{m,1:k-1}\right\}$$

$$= \mathbb{E}\left\{\mathcal{A}_{k-1}\xi_{k-1} + \mathcal{B}_{k-1}w_{k-1} \mid y_{m,1:k-1}\right\}$$

$$= \mathbb{E}\left\{\mathcal{A}_{k-1}\xi_{k-1} \mid y_{m,1:k-1}\right\}$$

$$= \mathcal{A}_{k-1}\hat{\xi}_{m,k-1},$$

which is exactly the same as (7.28), where the third equality holds from $\mathbb{E}\{w_{k-1}\} = 0$.

Subtracting (7.28) from (7.10) brings

$$\tilde{\xi}_{m,k}^- = \mathcal{A}_{k-1}\tilde{\xi}_{m,k-1}^- + \mathcal{B}_{k-1}w_{k-1}.$$

Inserting the expression of $\tilde{\xi}_{m,k}^{-}$ into

$$P_{\tilde{\xi}_{m,k}^{-}} \triangleq \mathbb{E}\left\{\tilde{\xi}_{m,k}^{-}\left(\tilde{\xi}_{m,k}^{-}\right)^{T}\right\}$$

gives

$$\begin{aligned}
P_{\tilde{\xi}_{m,k}^{-}} &= \mathbb{E}\left\{\left(\mathcal{A}_{k-1}\tilde{\xi}_{m,k-1}^{-} + \mathcal{B}_{k-1}w_{k-1}\right)\left(\mathcal{A}_{k-1}\tilde{\xi}_{m,k-1}^{-} + \mathcal{B}_{k-1}w_{k-1}\right)^{T}\right\} \\
&= \mathcal{A}_{k-1}P_{\tilde{\xi}_{m,k-1}^{-}}\mathcal{A}_{k-1}^{T} + \mathcal{B}_{k-1}\mathcal{Q}_{k-1}\mathcal{B}_{k-1}^{T},
\end{aligned}$$

which is precisely the same as (7.29).

By resorting to Lemma 7.1, we have

$$\begin{aligned}
\hat{y}_{\bar{m},k} &= \Phi\left(\bar{\vartheta}_{m,k}^{l}\right)\tau_{m}^{l} + \left(1 - \Phi\left(\bar{\vartheta}_{m,k}^{r}\right)\right)\tau_{m}^{r} \\
&\quad + \left(\Phi\left(\bar{\vartheta}_{m,k}^{r}\right) - \Phi\left(\bar{\vartheta}_{m,k}^{l}\right)\right)\left[\hat{\zeta}_{\bar{m},k} + \sqrt{\mathcal{R}_{m,k}}\lambda\left(\bar{\vartheta}_{m,k}^{r} - \bar{\vartheta}_{m,k}^{l}\right)\right] \\
&= \bar{\alpha}_{m,k}\tau_{m}^{l} + \bar{\beta}_{m,k}\tau_{m}^{r} + (1 - \bar{\alpha}_{m,k} - \bar{\beta}_{m,k})\left[\hat{\zeta}_{\bar{m},k} + \sqrt{\mathcal{R}_{m,k}}\lambda\left(\bar{\vartheta}_{m,k}^{r} - \bar{\vartheta}_{m,k}^{l}\right)\right],
\end{aligned}$$

which is precisely the same as (7.32).

Subtracting (7.32) from

$$\bar{y}_{m,k} \triangleq \zeta_{m,k} + v_{m,k}$$

leads to

$$\tilde{y}_{m,k}^{-} = \left(1 - \alpha_{m,k} - \beta_{m,k}\right)\bar{y}_{m,k} + \alpha_{m,k}\tau_{m}^{l} + \beta_{m,k}\tau_{m}^{r} - \hat{y}_{m,k}^{-}. \tag{7.36}$$

Bearing (7.36) in mind, we have

$$\begin{aligned}
P_{\tilde{\xi}_{m,k}^{-}\tilde{y}_{m,k}^{-}} &= \mathbb{E}\left\{\left(\xi_{k} - \hat{\xi}_{k}^{-}\right)\left((1 - \alpha_{m,k} - \beta_{m,k})\bar{y}_{m,k} + \alpha_{m,k}\tau_{m}^{l}\right.\right. \\
&\qquad\left.\left. + \beta_{m,k}\tau_{m}^{r} - \hat{y}_{\bar{m},k}\right)^{T}\right\} \\
&= \mathbb{E}\left\{\xi_{k}\left(\sum_{l=0}^{p-1}\Gamma_{m,h_{k-l}}C_{m,k-l}\xi_{k-l}\right)^{T}\left(1 - \alpha_{m,k} - \beta_{m,k}\right)\right\} \\
&\quad - \mathbb{E}\left\{\hat{\xi}_{k}^{-}\left(\sum_{l=0}^{p-1}\Gamma_{m,h_{k-l}}C_{m,k-l}\xi_{k-l}\right)^{T}\left(1 - \alpha_{m,k} - \beta_{m,k}\right)\right\} \\
&= \left(1 - \bar{\alpha}_{m,k} - \bar{\beta}_{m,k}\right)\sum_{l=0}^{p-1}P_{\xi_{m,k}\xi_{m,k-l}}\left(\Gamma_{m,h_{k-l}}C_{m,k-l}\right)^{T} \\
&\quad - \left(1 - \bar{\alpha}_{m,k} - \bar{\beta}_{m,k}\right)\hat{\xi}_{k}^{-}\sum_{l=0}^{p-1}\hat{\xi}_{k-l}^{-}\left(\Gamma_{m,h_{k-l}}C_{m,k-l}\right)^{T}.
\end{aligned} \tag{7.37}$$

Paying attention to the relationships among $\xi_{m,k}$, $\hat{\xi}_{m,k}$ and $\tilde{\xi}_{m,k}$, we have

$$
\begin{aligned}
P_{\xi_{m,k}\xi_{m,k-l}} &= \mathbb{E}\left\{ \left(\tilde{\xi}_{\bar{m},k} + \hat{\xi}_{\bar{m},k} \right)\left(\tilde{\xi}_{\bar{m},k-l} + \hat{\xi}_{\bar{m},k-l} \right)^T \right\} \\
&= P_{\tilde{\xi}^-_{m,k}\tilde{\xi}^-_{m,k-l}} + \hat{\xi}^-_{m,k}\left(\hat{\xi}^-_{m,k-l} \right)^T,
\end{aligned}
\tag{7.38}
$$

where the second equality holds from

$$
\mathbb{E}\left\{ \tilde{\xi}^-_k \left(\hat{\xi}^-_{k-l} \right)^T \right\} = 0, \ \mathbb{E}\left\{ \hat{\xi}^-_k \left(\tilde{\xi}^-_{k-l} \right)^T \right\} = 0.
$$

Substituting (7.38) into (7.37) produces

$$
\begin{aligned}
P_{\tilde{\xi}^-_{m,k}\tilde{y}^-_{m,k}} &= \left(1 - \bar{\alpha}_{m,k} - \bar{\beta}_{m,k} \right) \sum_{l=0}^{p-1} P_{\tilde{\xi}^-_k \tilde{\xi}^-_{k-l}} \left(\Gamma_{m,h_{k-l}} C_{m,k-l} \right)^T \\
&= \left(1 - \bar{\alpha}_{m,k} - \bar{\beta}_{m,k} \right) P_{\tilde{\xi}^-_k} \left(\Gamma_{m,h_k} C_{m,k} \right)^T,
\end{aligned}
\tag{7.39}
$$

where the last equality holds from

$$
P_{\tilde{\xi}^-_k \tilde{\xi}^-_t} \neq 0 \ for \ k \neq t, k,t = 0,1,2,\ldots.
$$

(7.39) is precisely the same as (7.32).

Retrospecting the expressions of $\zeta_{m,k}$, $\hat{\zeta}^-_{m,k}$ and $\tilde{\zeta}^-_{m,k}$ leads to

$$
\begin{aligned}
\zeta_{\bar{m},k} &= \sum_{l=0}^{p-1} \Gamma_{m,h_{k-l}} C_{m,k-l}\xi_{k-l} - \Gamma_{m,h_k} C_{m,k}\hat{\xi}^-_k - \sum_{l=1}^{p-1} \Gamma_{m,h_{k-l}} C_{m,k-l}\hat{\xi}_{k-l} \\
&= \Gamma_{m,h_k} C_{m,k}\tilde{\xi}^-_k + \sum_{l=1}^{p-1} \Gamma_{m,h_{k-l}} C_{m,k-l}\tilde{\xi}_{k-l},
\end{aligned}
$$

which indicates

$$
\begin{aligned}
\mathbb{E}\left\{ \tilde{\zeta}_{\bar{m},k} \left(\tilde{\zeta}_{\bar{m},k} \right)^T \right\} &= \Gamma_{m,h_k} C_{m,k} P_{\tilde{\xi}^-_k} C^T_{m,k} \Gamma_{m,h_k} \\
&\quad + \sum_{l=1}^{p-1} \Gamma_{m,h_{k-l}} C_{m,k-l} P_{\tilde{\xi}_{k-l}} C^T_{m,k-l} \Gamma_{m,h_{k-l}}.
\end{aligned}
\tag{7.40}
$$

Akin to the derivation of $P_{\tilde{\xi}^-_{m,k}\tilde{y}^-_{m,k}}$, one has

$$
\begin{aligned}
P_{\tilde{y}^-_{m,k}} &= \mathbb{E}\left\{ \tilde{y}^-_{m,k} \left(\tilde{y}^-_{m,k} \right)^T \right\} \\
&= \left(1 - \bar{\alpha}_{m,k} - \bar{\beta}_{m,k} \right)^2 \mathbb{E}\left\{ \tilde{\zeta}^-_{m,k} \left(\tilde{\zeta}^-_{m,k} \right)^T \right\} \\
&\quad + \mathbb{E}\left\{ \left(1 - \alpha_{m,k} - \beta_{m,k} \right) \tilde{v}_{m,k} \tilde{v}^T_{m,k} \left(1 - \alpha_{m,k} - \beta_{m,k} \right) \right\},
\end{aligned}
\tag{7.41}
$$

where

$$\tilde{v}_{m,k} \triangleq v_{m,k} - \sqrt{\mathcal{R}_{m,k}}\,\lambda\left(\bar{\vartheta}^{r}_{m,k} - \bar{\vartheta}^{l}_{m,k}\right).$$

Accordingly, we have

$$\mathbb{E}\left\{\left(1-\alpha_{m,k}-\beta_{m,k}\right)\tilde{v}_{m,k}\tilde{v}^{T}_{m,k}\left(1-\alpha_{m,k}-\beta_{m,k}\right)\right\}$$
$$= var\left\{y_{m,k} \mid x_{k}, \mathcal{R}_{m,k}, y_{m,1:k-1}\right\} \tag{7.42}$$
$$= \mathcal{R}_{m,k}\left[1+\varphi\left(\bar{\vartheta}^{r}_{m,k},\bar{\vartheta}^{l}_{m,k}\right)\right].$$

Inserting (7.40) and (7.42) into (7.41) results in

$$P_{\bar{y}_{m,k}} = \mathbb{E}\left\{\tilde{y}^{-}_{m,k}\left(\tilde{y}^{-}_{m,k}\right)^{T}\right\}$$
$$= \left(1-\bar{\alpha}_{m,k}-\bar{\beta}_{m,k}\right)^{2}\Gamma_{m,h_{k}}C_{m,k}P_{\bar{\xi}^{-}_{k}}C^{T}_{m,k}\Gamma_{m,h_{k}}$$
$$+ \left(1-\bar{\alpha}_{m,k}-\bar{\beta}_{m,k}\right)^{2}\sum_{l=1}^{p-1}\Gamma_{m,h_{k-l}}C_{m,k-l}P_{\bar{\xi}^{-}_{k-l}}C^{T}_{m,k-l}\Gamma_{m,h_{k-l}}$$
$$+ \mathcal{R}_{m,k}\left[1+\varphi\left(\bar{\vartheta}^{r}_{m,k},\bar{\vartheta}^{l}_{m,k}\right)\right].l$$

which is precisely the same as (7.33).

Remark 7.4 When making comparison between the proposed LTKF in Theorem 7.1 and the conventional TKF, we observe three major differences. The first difference is the replacement of the $C_{k}\hat{x}^{-}_{k}$ (which is the product of the original observation coefficient C_{k} and state prediction $\hat{x}^{-}_{k}$) by $\hat{\zeta}^{-}_{m,k} \triangleq \sum_{l=0}^{p-1}\Gamma_{m,h_{k-l}}C_{m,k-l}\hat{\xi}^{-}_{k-l}$ (which is the sum of p products in regard to the update coefficient $\Gamma_{m,h_{k-l}}$, observation coefficient $C_{m,k-l}$ and augmented state ξ_{k-l}) in all observation-related terms. The second difference is the substitution of the original observation noise covariance $R_{m,k}$ by $\mathcal{R}_{m,k} \triangleq \sum_{l=0}^{p-1}\Gamma^{2}_{m,h_{k-l}}R_{m,k-l}$ (which is the sum of p delayed noise covariances) in all terms relevant to the observation prediction, filter gain as well as error covariances. The last difference is the presence of several new terms including the augmentation-induced terms $\hat{\xi}^{-}_{m,k}$, $\hat{\xi}_{m,k}$, $P_{\bar{\xi}^{-}_{m,k}}$, $P_{\bar{\xi}_{m,k}}$, $P_{\bar{\xi}^{-}_{m,k}\bar{y}_{m,k}}$ and the protocol-induced terms $\bar{\gamma}_{m,k}\sum_{l=1}^{p-1}C_{m,k-l}P_{\bar{\xi}_{m,k-l}}C^{T}_{m,k-l}\bar{\gamma}_{m,k}$. These differences obviously reflect the influences from the dynamical bias and the RRP on the development of the filtering fusion algorithm.

Theorem 7.2 Suppose that $\hat{\xi}_{m,k}$ $(m=1,2,\ldots,p)$ and $P_{\bar{\xi}_{m,k}}$ are, respectively, the state estimate and the error covariance produced by the mth LTKF. Let $\hat{\xi}_{k}$ and $P_{\bar{\xi}_{k}}$ be, respectively, the global optimal estimate and covariance obtained at the fusion center under the federated fusion rule. Then, the FTKF for system (7.10)–(7.11) is expressed as

$$\begin{cases} P_{\bar{\xi}_{k}} = \left(\displaystyle\sum_{m=1}^{p}P^{-1}_{\bar{\xi}_{m,k}}\right)^{-1}, \\[4mm] \hat{\xi}_{k} = P_{\bar{\xi}_{k}}\left(\displaystyle\sum_{i=1}^{p}P^{-1}_{\bar{\xi}_{m,k}}\hat{\xi}_{m,k}\right). \end{cases} \tag{7.43}$$

where

$$\begin{cases} Q_{m,k-1} = \epsilon_m^{-1} Q_{k-1}, \\ P_{\tilde{\xi}_{m,k-1}} = \epsilon_m^{-1} P_{\tilde{\xi}_{k-1}}, \\ \hat{\xi}_{m,k-1} = \hat{\xi}_{k-1}. \end{cases} \tag{7.44}$$

Denoting

$$\Gamma_1 \triangleq \begin{bmatrix} I & 0 \end{bmatrix},$$

$$\Gamma_2 \triangleq \begin{bmatrix} 0 & I \end{bmatrix},$$

the next theorem presents the FTKF for system (7.1)–(7.8).

Theorem 7.3 Given system (7.1)–(7.8), its FTKF is as follows:

$$\begin{cases} \hat{x}_k = \Gamma_1 \hat{\xi}_k, \\ \hat{b}_k = \Gamma_2 \hat{\xi}_k, \\ P_{\tilde{x}_k} = \Gamma_1 P_{\tilde{\xi}_k} \Gamma_1^T, \\ P_{\tilde{b}_k} = \Gamma_2 P_{\tilde{\xi}_k} \Gamma_2^T. \end{cases} \tag{7.45}$$

Proof Theorem 7.3 follows readily from Theorem 7.1 by noting the correlation between system (7.1)–(7.8) and system (7.10)–(7.11).

Theorems 7.1–7.2, together with Lemma 7.1, constitute the FTKF algorithm with its pseudocode outlined in Table 7.1.

Remark 7.5 In accordance with Lemma 7.1 and Theorems 7.1–7.3, a protocol-based FTKF mechanism is formalized to settle the federated filtering fusion problem in the presence of the dead-zone-like censoring, dynamical bias and RRP. The investigated model (7.1)–(7.8) is generic for its inclusion of not only the multi-sensor sampling nature but also the modeling and observation uncertainties (i.e. the dynamical bias and dead-zone-like censoring), which are universally confronted in engineering ranges such as image processing, target tracking, fault diagnosis etc. These uncertainties are propitiously tackled in a holistic yet efficient framework under the RRP. It is noteworthy that the desired FTKF is, in essence, a distributed extension of the conventional TKF. In the event that the dynamical bias and the RRP are disregarded (i.e. $B_k = 0$ and $\bar{y}_{m,k} = z_{m,k}$), the LTKF in Theorem 7.1 will degrade to the standard TKF. Consequently, the FTKF in Theorem 7.2 will degenerate to the distributed generalization of the conventional TKF.

Next, we move forward to assess the performance of the designed FTKF. Due to the time-varying nature of the protocol-induced measurement update coefficient Γ_{m,h_k}, the convergence of the developed federated Tobit Kalman filtering algorithm cannot be guaranteed in general. Thus, we turn to pursue the boundedness of the developed algorithm where the upper and lower bounds of the estimation error

TABLE 7.1

FTKF Pseudocode

Algorithm: FTKF

Input: $\bar{x}_0, \bar{b}_0, P_{\tilde{x}_0}, P_{\tilde{b}_0}, y_{m,1:k}$

Output: $\hat{x}_k, \hat{b}_k, P_{\tilde{x}_k}$.

1: let $\hat{\xi}_0 = \left[\bar{x}_0^T \quad \bar{b}_0^T\right]^T, \hat{\xi}_{m,0} = \bar{\xi}_0, P_{\tilde{\xi}_{m,0}} = \epsilon_m^{-1} P_{\tilde{\xi}_0},$

$Q_{m,0} = \epsilon_m^{-1} Q_0, P_{\tilde{\xi}_0} = \text{diag}\left\{P_{\tilde{x}_0}, P_{\tilde{b}_0}\right\}.$

2: **for** $k = 1 : N$ **do**

3:       calculate the local predicted estimate $\hat{\xi}_{m,k}^-$ and

      covariance $P_{\tilde{\xi}_{m,k}^-}$ by (7.28)–(7.29);

4:       calculate the local gain matrix $K_{m,k}$ by (7.32);

5:       calculate the local updated estimate $\hat{\xi}_{m,k}$ and

      covariance $P_{\tilde{\xi}_{m,k}}$ by (7.30)–(7.31);

6:       calculate the optimal fused estimate $\hat{\xi}_k$ and

      covariance $P_{\tilde{\xi}_k}$ by (7.43);

7:       calculate the optimal fused estimates $\hat{x}_k$ and $\hat{b}_k$

      and associate covariances $P_{\tilde{x}_k}$ and $P_{\tilde{b}_k}$ by (7.45)

8:       reallocate $\hat{\xi}_{m,k}, P_{\tilde{\xi}_{m,k}}$ and $Q_{m,k}$ by (7.44);

9: **end for**

covariance are explored. The pursuit of such bounds can be carried out based on three principles: (1) the optimality of the developed LTKF motivates us to construct a set of suboptimal filters whose estimation error covariances are envisioned to be the upper bounds on $P_{\tilde{\xi}_{m,k}^-}$; (2) the semi-positive definiteness of $P_{\tilde{\xi}_{m,k}^-}, Q_k$ and $R_{m,k}$ paves the way for us to envisage the lower bound on $P_{\tilde{\xi}_{m,k}^-}$ via some subtle matrix manipulations; and (3) the federated fusion rule provides us with the correlation between the bounds of the LTKF and that of the fused filter.

Recalling that the optimal protocol-based LTKF derived in Theorem 7.2 has an optimal filtering gain $K_{m,k}$ (which depends on both left- and right-censoring probabilities $\bar{\alpha}_{m,k}$ and $\bar{\beta}_{m,k}$), the following theorem is dedicated to the derivation of a self-propagating upper bound on $P_{\tilde{\xi}_{m,k}^-}$ via constructing suboptimal protocol-based LTKFs with suboptimal filtering gains $K_{m,k}^u$ (which depends only on the left-censoring probabilities $\bar{\alpha}_{m,k}$). As to the lower bound on $P_{\tilde{\xi}_{m,k}}$, it can be acquired by making use of the semi-positive definiteness of matrices $P_{\tilde{\xi}_{m,k}^-}, Q_k$ and $R_{m,k}$.

Theorem 7.4 Let the initial condition $P^u_{\check{\xi}^-_{m,0}} > 0$ and $P^l_{\check{\xi}^-_{m,0}} > 0$ be given. Calculate the matrix sequences $\left\{ P^u_{\check{\xi}^-_{m,k+1}} \right\}_{k \geq 0}$ and $\left\{ P^l_{\check{\xi}^-_{m,k+1}} \right\}_{k \geq 0}$ according to the following difference equations:

$$P^u_{\check{\xi}^-_{m,k+1}} = A_k P^u_{\check{\xi}^-_k} A_k^T + B_k Q_k B_k^T - A_k \left(1 - \bar{\alpha}_{m,k} \right)^2 P_{\check{\xi}^-_{m,k}}$$

$$\times \left(\Gamma_{m,h_k} C_{m,k} \right)^T \left\{ \left(1 - \bar{\alpha}_{m,k} \right)^2 \Gamma_{m,h_k} C_{m,k} P_{\check{\xi}^-_{m,k}} C_{m,k}^T \Gamma_{m,h_k}^T \right.$$

$$+ \left(1 - \bar{\alpha}_{m,k} \right)^2 \sum_{l=1}^{p-1} \Gamma_{m,h_{k-l}} C_{m,k-l} P_{\check{\xi}^-_{m,k-l}} C_{m,k-l}^T \Gamma_{m,h_{k-l}}^T$$

$$\left. + \mathcal{R}_{m,k} \left[1 + \varphi \left(\bar{\vartheta}^l_{m,k} \right) \right] \right\}^{-1} \Gamma_{m,h_k} C_{m,k} P_{\check{\xi}^-_{m,k}} A_k^T,$$

$$P^l_{\check{\xi}^-_{m,k+1}} = \left(1 - \left(1 - \bar{\alpha}_{m,k} - \bar{\beta}_{m,k} \right)^2 \right) A_k P^l_{\check{\xi}^-_{m,k}} A_k^T + B_k Q_k B_k^T.$$

Then, the calculated matrices $P^u_{\check{\xi}^-_{m,k+1}}$ and $P^l_{\check{\xi}^-_{m,k+1}}$ satisfy

$$P^l_{\check{\xi}^-_{m,k+1}} \leq P_{\check{\xi}^-_{m,k+1}} \leq P^u_{\check{\xi}^-_{m,k+1}},$$

i.e. $P^u_{\check{\xi}^-_{m,k+1}}$ and $P^l_{\check{\xi}^-_{m,k+1}}$ are, respectively, the self-propagating upper and lower bounds on $P_{\check{\xi}^-_{m,k+1}}$, where $\varphi \left(\bar{\vartheta}^l_{m,k} \right) = \lambda \left(\bar{\vartheta}^l_{m,k} \right) \left[\lambda(\bar{\vartheta}^l_{m,k}) - \bar{\vartheta}^l_{m,k} \right]$ and $\lambda \left(\bar{\vartheta}^l_{m,k} \right) = \phi \left(\bar{\vartheta}^l_{m,k} \right) / \left[1 - \Phi \left(\bar{\vartheta}^l_{m,k} \right) \right]$.

Proof First, let us concentrate on designing a suboptimal protocol-based LTKF whose gain matrix $K^u_{m,k}$ only hinges on the left-censoring probability $\bar{\alpha}_{m,k}$. Recalling the LTKF designed in Theorem 7.2, its error covariance $P_{\check{\xi}^-_{m,k+1}}$ in (7.29) is rearranged into

$$P_{\check{\xi}^-_{m,k+1}}$$

$$= A_k P_{\check{\xi}^-_{m,k}} A_k^T + B_k Q_k B_k^T - \left(1 - \bar{\alpha}_{m,k} - \bar{\beta}_{m,k} \right)^2 A_k P_{\check{\xi}^-_{m,k}}$$

$$\times \left(\Gamma_{m,h_k} C_{m,k} \right)^T \left\{ \left(1 - \bar{\alpha}_{m,k} - \bar{\beta}_{m,k} \right)^2 \Gamma_{m,h_k} C_{m,k} P_{\check{\xi}^-_{m,k}} \right.$$

$$\times C_{m,k}^T \Gamma_{m,h_k}^T + \left(1 - \bar{\alpha}_{m,k} - \bar{\beta}_{m,k} \right)^2 \sum_{l=1}^{p-1} \Gamma_{m,h_{k-l}} C_{m,k-l} \tag{7.46}$$

$$\times P_{\check{\xi}^-_{m,k-l}} C_{m,k-l}^T \Gamma_{m,h_{k-l}}^T$$

$$\left. + \mathcal{R}_{m,k} \left[1 + \varphi \left(\bar{\vartheta}^r_{m,k}, \bar{\vartheta}^l_{m,k} \right) \right] \right\}^{-1} \Gamma_{m,h_k} C_{m,k} P_{\check{\xi}^-_{m,k}} A_k^T.$$

Accordingly, the error covariance $P^u_{\check{\xi}^-_{m,k+1}}$ with regard to the suboptimal LTKF can be easily obtained as the one in Theorem 7.4.

For k = 0, by setting $P^u_{\check{\xi}^-_{m,0}} = P_{\check{\xi}^-_{m,0}} > 0$, we easily have $P_{\check{\xi}^-_{m,0}} \leq P^u_{\check{\xi}^-_{m,0}}$. For k > 0, since the performance of the optimal protocol-based LTKF must be no less than any of its suboptimal counterparts, it can be concluded that $P_{\tilde{\xi}_{m,k}} \leq P^{u}_{\tilde{\xi}_{m,k}}$. As such, $P^{u}_{\tilde{\xi}_{m,k}}$ in Theorem 7.4 is an upper bound on $P_{\tilde{\xi}_{m,k}}$ for all $k \geq 0$.

Next, we move forward to seek a lower bound on $P_{\tilde{\xi}_{m,k}}$. Denoting $\mathfrak{A}_{m,k} \triangleq \mathcal{A}_k + \left(1-\bar{\alpha}_{m,k}-\bar{\beta}_{m,k}\right)^2 \mathfrak{K}_{m,k}\Gamma_{m,h_k}C_{m,k}$ and $\mathfrak{K}_{m,k} \triangleq -\mathcal{A}_{m,k}K_{m,k}$, we have

$$\mathfrak{A}_{m,k}P_{\tilde{\xi}_{m,k}}\left(\Gamma_{m,h_k}C_{m,k}\right)^T + \mathfrak{K}_{m,k}\left\{\left(1-\bar{\alpha}_{m,k}-\bar{\beta}_{m,k}\right)^2\sum_{l=1}^{p-1}\Gamma_{m,h_{k-l}}C_{m,k-l}\right.$$

$$\times P_{\tilde{\xi}_{m,k-l}}C_{m,k-l}^T\Gamma_{m,h_{k-l}}^T + \mathcal{R}_{m,k}\left[1+\varphi(\bar{\vartheta}_{m,k}^{r},\bar{\vartheta}_{m,k}^{l})\right]\right\}$$

$$=\left(\mathcal{A}_k + \left(1-\bar{\alpha}_{m,k}-\bar{\beta}_{m,k}\right)^2\mathfrak{K}_{m,k}\Gamma_{m,h_k}C_{m,k}\right)P_{\tilde{\xi}_{m,k}}\left(\Gamma_{m,h_k}C_{m,k}\right)^T$$

$$+\mathfrak{K}_{m,k}\left\{\left(1-\bar{\alpha}_{m,k}-\bar{\beta}_{m,k}\right)^2\sum_{l=1}^{p-1}\Gamma_{m,h_{k-l}}C_{m,k-l}P_{\tilde{\xi}_{m,k-l}}C_{m,k-l}^T\Gamma_{m,h_{k-l}}^T\right. \tag{7.47}$$

$$+\mathcal{R}_{m,k}\left[1+\varphi\left(\bar{\vartheta}_{m,k}^{r},\bar{\vartheta}_{m,k}^{l}\right)\right]\right\}$$

$$=\mathcal{A}_k P_{\tilde{\xi}_{m,k}}\left(\Gamma_{m,h_k}C_{m,k}\right)^T + \mathfrak{K}_{m,k}P_{\bar{y}_{m,k}^-}$$

$$=0.$$

It follows from (7.46) that

$$P_{\tilde{\xi}_{m,k+1}} = \left(1-\left(1-\bar{\alpha}_{m,k}-\bar{\beta}_{m,k}\right)^2\right)\left(\mathcal{A}_k P_{\tilde{\xi}_{m,k}}\mathcal{A}_k^T + \mathcal{B}_k Q_k \mathcal{B}_k^T\right)$$

$$+\left(1-\bar{\alpha}_{m,k}-\bar{\beta}_{m,k}\right)^2\left(\mathcal{A}_k P_{\tilde{\xi}_{m,k}}\mathcal{A}_k^T + \mathcal{B}_k Q_k \mathcal{B}_k^T\right)$$

$$-\left(1-\bar{\alpha}_{m,k}-\bar{\beta}_{m,k}\right)^2 \mathcal{A}_k P_{\tilde{\xi}_k}\left(\Gamma_{m,h_k}C_{m,k}\right)^T P_{\bar{y}_{m,k}^-}^{-1}\Gamma_{m,h_k}C_{m,k}P_{\tilde{\xi}_k}\mathcal{A}_k^T$$

$$=\left(1-\left(1-\bar{\alpha}_{m,k}-\bar{\beta}_{m,k}\right)^2\right)\left(\mathcal{A}_k P_{\tilde{\xi}_{m,k}}\mathcal{A}_k^T + \mathcal{B}_k Q_k \mathcal{B}_k^T\right)$$

$$+\left(1-\bar{\alpha}_{m,k}-\bar{\beta}_{m,k}\right)^2\left(\mathcal{A}_k P_{\tilde{\xi}_{m,k}}\mathcal{A}_k^T + \mathcal{B}_k Q_k \mathcal{B}_k^T\right) \tag{7.48}$$

$$+\left(1-\bar{\alpha}_{m,k}-\bar{\beta}_{m,k}\right)^2 \mathfrak{K}_{m,k}\Gamma_{m,h_k}C_{m,k}P_{\tilde{\xi}_{m,k}}\mathcal{A}_k^T$$

$$=\left(1-\left(1-\bar{\alpha}_{m,k}-\bar{\beta}_{m,k}\right)^2\right)\left(\mathcal{A}_k P_{\tilde{\xi}_{m,k}}\mathcal{A}_k^T + \mathcal{B}_k Q_k \mathcal{B}_k^T\right)$$

$$+\left(1-\bar{\alpha}_{m,k}-\bar{\beta}_{m,k}\right)^2\left(\mathfrak{A}_{m,k}P_{\tilde{\xi}_{m,k}}\mathcal{A}_k^T + \mathcal{B}_k Q_k \mathcal{B}_k^T\right).$$

Inserting (7.47) into (7.48) generates

$$
\begin{aligned}
P_{\hat{\xi}^-_{\varsigma m,k+1}} &= \left(1-\left(1-\bar{\alpha}_{m,k}-\bar{\beta}_{m,k}\right)^2\right)\left(\mathcal{A}_k P_{\hat{\xi}^-_{\varsigma m,k}} \mathcal{A}_k^T + \mathcal{B}_k Q_k \mathcal{B}_k^T\right) \\
&\quad + (1-\bar{\alpha}_{m,k}-\bar{\beta}_{m,k})^2\left(\mathfrak{A}_{m,k} P_{\hat{\xi}^-_{\varsigma m,k}} \mathfrak{A}_k^T + \mathcal{B}_k Q_k \mathcal{B}_k^T\right) \\
&\quad + \left(1-\bar{\alpha}_{m,k}-\bar{\beta}_{m,k}\right)^2 \mathfrak{K}_{m,k} P_{\tilde{y}^-_{m,k}} \mathfrak{K}^T_{m,k} \\
&\geq \left(1-\left(1-\bar{\alpha}_{m,k}-\bar{\beta}_{m,k}\right)^2\right)\mathcal{A}_k P_{\hat{\xi}^-_{\varsigma m,k}} \mathcal{A}_k^T \\
&\quad + \left(1-\bar{\alpha}_{m,k}-\bar{\beta}_{m,k}\right)^2 \mathcal{B}_k Q_k \mathcal{B}_k^T.
\end{aligned} \tag{7.49}
$$

Inspired by (7.49), let us define

$$
P^l_{\hat{\xi}^-_{\varsigma m,k+1}} \triangleq \left(1-\left(1-\bar{\alpha}_{m,k}-\bar{\beta}_{m,k}\right)^2\right)\mathcal{A}_k P^l_{\hat{\xi}^-_{\varsigma m,k}} \mathcal{A}_k^T + \left(1-\bar{\alpha}_{m,k}-\bar{\beta}_{m,k}\right)^2 \mathcal{B}_k Q_k \mathcal{B}_k^T,
$$

which is initialized at $P^l_{\hat{\xi}^-_{\varsigma m,0}}=0$. Now, let us prove $P^l_{\hat{\xi}^-_{\varsigma m,k+1}} \leq P_{\hat{\xi}^-_{\varsigma m,k+1}}$ via mathematical induction. Noticeably, at the initial time k = 0, we have $P^l_{\hat{\xi}^-_{\varsigma m,0}}=0 \leq P_{\hat{\xi}^-_{\varsigma m,0}}$. Supposing that $P^l_{\hat{\xi}^-_{\varsigma m,k}} \leq P_{\hat{\xi}^-_{\varsigma m,k}}$ holds at time k, we have

$$
\begin{aligned}
P^l_{\hat{\xi}^-_{\varsigma m,k+1}} &= \left(1-\left(1-\bar{\alpha}_{m,k}-\bar{\beta}_{m,k}\right)^2\right)\mathcal{A}_k P^l_{\hat{\xi}^-_{\varsigma m,k}} \mathcal{A}_k^T + \left(1-\bar{\alpha}_{m,k}-\bar{\beta}_{m,k}\right)^2 \mathcal{B}_k Q_k \mathcal{B}_k^T \\
&\leq \left(1-\left(1-\bar{\alpha}_{m,k}-\bar{\beta}_{m,k}\right)^2\right)\mathcal{A}_k P_{\hat{\xi}^-_{\varsigma m,k}} \mathcal{A}_k^T + \left(1-\bar{\alpha}_{m,k}-\bar{\beta}_{m,k}\right)^2 \mathcal{B}_k Q_k \mathcal{B}_k^T \\
&\leq P_{\hat{\xi}^-_{\varsigma m,k+1}},
\end{aligned}
$$

where the last inequality holds from (7.49). As a result, it can be concluded that $P^l_{\hat{\xi}^-_{\varsigma m,k+1}} \leq P_{\hat{\xi}^-_{\varsigma m,k+1}}$ holds for all $k \geq 0$, i.e. $P^l_{\hat{\xi}^-_{\varsigma m,k+1}}$ given by (7.19) is a lower bound on $P_{\hat{\xi}^-_{\varsigma m,k+1}}$. As such, we have $P^l_{\hat{\xi}^-_{\varsigma m,k+1}} \leq P_{\hat{\xi}^-_{\varsigma m,k+1}} \leq P^u_{\hat{\xi}^-_{\varsigma m,k+1}}$ for all $k \geq 0$. This completes the proof.

Theorem 7.5 Letting the initial condition $P^u_{\hat{\xi}^-_{\varsigma m,0}} > 0$ and $P^l_{\hat{\xi}^-_{\varsigma m,0}} > 0$ be given, we have

$$
P^l_{\tilde{x}^-_{k+1}} \leq P_{\tilde{x}^-_{k+1}} \leq P^u_{\hat{\xi}^-_{k+1}},
$$

i.e. $P^u_{\tilde{x}^-_{k+1}}$ and $P^l_{\tilde{x}^-_{k+1}}$ are, respectively, the self-propagating upper and lower bounds on $P_{\tilde{x}^-_{k+1}}$, where

$$
P^u_{\tilde{x}^-_{k+1}} = \Gamma_1 \left[\sum_{m=1}^{p}\left(P^u_{\hat{\xi}^-_{m,k}}\right)^{-1}\right]^{-1} \Gamma_1^T,
$$

$$
P^l_{\tilde{x}^-_{k+1}} = \Gamma_1 \left[\sum_{m=1}^{p}\left(P^l_{\hat{\xi}^-_{m,k}}\right)^{-1}\right]^{-1} \Gamma_1^T.
$$

Proof Theorem 7.5 follows directly from Theorems 7.2–7.4.

Remark 7.6 It should be noted that although the main results in this chapter are obtained based on scalar sensor observation models for simplicity, they can be directly generalized to the case of multi-dimensional observation models. This generalization would bring in changes that include (1) the transformation of the update coefficient from $\Gamma_{m,h_k} \triangleq \delta(\hbar_k - m)$ (which is a scalar) to $\Gamma_{m,h_k} \triangleq \delta(\hbar_k - m)I$ (which is now a matrix); and (2) the introduction of a few more Bernoulli random variables $\alpha_{m,k}^i$ $(i = 1, 2, \ldots, n_{z_m})$ and $\beta_{m,k}^i$ that are adopted to regulate the left and right censoring of $y_{m,k}^i$ (which is the ith entry of $y_{m,k}$). Fortunately, the observation model can still be written in a form similar to (7.8), and this makes it possible for us to easily extend our main results in this chapter to the case of multi-dimensional sensor observations.

7.3 ILLUSTRATIVE EXAMPLES

In this section, the oscillator example and distributed target tracking example are leveraged to elucidate the usefulness of the presented filter design approach and associated filtering performance. Denote the first and second dimensions of x_k, respectively, as $x_{1,k}$ and $x_{2,k}$, and the root mean-squared errors (RMSEs) and average root mean-squared errors (ARMSEs) of $x_{1,k}$, $x_{2,k}$ and b_k as

$$\text{RMSE1} \triangleq \sqrt{\frac{1}{M} \sum_{i=1}^{M} \left(x_{1,k}^{(i)} - \hat{x}_{1,k}^{(i)} \right)^2},$$

$$\text{RMSE2} \triangleq \sqrt{\frac{1}{M} \sum_{i=1}^{M} \left(x_{2,k}^{(i)} - \hat{x}_{2,k}^{(i)} \right)^2},$$

$$\text{RMSE3} \triangleq \sqrt{\frac{1}{M} \sum_{i=1}^{M} \left(b_k^{(i)} - \hat{b}_k^{(i)} \right)^2},$$

$$\text{ARMSE1} \triangleq \frac{1}{N} \sum_{k=1}^{N} \sqrt{\frac{1}{M} \sum_{i=1}^{M} \left(x_{1,k}^{(i)} - \hat{x}_{1,k}^{(i)} \right)^2},$$

$$\text{ARMSE2} \triangleq \frac{1}{N} \sum_{k=1}^{N} \sqrt{\frac{1}{M} \sum_{i=1}^{M} \left(x_{2,k}^{(i)} - \hat{x}_{2,k}^{(i)} \right)^2},$$

$$\text{ARMSE3} \triangleq \frac{1}{N} \sum_{k=1}^{N} \sqrt{\frac{1}{M} \sum_{i=1}^{M} \left(b_k^{(i)} - \hat{b}_k^{(i)} \right)^2},$$

Where M is the number of Monte Carlo trials and N is the number of time steps in each trial.

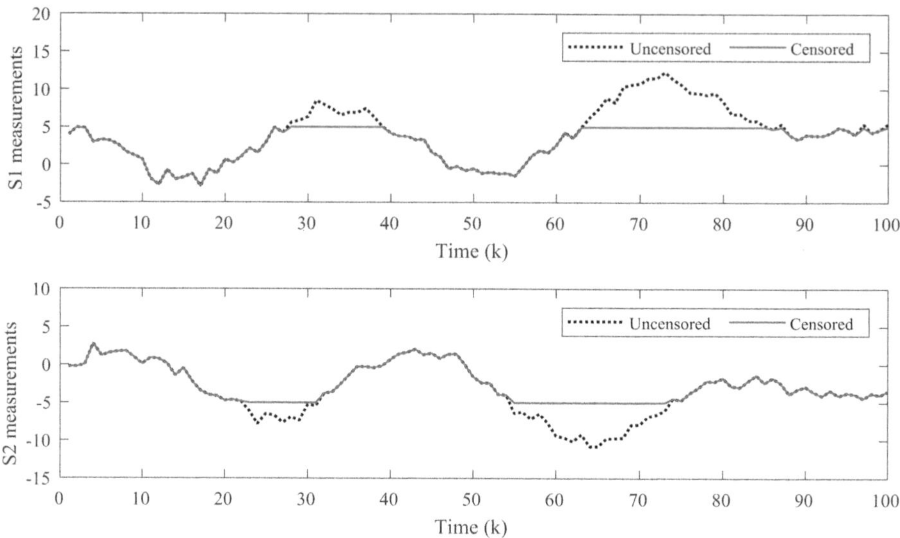

FIGURE 7.2 $\bar{y}_{1,k}$, $\bar{y}_{2,k}$ (uncensored), and $y_{1,k}$ and $y_{2,k}$ (censored).

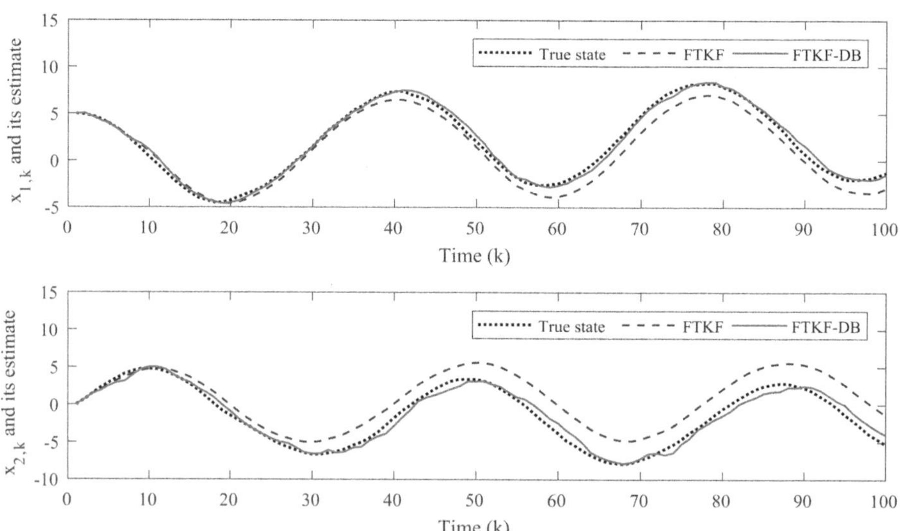

FIGURE 7.3 States and estimates.

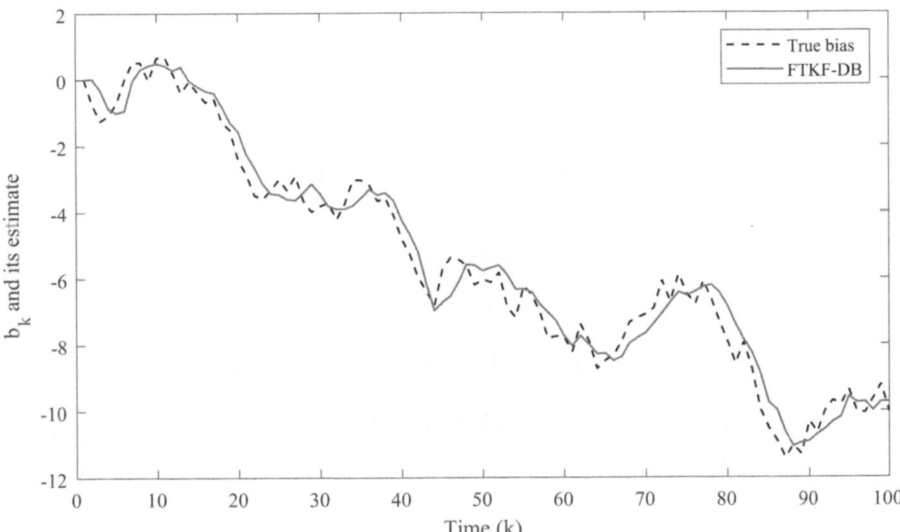

FIGURE 7.4 The bias and its estimate.

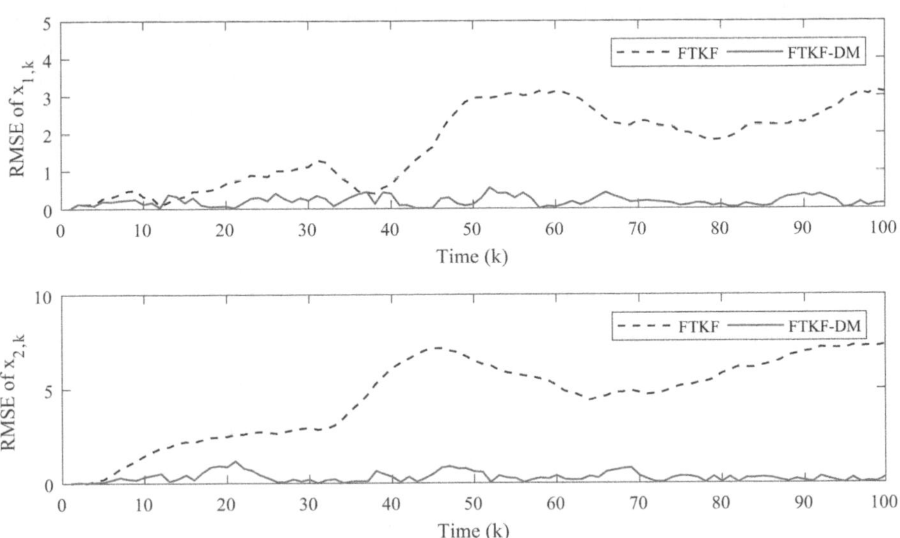

FIGURE 7.5 Comparison in RMSE1 and RMSE2 between the FTKF and FTKF-DB.

7.3.1 Oscillator Example

Let system (7.1)–(7.8) have the following parameters

$$A_k = \begin{bmatrix} \cos(\omega) & -\sin(\omega) \\ \sin(\omega) & \cos(\omega) \end{bmatrix}, B_k = \begin{bmatrix} 0.1 \\ 0.15 \end{bmatrix},$$

$$\Xi_k = 0.25, C_{1,k} = \begin{bmatrix} 1 & 0 \end{bmatrix}, C_{2,k} = \begin{bmatrix} 0 & 1 \end{bmatrix},$$

$$Q_k = \mathrm{diag}\{0.0025, 0.0025\}, R_{1,k} = R_{2,k} = 0.6,$$

$$H_k = 1, \omega = 0.052\pi, \tau_1^l = \tau_2^l = -5, \tau_1^r = \tau_2^r = 5,$$

$$P_{\tilde{x}_0} = 0.1 I_2, \bar{x}_0 = \begin{bmatrix} 5 & 0 \end{bmatrix}^T, \bar{b}_0 = 0, P_{\tilde{b}_0} = 0.1.$$

Note that the dynamic model is corrupted by ambient disturbances entering the system via the process noise ω_k and dynamical bias b_k. The magnetometer sampling is subject to the dead-zone-like censoring, and the data transmission is scheduled by the RRP. In the sequel, in order to testify the effectiveness of our filtering approach in appropriately addressing the dynamical bias issue, we first make a comparison between the proposed FTKF (which is barely capable of addressing the multi-sensor dead-zone-like censored observations in case of the RRP) and the FTKF with dynamical bias (FTKF-DB) (which is capable of simultaneously addressing the bias-corrupted dynamic model and the multi-sensor dead-zone-like censored observations in case of the RRP). From the following demonstrated figures, one can apparently observe the superiority of our FTKF-DB over the FTKF in accurately estimating the target state.

For Sensor 1 (S1) and Sensor 2 (S2), Figure 7.2 sketches the corresponding uncensored observations ($\bar{y}_{1,k}$ and $\bar{y}_{2,k}$) and censored observations ($y_{1,k}$ and $y_{2,k}$). Figure 7.3 depicts the true state values and their estimates provided by the FTKF and FTKF-DB, and Figure 7.4 illustrates the bias estimates produced by our FTKF-DB. It is seen from Figures 7.3–7.4 that our FTKF-DB manages to track the state and bias values precisely, whilst the FTKF has considerable deviations from true state values and, moreover, is unable to track the bias. Figures 7.5–7.6 plot the RMSE curves of the FTKF and FTKF-DB after 1000 independent Monte Carlo trials. It is spotted from Figures 7.5–7.6 that the RMSE1 and RMSE2 curves of our FTKF-DB always reside lower than that of the FTKF, whilst the RMSE3 curve of our FTKF-DB is within a satisfactory range. This is because both state and bias estimation issues are well settled in our FTKF-DB, whereas the bias estimation problem is not disposed of in the FTKF.

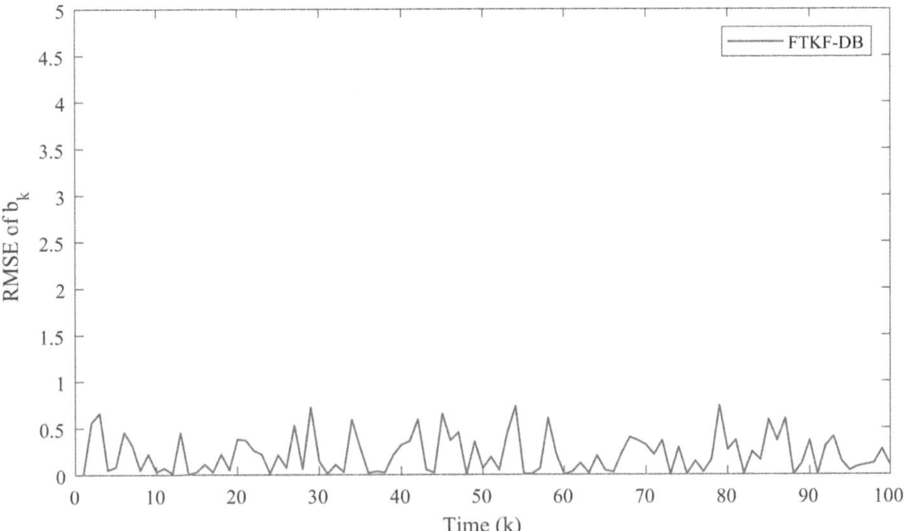

FIGURE 7.6   RMSE3 produced by FTKF-DB.

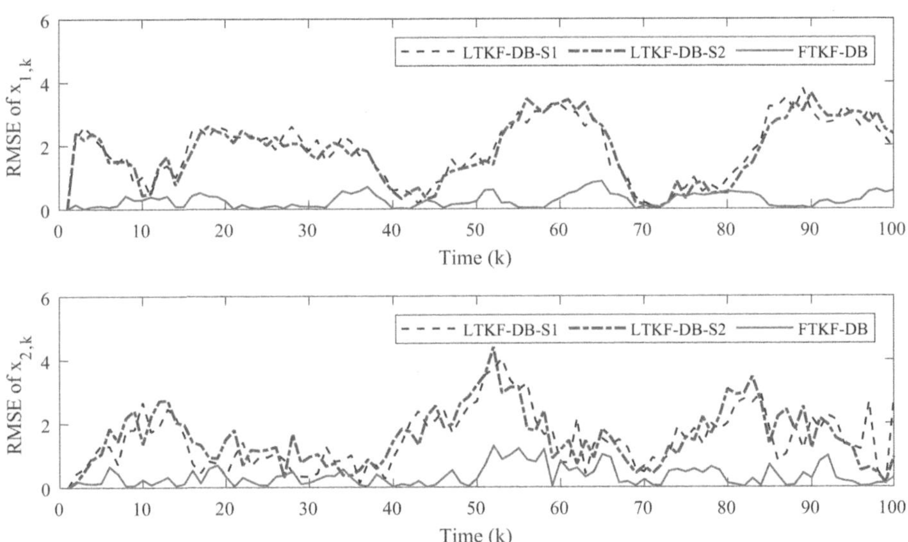

FIGURE 7.7   Comparison in RMSE1 and RMSE2 between the local and fused filters.

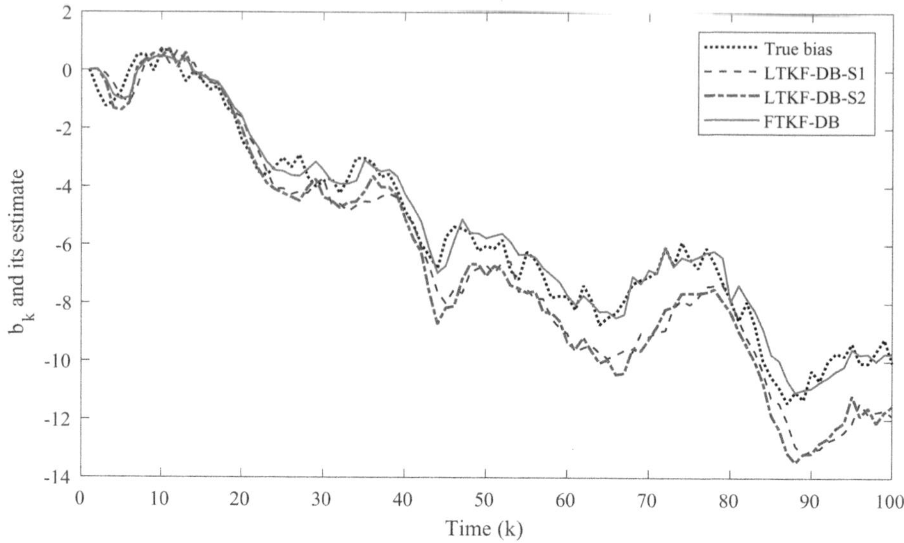

FIGURE 7.8 Comparison in $\hat{b}_k$ between the local and fused filters.

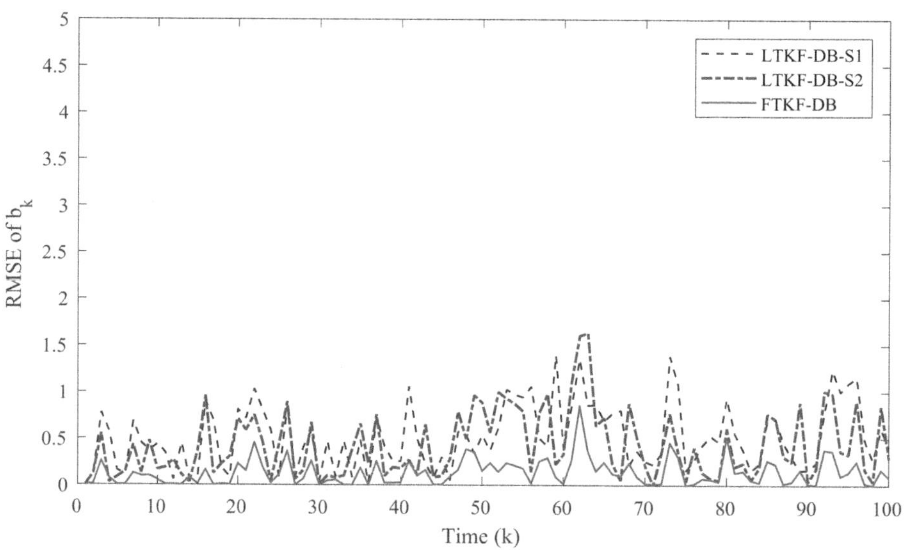

FIGURE 7.9 Comparison in RMSE3 between the local and fused filters.

TABLE 7.2

Comparison among ARMSEs for Different Censoring Regions $[\tau_m^l, \tau_m^r]$

$[\tau_m^l, \tau_m^r]$	ARMSE1	ARMSE2	ARMSE3
[−5, 5]	0.2329	0.4768	0.2515
[−1, 1]	4.4252	4.1257	3.6580
[−0.5, 0.5]	7.1406	6.7888	5.7826

TABLE 7.3

Comparison among ARMSEs for Different Noise Covariances $R_{m,k}$

$R_{m,k}$	ARMSE1	ARMSE2	ARMSE3
0.6	0.2329	0.4768	0.2515
0.1	0.0948	0.3707	0.2082
0.06	0.0742	0.2924	0.1601

Besides, to further elaborate the superiority of our FTKF-DB over its local counterparts, the performance comparison is made among these filters, where the LTKFs based on S1 and S2 observations are, respectively, named LTKF-DB-S1 and LTKF-DB-S2. After 1000 times of Monte Carlo trials, Figure 7.7 draws the RMSE curves in regard to the state tracking, whilst Figures 7.8–7.9 demonstrate both estimation and RMSE curves with respect to the bias tracking. It is evidently seen from Figures 7.7–7.9 that the performance of our FTKF-DB outperforms that of the LTKF-DB-S1 and LTKF-DB-S2 in both state and bias tracking. This is because our FTKF-DB properly integrates all available sensor information in accomplishing the task of target state and bias tracking, whilst both LTKF-DB-S1 and LTKF-DB-S2 produce their tracking results based at partial sensor observations.

At last, we show the influences from the censoring threshold and noise covariance on the filtering algorithm of our FTKF-DB. For different censoring regions $[\tau_m^l, \tau_m^r]$ and noise covariances $R_{m,k}$, the associated ARMSEs of our developed filtering fusion algorithm based on 1000 Monte Carlo trials (with each trial comprised of 100 time steps) are demonstrated in Tables 7.2–7.3. Looking at the two tables, we can draw the following conclusions: (1) as the censoring region $[\tau_m^l, \tau_m^r]$ narrows, the filtering accuracy with respect to the system state and dynamical bias deteriorates; and (2) as the covariance $R_{m,k}$ decreases, the filtering accuracy in regard to the system state and dynamical bias improves.

7.3.2 DISTRIBUTED TARGET TRACKING EXAMPLE

Consider an example of distributive target tracking with two sensors. According to Newton's force principle, the movement of the target can be modeled as a constant-velocity nominal system, and the target manoeuver is modeled as an unknown acceleration bias to the constant velocity. Meanwhile, the nominal system is corrupted by ambient disturbances entering the system via process noises. The two sensors sample the target position (range) and velocity (Doppler), respectively, with the same sampling period $T = 0.1s$. The measurement sampling is subject to the dead-zone-like censoring, and the data transmission is scheduled by the RRP. This situation can be modeled by (7.1)–(7.8) with the following parameters.

$$A_k = \begin{bmatrix} 1 & T \\ 0 & 1 \end{bmatrix}, B_k = \begin{bmatrix} 0.5T^2 \\ T \end{bmatrix}, \bar{x}_0 = \begin{bmatrix} 10 & 0 \end{bmatrix}^T, \bar{b}_0 = 0,$$

$$\Xi_k = 0.25, C_{1,k} = \begin{bmatrix} 1 & 0 \end{bmatrix}, C_{2,k} = \begin{bmatrix} 0 & 1 \end{bmatrix}, P_{\bar{x}_0} = I_2,$$

$$H_k = 1, \tau_1^l = -1000, \tau_2^l = -200, \tau_1^r = 1000, \tau_2^r = 200,$$

$$Q_k = 10^{-4} \operatorname{diag}\{25, 25\}, R_{1,k} = 0.1, R_{2,k} = 0.3, P_{\bar{b}_0} = 0.1.$$

Figures 7.10–7.11 plot the RMSE curves of the FTKF and FTKF-DB after 1000 independent Monte Carlo trials. It is spotted from Figures 7.10–7.11 that the RMSE curves with respect to the target position and velocity produced by our FTKF-DB always locate lower than that produced by the FTKF, whilst the RMSE curve in regard to the dynamical bias generated by our FTKF-DB is within a satisfactory range. These demonstration results obviously showcase the effectiveness and applicability of the proposed FTKF-DB in addressing the distributive target-tracking problem subject to the dead-zone-like censoring, dynamical biases and RRP.

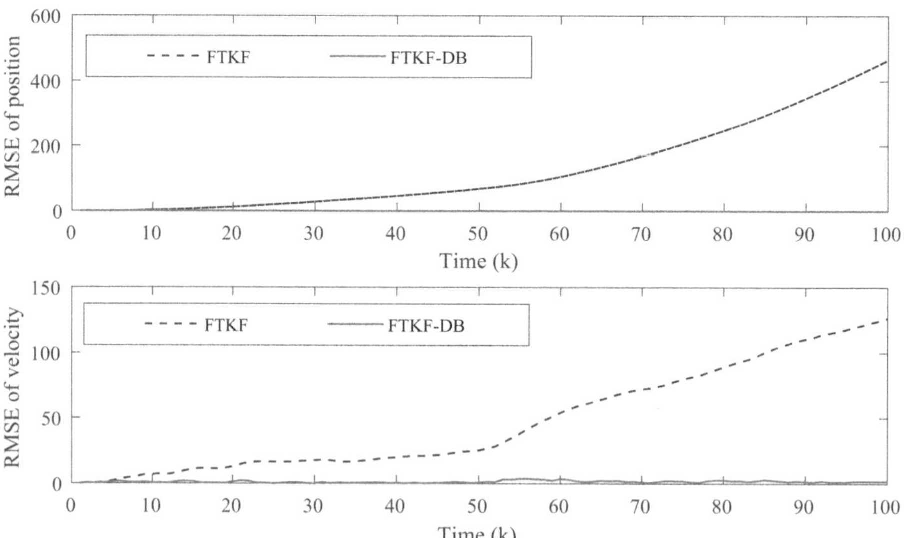

FIGURE 7.10 Comparison in RMSEs of target position and velocity between the FTKF and FTKF-DB.

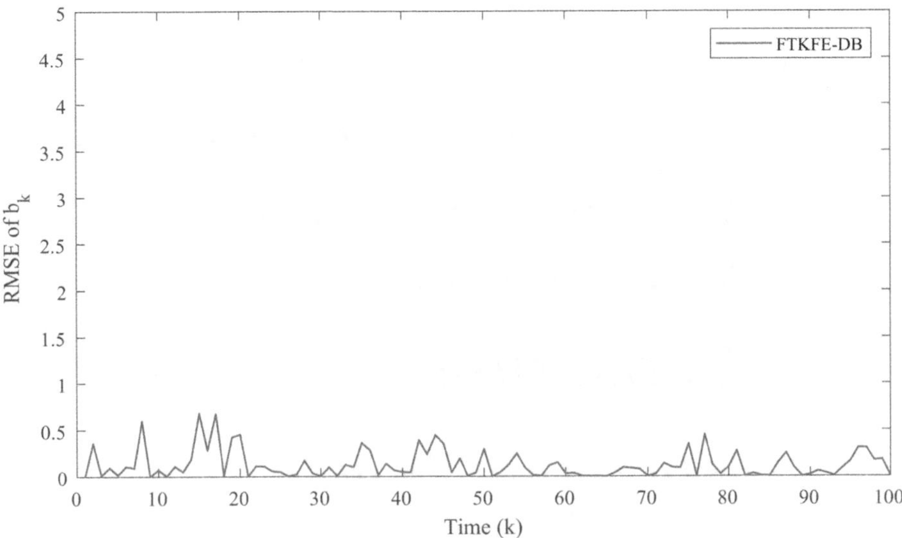

FIGURE 7.11   RMSE of dynamical bias produced by FTKF-DB.

7.4   SUMMARY

In this chapter, we have dealt with the filtering fusion problem for a class of multi-sensor systems with the concurrence of dead-zone-like censoring and dynamical biases under the RRP. The censoring phenomenon is depicted by the two-side Tobit observation model to manifest the dead-zone-like feature of the observation. The bias takes the form of a dynamic model driven by Gaussian white noises with known means and variances. For each sensor, taking into consideration the effects incurred by the censored observations, dynamic bias as well as RRP, an enhanced regression model has been constructed, based on which a unified filtering fusion framework has been formulated via the proposed LTKF. Thanks to the federated fusion rule, local estimates generated by all the LTKFs have been well incorporated to establish the desired FTKF. Finally, the applicability of the presented FTKF has been validated, and its superiority over the local counterpart has also been certified by simulation experiments.

8 Multi-Sensor Filtering Fusion with Parametric Uncertainties and Measurement Censoring
Monotonicity and Boundedness

In multi-sensor fusion estimation (MSFE) scenarios, the target plant is inherently subject to various uncertainties, e.g. parametric uncertainties due to modeling errors, unmodeled dynamics and parameter perturbations and measurement uncertainties due to ambient disturbances, inherent sensor limitations and network constraints. It has been fully acknowledged that these parametric and measurement uncertainties, if not adequately dealt with, could lead to significant deterioration or even divergence of the system performance. Consequently, the MSFE problems contaminated by different uncertainties have gained a recurring research interest in applications ranging from communication, image processing and ocean circulation to machine intelligence, and a great many results have been reported in literature, where the overwhelming majority build themselves on mature estimation techniques. Among various filters proposed so far, we highlight the Kalman filter (KF), least-square filter, H_∞ filter and variance-constrained filter that have proven to be rather popular. Note that the KF has been envisioned as a particularly effective tool in handling multiplicative noises because of its recursive structure and concise implementation. Nevertheless, most of the foregoing work is restrained to the setting where the concerned system is subject to either parametric uncertainty or measurement uncertainty, whereas systems might be corrupted by not only parametric but also measurement uncertainties in case of serious internal/external disturbances and constraints.

For MSFE problems with both parametric and measurement uncertainties, some initial attempts have been made in the past decade under the assumption that the parametric uncertainty (e.g. multiplicative noises) and the measurement uncertainty (e.g. packet dropouts, cyber-attacks and censored measurements) are uncorrelated with each other, and thus, their influences on the filter design and performance analysis are tackled separately via different handling techniques. To be specific, the

DOI: 10.1201/9781003461623-8

MSFE problem has been resolved in [180] for multi-sensor systems (MSSs) with mixed uncertainties, where the state augmentation and stochastic parameter matrix techniques have been, respectively, leveraged to tackle the multiplicative noise and packet dropout so as to develop a robust Kalman fusion estimator. The variance-constrained distributed filtering problem has been studied in [181] for sensor networks with multiplicative noises and deception attacks, where the sufficient condition has been established for the existence of the required estimator that satisfies the estimation error variance constraints by means of the recursive linear matrix inequality.

It should be noted that one obvious drawback with these estimators is that the use of the augmentation method in handling multiplicative noises often leads to heavy computational burdens which are not favorable in real-time applications. Moreover, these estimators are built under the assumption that the parametric uncertainty (e.g. the multiplicative noise) is required to be uncorrelated with the measurement uncertainty (e.g. the packet dropout and deception attack). As a consequence, these estimators are sensitive to the accuracy of this independence assumption, and any deviation from such an assumption might lead to a great degradation of the estimation performance. As such, it would be an interesting topic to develop a novel MSFE approach for multi-sensor systems under the Kalman filtering framework while abandoning the rather strict assumption on the independence between the parametric and measurement uncertainties.

Summarizing these discussions, it can be seen that the simultaneous presence of parametric and measurement uncertainties has given rise to challenges for the filter design, and it becomes necessary to evaluate their impacts on the filtering performance. Moreover, to the best of the authors' knowledge, only scattered results have been available in the literature on the filtering problem for multi-sensor systems subject to correlated parametric and measurement uncertainties. It is, therefore, of particular significance to shorten such a gap by initiating a study on the corresponding MSFE filtering problem, where the parametric uncertainty takes the form of the multiplicative noise and has an inherent impact on the measurement uncertainty that takes the form of measurement censoring. In this chapter, our aim is to derive a federated Tobit Kalman fusion estimator (FTKFE) and analyze its performance by studying the monotonicity of the estimation error covariance (EEC) with respect to the censoring threshold and investigating the boundedness of the EEC.

8.1　PROBLEM FORMULATION

Consider a multi-sensor system with multiplicative noises:

$$x_{k+1} = \left(A_k + F_k \alpha_k \right) x_k + \omega_k, \tag{8.1}$$

$$z_{i,k} = \left(C_{i,k} + G_{i,k} \beta_{i,k} \right) x_k + \upsilon_{i,k}, \, i = 1, 2, \ldots, p, \tag{8.2}$$

where $x_k \in \mathbb{R}^{n_x}$ is the state vector and $z_{i,k} \in \mathbb{R}$ is the uncensored observation of the ith sensor. A_k, F_k, $C_{i,k}$ and $G_{i,k}$ are known matrices. $\omega_k \in \mathbb{R}^{n_x}$ and $\upsilon_{i,k} \in \mathbb{R}$ are zero-mean white Gaussian noises in the process, and the measurement with covariances Q_k and $R_{i,k}$, respectively. $\alpha_k \in \mathbb{R}^{n_\alpha}$ and $\beta_k \in \mathbb{R}^{n_\beta}$ are multiplicative noises.

The Tobit observation model is given as follows:

$$y_{i,k} = \begin{cases} z_{i,k}, & z_{i,k} > \mathcal{I}_i, \\ \mathcal{I}_i, & z_{i,k} \le \mathcal{I}_i, \end{cases} \tag{8.3}$$

where $y_{i,k}$ $(i = 1, 2, \ldots, p)$ are output observations of the ith sensor with censoring thresholds $\mathcal{I}_i$.

Based on this Tobit observation model, let us define Bernoulli random variables $\gamma_{i,k}$ to regulate the censoring phenomenon of $y_{i,k}$ as follows:

$$\gamma_{i,k} = \begin{cases} 1, & z_{i,k} > \mathcal{I}_i, \\ 0, & z_{i,k} > \mathcal{I}_i, \end{cases} \tag{8.4}$$

with the following probability distributions: $\mathrm{Prob}\{\gamma_{ki,} = 1\} = \overline{\gamma}_{i,k}$, $\mathrm{Prob}\{\gamma_{i,k} = 0\} = 1 - \overline{\gamma}_{i,k}$. Here, $\overline{\gamma}_{i,k}$ are censoring probabilities whose values are given later. Taking advantage of $\gamma_{i,k}$, $y_{i,k}$ given by (8.3) can be rewritten as follows:

$$y_{i,k} = \gamma_{i,k} z_{i,k} + (1 - \gamma_{i,k}) \mathcal{I}_i. \tag{8.5}$$

Assumption 8.1 (1) x_0, ω_k and $\upsilon_{i,k}$ are mutually independent. (2) The multiplicative noises α_k and $\beta_{i,k}$ are mutually independent, zero-mean and white with covariances Σ_{α_k} and $\Sigma_{\beta_{i,k}}$ and are uncorrelated with x_0, ω_k and $\upsilon_{i,k}$.

Remark 8.1 It is observed from (8.5) that the random variables $\gamma_{i,k}$ $(i = 1, 2, \ldots, p)$ are employed to describe the censoring phenomena of $y_{i,k}$. Letting $z_{i,k} \triangleq \zeta_{i,k} + \upsilon_{i,k}$ and $\zeta_{i,k} \triangleq (C_{i,k} + G_{i,k} \beta_{i,k}) x_k$, $\zeta_{i,k}$ can be regarded as a pseudo-state of measurement $z_{i,k}$. $\overline{\gamma}_{i,k}$ can be approximated by

$$\overline{\gamma}_{i,k} \approx \Phi\left(\frac{\hat{\zeta}_{i,k}^- - \mathcal{I}_i}{\sqrt{P_{\tilde{z}_{i,k}^-}}} \right), \tag{8.6}$$

where $\hat{\zeta}_{i,k}^-$ is the one-step prediction of $\zeta_{i,k}$ with the prediction error covariance $P_{\tilde{z}_{i,k}^-}$ to be determined later.

The objectives of this chapter are to (i) design the federated TKF for the dynamic model (8.1) and the measurement model (8.5) where the censoring probability is calculated via (8.6) and (ii) conduct the performance assessment by studying the monotonicity of the EEC with respect to the censoring threshold and investigating the boundedness property of the devised filtering algorithm.

8.2 DESIGN OF THE FUSION ESTIMATOR

In this section, we are dedicated to formulating a federated Tobit Kalman fusion estimation paradigm to surmount the manifold difficulties resulting from the concurrence of the MSS, censored observations and multiplicative noises. The fusion estimation paradigm is quite unique because of (1) its delicately devised LTKFs based on an ameliorated regression model that stems from the involvement of the multiplicative noises, where extra effort is required for calculating the measurement prediction, filtering gain and error covariance; (2) its skillfully selected fusion rule that effectively accommodates available information from all the LTKFs, where the global optimality of the fusion estimation is guaranteed.

Before procceding further, we put forward the following ameliorated regression model to accommodate the impact from the multiplicative noise.

Lemma 8.1 The expectation and variance of $y_{i,k}$ conditioned on the pseudo-state $\zeta_{i,k}$ are

$$
\mathbb{E}\left\{y_{i,k} \mid \zeta_{i,k}\right\} = \Phi\left(\frac{\mathcal{I}_i - \zeta_{i,k}}{\sqrt{R_{i,k}}}\right)\mathcal{I}_i
$$
$$
+ \Phi\left(\frac{\zeta_{i,k} - \mathcal{I}_i}{\sqrt{R_{i,k}}}\right)\left[\zeta_{i,k} + \sqrt{R_{i,k}}\,\lambda\left(\frac{\mathcal{I}_i - \zeta_{i,k}}{\sqrt{R_{i,k}}}\right)\right], \tag{8.7}
$$

and

$$
Var\left\{y_{i,k} \mid \zeta_{i,k}\right\}
$$
$$
= R_{i,k}\left[1 - \varphi\left(\frac{\mathcal{I}_i - \zeta_{i,k}}{\sqrt{R_{i,k}}}\right)\right]\Phi\left(\frac{\zeta_{i,k} - \mathcal{I}_i}{\sqrt{R_{i,k}}}\right)
$$
$$
+ \left[\zeta_{i,k} + \sqrt{R_{i,k}}\,\lambda\left(\frac{\mathcal{I}_i - \zeta_{i,k}}{\sqrt{R_{i,k}}}\right)\right]^2
$$
$$
\times \Phi\left(\frac{\zeta_{i,k} - \mathcal{I}_i}{\sqrt{R_{i,k}}}\right)\Phi\left(\frac{\mathcal{I}_i - \zeta_{i,k}}{\sqrt{R_{i,k}}}\right) \tag{8.8}
$$
$$
- 2\mathcal{I}\Phi\left(\frac{\zeta_{i,k} - \mathcal{I}_i}{\sqrt{R_{i,k}}}\right)\left[1 - \Phi\left(\frac{\zeta_{i,k} - \mathcal{I}_i}{\sqrt{R_{i,k}}}\right)\right]
$$
$$
\times \left[\zeta_{i,k} + \sqrt{R_{i,k}}\,\lambda\left(\frac{\mathcal{I}_i - \zeta_{i,k}}{\sqrt{R_{i,k}}}\right)\right]
$$
$$
+ \Phi\left(\frac{\mathcal{I}_i - \zeta_{i,k}}{\sqrt{R_{i,k}}}\right)\mathcal{I}_i^2\left[1 - \Phi\left(\frac{\mathcal{I}_i - \zeta_{i,k}}{\sqrt{R_{i,k}}}\right)\right],
$$

where

$$\lambda\left(\frac{\mathcal{I}_i-\zeta_{i,k}}{\sqrt{R_{i,k}}}\right)=\frac{\phi\left(\frac{\mathcal{I}_i-\zeta_{i,k}}{\sqrt{R_{i,k}}}\right)}{1-\Phi\left(\frac{\mathcal{I}_i-\zeta_{i,k}}{\sqrt{R_{i,k}}}\right)}, \tag{8.9}$$

$$\varphi\left(\frac{\mathcal{I}_i-\zeta_{i,k}}{\sqrt{R_{i,k}}}\right)=\lambda\left(\frac{\mathcal{I}_i-\zeta_{i,k}}{\sqrt{R_{i,k}}}\right)\left[\lambda\left(\frac{\mathcal{I}_i-\zeta_{i,k}}{\sqrt{R_{i,k}}}\right)-\left(\frac{\mathcal{I}_i-\zeta_{i,k}}{\sqrt{R_{i,k}}}\right)\right]. \tag{8.10}$$

and $\lambda(\cdot)$ is called the inverse Mills ratio. Here, $\phi(\cdot)$ and $\Phi(\cdot)$ are, respectively, the probability density function (PDF) and CDF of the Gaussian random variable $z_{i,k}$ of following forms:

$$\phi\left(\frac{z_{i,k}-\zeta_{i,k}}{\sqrt{R_{i,k}}}\right)=\frac{1}{\sqrt{2\pi R_{i,k}}}e^{-\frac{(z_{i,k}-\zeta_{i,k})^2}{2R_{i,k}}}, \tag{8.11}$$

$$\Phi\left(\frac{z_{i,k}-\zeta_{i,k}}{\sqrt{R_{i,k}}}\right)=\int_{-\infty}^{z_{i,k}}\frac{1}{\sqrt{2\pi R_{i,k}}}e^{-\frac{(\eta_{i,k}-\zeta_{i,k})^2}{2R_{i,k}}}\,d_{\eta_{i,k}}. \tag{8.12}$$

Proof Recalling $z_{i,k}\triangleq\left(C_{i,k}+G_{i,k}\beta_{i,k}\right)x_k+\upsilon_{i,k}$ and $\zeta_{i,k}\triangleq\left(C_{i,k}+G_{i,k}\beta_{i,k}\right)$, we know that the latent measurement $z_{i,k}$ obeys a normal distribution with mean $\zeta_{i,k}$ and variance $R_{i,k}$. In line with both latent and censored measurement models, the PDF of $y_{i,k}$ is

$$f\left(y_{i,k}\mid\zeta_{i,k}\right)=\phi\left(\frac{y_{i,k}-\zeta_{i,k}}{\sqrt{R_{i,k}}}\right)u\left(y_{i,k}-\mathcal{I}_i\right)+\delta\left(\mathcal{I}_i-y_{i,k}\right)\Phi\left(\frac{y_{i,k}-\zeta_{i,k}}{\sqrt{R_{i,k}}}\right), \tag{8.13}$$

where $\phi\left(\frac{y_{i,k}-\zeta_{i,k}}{\sqrt{R_{i,k}}}\right)$ and $\Phi\left(\frac{y_{i,k}-\zeta_{i,k}}{\sqrt{R_{i,k}}}\right)$ are calculated via (8.11)–(8.12), and $u\left(y_{i,k}-\mathcal{I}_i\right)$ is the unit step function.

With the help of (8.13), we arrive at

$$\begin{aligned}\mathbb{E}\{y_{i,k}\mid\zeta_{i,k}\}=&\ Prob\{y_{i,k}>\mathcal{I}_i\mid\zeta_{i,k}\}\mathbb{E}\{y_{i,k}\mid y_{i,k}>\mathcal{I}_i,\zeta_{i,k}\}\\&+Prob\{y_{i,k}=\mathcal{I}_i\mid\zeta_{i,k}\}\mathbb{E}\{y_{i,k}\mid y_{i,k}=\mathcal{I}_i,\zeta_{i,k}\},\end{aligned} \tag{8.14}$$

from which one can see that the calculation of all probabilities and expectations in (2.1) are required prior to the calculation of $\mathbb{E}\{y_{i,k}|\zeta_{i,k}\}$.

Keeping in mind that $z_{i,k}$ obeys a normal distribution with mean $\zeta_{i,k}$ and variance $R_{i,k}$, we have

$$Prob\{y_{i,k} > \mathcal{I}_i \mid \zeta_{i,k}\} = Prob\{z_{i,k} > \mathcal{I}_i \mid \zeta_{i,k}\}$$

$$= \Phi\left(\frac{\zeta_{i,k} - \mathcal{I}_i}{\sqrt{R_{i,k}}}\right). \tag{8.15}$$

Besides, denoting $\lambda\left(\dfrac{\mathcal{I}_i - \zeta_{i,k}}{\sqrt{R_{i,k}}}\right) \triangleq \dfrac{\phi\left(\dfrac{\mathcal{I}_i - \zeta_{i,k}}{\sqrt{R_{i,k}}}\right)}{1 - \Phi\left(\dfrac{\mathcal{I}_i - \zeta_{i,k}}{\sqrt{R_{i,k}}}\right)}$, it follows from (8.13) that

$$\mathbb{E}\{y_{i,k} \mid y_{i,k} > \mathcal{I}_i, \zeta_{i,k}\} = \frac{1}{1 - \Phi\left(\dfrac{\mathcal{I}_i - \hat{\zeta}_{i,k}^-}{\sqrt{P_{\tilde{z}_{i,k}^-}}}\right)} \int_{\mathcal{I}_i}^{+\infty} y_{i,k} \phi\left(\frac{y_{i,k} - \hat{\zeta}_{i,k}^-}{\sqrt{P_{\tilde{z}_{i,k}^-}}}\right) dy_{i,k}$$

$$= \frac{\hat{\zeta}_{i,k}^-}{1 - \Phi\left(\dfrac{\mathcal{I}_i - \hat{\zeta}_{i,k}^-}{\sqrt{P_{\tilde{z}_{i,k}^-}}}\right)} \Phi\left(\frac{\hat{\zeta}_{i,k}^- - \mathcal{I}_i}{\sqrt{P_{\tilde{z}_{i,k}^-}}}\right) - \frac{P_{\tilde{z}_{i,k}^-}}{1 - \Phi\left(\dfrac{\mathcal{I}_i - \hat{\zeta}_{i,k}^-}{\sqrt{P_{\tilde{z}_{i,k}^-}}}\right)} \tag{8.16}$$

$$\times \left(\phi\left(\frac{+\infty - \hat{\zeta}_{i,k}^-}{\sqrt{P_{\tilde{z}_{i,k}^-}}}\right) - \phi\left(\frac{\mathcal{I}_i - \hat{\zeta}_{i,k}^-}{\sqrt{P_{\tilde{z}_{i,k}^-}}}\right)\right)$$

$$= \hat{\zeta}_{i,k}^- + P_{\tilde{z}_{i,k}^-} \lambda\left(\frac{\mathcal{I}_i - \zeta_{i,k}}{\sqrt{R_{i,k}}}\right).$$

Parallel to (8.15)–(8.16), we have

$$\begin{cases} Prob\{y_{i,k} = \mathcal{I}_i \mid \zeta_{i,k}\} = \Phi\left(\dfrac{\mathcal{I}_i - \zeta_{i,k}}{\sqrt{R_{i,k}}}\right), \\[2ex] \mathbb{E}\{y_{i,k} \mid y_{i,k} = \mathcal{I}_i, \zeta_{i,k}\} = \mathcal{I}_i. \end{cases}$$

Inserting the above equation and (8.15)–(8.16) into (8.14) yields (8.7). It follows from Lemma 1 in [51] that

$$
\mathbb{E}\left\{y_{i,k}^2 \mid \zeta_{i,k}\right\} = \Phi\left(\frac{\mathcal{I}_i - \zeta_{i,k}}{\sqrt{R_{i,k}}}\right)\mathcal{I}_i^2 + \Phi\left(\frac{\zeta_{i,k} - \mathcal{I}_i}{\sqrt{R_{i,k}}}\right)
$$
$$
\times \mathbb{E}\left\{z_{i,k}^2 \mid z_{i,k} > \mathcal{I}_i, \zeta_{i,k}\right\},
\tag{8.17}
$$

where

$$
\mathbb{E}\left\{z_{i,k}^2 \mid z_{i,k} > \mathcal{I}_i, \zeta_{i,k}\right\} = R_{i,k}\left[1 - \varphi\left(\frac{\mathcal{I}_i - \zeta_{i,k}}{\sqrt{R_{i,k}}}\right)\right]
$$
$$
+ \left[\zeta_{i,k} + \sqrt{R_{i,k}}\,\lambda\left(\frac{\mathcal{I}_i - \zeta_{i,k}}{\sqrt{R_{i,k}}}\right)\right]^2,
\tag{8.18}
$$

can be obtained along the similar derivation of $\mathbb{E}\left\{y_{i,k} \mid \zeta_{i,k}\right\}$, and $\varphi\left(\dfrac{\mathcal{I}_i - \zeta_{i,k}}{\sqrt{R_{i,k}}}\right)$ is defined by (8.10).

Putting (8.17)–(8.18) and (8.7) into the expression of $Var\left\{y_{i,k} \mid \zeta_{i,k}\right\}$ further leads to (8.8). This completes the proof.

Remark 8.2 Compared to its conventional counterpart in the standard TKF, the regression model in Lemma 8.1 has two advantages in the simultaneous handling of the measurement censoring and multiplicative noises. The first advantage is the introduction of the pseudo-state $z_{i,k}$, which noticeably characterizes the impact from the multiplicative noise. The second advantage is the elimination of the incorrect assumption that the measurement variance can be split into the sum of the variances of the censored and uncensored measurements, which obviously improves the calculation accuracy of the resultant algorithm. In a word, our newly derived model has the advantages of better calculation accuracy (of the measurement variance) and stronger robustness (against multiplicative noises), although at the cost of adding further intricacy to the follow-up filter design and performance analysis.

Let

$$
P_{\tilde{x}_{i,k}\tilde{y}_{i,k}} \triangleq \mathbb{E}\left\{\tilde{x}_{i,k}^-(\tilde{y}_{i,k}^-)^T \mid y_{i,1:k-1}\right\},
$$
$$
P_{\tilde{x}_{i,k}} \triangleq \mathbb{E}\left\{\tilde{x}_{i,k}\tilde{x}_{i,k}^T \mid y_{i,1:k}\right\}, P_{x_k} \triangleq \mathbb{E}\left\{x_k x_k^T\right\},
$$
$$
\hat{x}_{i,k} \triangleq \mathbb{E}\left\{x_k \mid y_{i,1:k}\right\}, \hat{y}_{i,k}^- \triangleq \mathbb{E}\left\{y_{i,k} \mid y_{i,1:k-1}\right\},
$$
$$
P_{\tilde{x}_{i,k}^-} \triangleq \mathbb{E}\left\{\tilde{x}_{i,k}^-\left(\tilde{x}_{i,k}^-\right)^T \mid y_{i,1:k-1}\right\}, \tilde{y}_{i,k}^- \triangleq y_{i,k} - \hat{y}_{i,k}^-,
$$
$$
y_{i,1:k} \triangleq \left\{y_{i,1}, y_{i,2}, \ldots, y_{i,k}\right\}, \hat{\zeta}_{i,k}^- \triangleq \mathbb{E}\left\{\zeta_{i,k} \mid y_{i,1:k-1}\right\},
$$

$$\hat{x}_{i,k}^- \triangleq \mathbb{E}\left\{x_k \mid y_{i,1:k-1}\right\}, \tilde{x}_{i,k}^- \triangleq x_k - \hat{x}_{i,k}^-, \tilde{x}_{i,k} \triangleq x_k - \hat{x}_{i,k}.$$

Let $\hat{x}_{k-1}$ and $P_{\tilde{x}_{k-1}}$ be, respectively, the fused estimate and EEC attained by the fusion center under the federated fusion rule [33] at time $k-1$. According to the information-sharing principle of the federated fusion rule, at time $k-1$, the LTKF is initialized as

$$\hat{x}_{i,k-1} \triangleq \hat{x}_{k-1}, \; P_{\tilde{x}_{i,k-1}} \triangleq \epsilon_i^{-1} P_{\tilde{x}_{k-1}} \; \text{and} \; Q_{i,k-1} \triangleq \epsilon_i^{-1} Q_k \; \text{where} \; \sum_{i=1}^{p} \epsilon_i \triangleq 1.$$

Here, $\epsilon_i \; (i = 1, 2, \ldots, p)$ are the information-sharing weights that are adopted to determine the distribution of the information among local filters. For simplicity, the information-sharing weights in this chapter are set as $1/p$. $Q_{i,k-1}$ are the reallocated noise covariances, and $P_{\tilde{x}_{k-1}}$ is the fused EEC at time $k-1$.

In the light of Lemma 8.1, the following theorem manifests the LTKF subject to multiplicative noises.

Theorem 8.1 The LTKF for system (8.1)–(8.4) is given by

$$\hat{x}_{i,k+1}^- = A_k \hat{x}_{i,k}, \tag{8.19}$$

$$P_{\tilde{x}_{i,k+1}^-} = A_k P_{\tilde{x}_{i,k}} A_k^T + F_k P_{x_k} F_k^T + Q_{i,k}, \tag{8.20}$$

$$\hat{x}_{i,k+1} = \hat{x}_{i,k+1}^- + K_{i,k+1}\left(y_{i,k+1} - \hat{y}_{i,k+1}^-\right), \tag{8.21}$$

$$P_{\tilde{x}_{i,k+1}} = P_{\tilde{x}_{i,k+1}^-} - K_{i,k+1} P_{\tilde{x}_{i,k+1}^- \tilde{y}_{i,k+1}^-}^T, \tag{8.22}$$

$$P_{x_{k+1}} = A_k P_{x_k} A_k^T + F_k P_{x_k} F_k^T + Q_k. \tag{8.23}$$

Furthermore, the local gain matrix $K_{i,k}$ and the measurement prediction $\hat{y}_{i,k}^-$ are

$$K_{i,k} = P_{\tilde{x}\tilde{y}_{i,k}^-} P_{\tilde{y}_{i,k}^-}^{-1}, \tag{8.24}$$

$$\hat{y}_{i,k}^- = \overline{\gamma}_{i,k}\left[\hat{\zeta}_{i,k}^- + P_{\tilde{z}_{i,k}^-} \lambda\left(\frac{\mathcal{I}_i - \hat{\zeta}_{i,k}^-}{\sqrt{P_{\tilde{z}_{i,k}^-}}}\right)\right] + (1 - \overline{\gamma}_{i,k})\mathcal{I}_i, \tag{8.25}$$

where

$$P_{\tilde{x}\tilde{y}_{i,k}^-} = P_{\tilde{x}_k^-} C_{i,k}^T \overline{\gamma}_{i,k}, \tag{8.26}$$

$$P_{\tilde{y}_{i,k}^-} = \overline{\gamma}_{i,k}^2 C_{i,k} P_{\tilde{x}_{i,k}^-} C_{i,k}^T + \overline{\gamma}_{i,k}^2 G_{i,k} P_{x_k} G_{i,k}^T + \mathcal{R}_{i,k}. \tag{8.27}$$

Here, $\hat{\zeta}_{i,k}^- = C_{i,k} \hat{x}_{i,k}^-$, $P_{\tilde{z}_{i,k}^-} = C_{i,k} P_{\tilde{x}_{i,k}^-} C_{i,k}^T + R_{i,k}$ and

$$
\begin{aligned}
\mathcal{R}_{i,k} = {}& P_{\tilde{z}_{i,k}^-}\left[1-\varphi\left(\frac{\mathcal{I}_i-\widehat{\zeta}_{i,k}^-}{\sqrt{P_{\tilde{z}_{i,k}^-}}}\right)\right]\Phi\left(\frac{\widehat{\zeta}_{i,k}^--\mathcal{I}_i}{\sqrt{P_{\tilde{z}_{i,k}^-}}}\right) \\[2mm]
& +\left[\widehat{\zeta}_{i,k}^-+\sqrt{P_{\tilde{z}_{i,k}^-}}\,\lambda\left(\frac{\mathcal{I}_i-\widehat{\zeta}_{i,k}^-}{\sqrt{P_{\tilde{z}_{i,k}^-}}}\right)\right]^2 \\[2mm]
& \times\Phi\left(\frac{\widehat{\zeta}_{i,k}^--\mathcal{I}_i}{\sqrt{P_{\tilde{z}_{i,k}^-}}}\right)\Phi\left(\frac{\mathcal{I}_i-\widehat{\zeta}_{i,k}^-}{\sqrt{P_{\tilde{z}_{i,k}^-}}}\right) \\[2mm]
& -2\mathcal{I}\Phi\left(\frac{\widehat{\zeta}_{i,k}^--\mathcal{I}_i}{\sqrt{P_{\tilde{z}_{i,k}^-}}}\right)\left[1-\Phi\left(\frac{\widehat{\zeta}_{i,k}^--\mathcal{I}_i}{\sqrt{P_{\tilde{z}_{i,k}^-}}}\right)\right] \\[2mm]
& \times\left[\widehat{\zeta}_{i,k}^-+\sqrt{P_{\tilde{z}_{i,k}^-}}\,\lambda\left(\frac{\mathcal{I}_i-\widehat{\zeta}_{i,k}^-}{\sqrt{P_{\tilde{z}_{i,k}^-}}}\right)\right] \\[2mm]
& +\Phi\left(\frac{\mathcal{I}_i-\widehat{\zeta}_{i,k}^-}{\sqrt{P_{\tilde{z}_{i,k}^-}}}\right)\mathcal{I}_i^2\left[1-\Phi\left(\frac{\mathcal{I}_i-\widehat{\zeta}_{i,k}^-}{\sqrt{P_{\tilde{z}_{i,k}^-}}}\right)\right].
\end{aligned}
\tag{8.28}
$$

Proof The plain employment of the orthogonality projection principle on system (8.1)–(8.4) yields (8.19) and (8.21). Subtracting (8.19) from (8.1) yields $\tilde{x}_{k+1}^- = A_k\tilde{x}_{i,k}+F_k\alpha_k x_k+\omega_k$ which gives rise to

$$
\begin{aligned}
P_{\tilde{x}_{i,k}^-} &\triangleq \mathbb{E}\left\{\tilde{x}_{i,k}^-\left(\tilde{x}_{i,k}^-\right)^T\mid y_{i,1:k-1}\right\} \\
&= A_{k-1}P_{\tilde{x}_{i,k-1}}A_{k-1}^T+F_{k-1}P_{x_{k-1}}F_{k-1}^T+Q_{i,k-1},
\end{aligned}
$$

which yields (8.20). (8.23) follows immediately from (8.1). The combination of (8.4) and (8.25) yields

$$
\tilde{y}_{i,k}^- = \gamma_{i,k}z_{i,k}+\left(1-\gamma_{i,k}\right)\mathcal{I}_i-\hat{y}_{i,k}^-.
\tag{8.29}
$$

Putting (8.1), (8.19) and (8.29) into $P_{\tilde{x}_{i,k}}$ leads to

$$
P_{\tilde{x}_{i,k}} = P_{\tilde{x}_{i,k}^-}-P_{\tilde{x}\tilde{y}_{i,k}^-}K_{i,k}^T-K_{i,k}P_{\tilde{x}\tilde{y}_{i,k}^-}^T+K_{i,k}P_{\tilde{y}_{i,k}^-}K_{i,k}^T,
\tag{8.30}
$$

and taking the trace on both sides of (8.30) gives

$$
trace\left\{P_{\tilde{x}_{i,k}}\right\} = trace\left\{P_{\tilde{x}_{i,k}^-}\right\}-2trace\left\{P_{\tilde{x}\tilde{y}_{i,k}^-}K_{i,k}^T\right\}+trace\left\{K_{i,k}P_{\tilde{y}_{i,k}^-}K_{i,k}^T\right\}.
\tag{8.31}
$$

Taking the partial derivative with respect to $K_{i,k}$ on both sides of (2.1) gives

$$\frac{\partial trace\{P_{\tilde{x}_{i,k}}\}}{\partial K_{i,k}} = -2trace\{P_{\tilde{x}\tilde{y}_{i,k}}\} + 2trace\{K_{i,k}P_{\tilde{y}_{i,k}}\}, \tag{8.32}$$

and setting (8.32) to zero, one has the optimal local gain matrix (8.24). Putting (8.24) into $P_{\tilde{x}_{i,k}}$ results in (8.22). Recalling (8.29) leads to

$$\begin{aligned}
P_{\tilde{x}\tilde{y}_{i,k}} &= \mathbb{E}\{x_k x_k^T C_{i,k}^T \gamma_{i,k}^T\} + \mathbb{E}\{x_k \upsilon_{i,k}^T \gamma_{i,k}^T\} \\
&\quad - \mathbb{E}\{\hat{x}_{i,k}^- x_k^T C_{i,k}^T \gamma_{i,k}^T\} - \mathbb{E}\{\hat{x}_{i,k}^- \upsilon_{i,k}^T \gamma_{i,k}^T\}.
\end{aligned} \tag{8.33}$$

As the approximation in (8.6) explains that x_k and $\gamma_{i,k}$ are uncorrelated, we further have

$$P_{\tilde{x}\tilde{y}_{i,k}} = P_{x_k} C_{i,k}^T \overline{\gamma}_{i,k}^T - \hat{x}_{i,k}^- \left(\hat{x}_{i,k}^-\right)^T C_{i,k}^T \overline{\gamma}_{i,k}^T. \tag{8.34}$$

Moving forward, we have

$$\begin{aligned}
P_{x_k} &= \mathbb{E}\left\{\left(\tilde{x}_k^- + \hat{x}_k^-\right)\left(\tilde{x}_k^- + \hat{x}_k^-\right)^T\right\} \\
&= P_{\tilde{x}_k^-} + \hat{x}_k^- \left(\hat{x}_k^-\right)^T.
\end{aligned} \tag{8.35}$$

Along the similar derivation line of $P_{\tilde{x}\tilde{y}_{i,k}}$, one has

$$\begin{aligned}
P_{\tilde{y}_{i,k}} &= \overline{\gamma}_{i,k} C_{i,k} P_{\tilde{x}_{i,k}^-} C_{i,k}^T \overline{\gamma}_{i,k}^T + \overline{\gamma}_{i,k} G_{i,k} P_{x_k} G_{i,k}^T \overline{\gamma}_{i,k}^T \\
&\quad + \mathbb{E}\left\{\gamma_{i,k}\left(\upsilon_{i,k} - \tilde{\upsilon}_{i,k}\right)\left(\upsilon_{i,k} - \tilde{\upsilon}_{i,k}\right)^T \gamma_{i,k}^T \,|\, y_{i,1:k-1}\right\},
\end{aligned}$$

where $\tilde{\upsilon}_{i,k} = \sqrt{P_{\tilde{z}_{i,k}}}\,\lambda\!\left(\dfrac{\mathcal{I}_i - \hat{\zeta}_{i,k}^-}{\sqrt{P_{\tilde{z}_{i,k}}}}\right)$. Referencing Lemma 8.1, we have (8.28), which,

together with the preceding equation, results in (8.27). This completes the proof.

Remark 8.3 Two noteworthy features can be witnessed by comparing our LTKF in Theorem 8.1 with its standard TKF counterpart. The first feature is the replacement of $C_k \hat{x}_k^-$ (which is the product of the observation coefficient C_k and state prediction $\hat{x}_k^-$) by $\hat{\zeta}_{i,k}^- = C_{i,k}\hat{x}_k^-$ (which is now the product of the sensor-based coefficient $C_{i,k}$ and state prediction $\hat{x}_{i,k}^-$). The second feature is the presence of a suite of new terms P_{x_k}, $F_k P_{x_k} F_k^T$ and $\overline{\gamma}_{i,k}^2 G_{i,k} P_{x_k} G_{i,k}^T$. It is notable that the first feature originates from the multi-sensor nature, whilst the second feature comes from the consideration of the multiplicative noises in both (8.1) and (8.2).

Based on Theorem 8.1, the final estimate and EEC attained by the fusion center under the federated fusion rule at time k are obtained as follows:

$$\begin{cases} P_{\tilde{x}_k} = \left(\sum_{i=1}^{p} P_{\tilde{x}_{i,k}}^{-1} \right)^{-1}, \\ \hat{x}_k = P_{\tilde{x}_k} \left(\sum_{i=1}^{p} P_{\tilde{x}_{i,k}}^{-1} \hat{x}_{i,k} \right). \end{cases} \tag{8.36}$$

It is pointed out in [33] that with the help of the upper-bounding technique and the information-sharing principle, the federated filtering fusion avoids the need to maintain and apply cross-covariances among the local filters by constructing the initial local filter covariances and per-step process noise covariances so as to satisfy the information sum in (8.36). In other words, under the federated fusion, the constructed LTKFs are independent from each other, and thus the local estimates can be directly combined to yield a fused solution via the relatively simple approach as shown in (8.36). Table 8.1 summarizes the implementation procedure of the proposed FTKFE.

TABLE 8.1

Implementation Procedure of the Proposed FTKFE

Step 1. Initiation

1. Let $k = 1$ and the initial values be $\hat{x}_0 = \overline{x}_0$, $P_{x_0} = \overline{x}_0 \overline{x}_0^T$,

   $P_{\tilde{x}_0} = P_0$.

Step 2. Local estimation

1. Calculate $P_{x_{k+1}}$ by (8.23);
2. Reallocate the information among local filters via computing

   $\hat{x}_{i,k}$, $P_{\tilde{x}_{i,k}}$ and $Q_{i,k}$ by the information-sharing principle.

3. Calculate the predicted state $\hat{x}_{i,k+1}^{-}$ by (8.19) and covariance

   $P_{\tilde{x}_{i,k+1}^{-}}$ by (8.20);

4. Calculate the predicted measurement $\hat{y}_{i,k+1}^{-}$ and covariance

   $P_{\tilde{y}_{i,k+1}^{-}}$ by Lemma 1;

5. Calculate the gain matrix $K_{i,k+1}$ by (8.24);

6. Calculate the state estimate $\hat{x}_{i,k+1}$ by (8.19) and covariance

   $P_{\tilde{x}_{i,k+1}}$ by (8.20);

Step 3. Federated fusion

1. Calculate the fused estimate $\hat{x}_{k+1}$ and covariance $P_{\tilde{x}_{k+1}}$
   by (8.36);

Step 4. Set $k = k + 1$ and go to Step 2.

Remark 8.4 In accordance with Lemma 8.1 and Theorem 8.1, a FTKFE framework is formulated to address the novel MSFE problem suffering simultaneously from censored observations and multiplicative noises. The concerned system model (8.1)–(8.4) is holistic, as it not only incorporates the MSS but also accounts for two significant observation uncertainties (i.e. censored observations and multiplicative noises), which are prevalently seen in scenarios ranging from image processing and target tracking to fault detection and cooperative localization. Thanks to the federated fusion rule, the correlations among these disparate uncertainties are prudently tackled within a unified yet valid prototype. It is noteworthy that the FTKFE presented by (8.28)–(8.36) is, in fact, the distributed generalization of the traditional TKF with multiplicative noises. In the case that the multiplicative noises are absent (i.e. $F_k = 0$ and $G_{i,k} = 0$), the LTKF in Theorem 8.1 would degrade to the conventional TKF, and as a result, the FTKFE would degenerate to the distributed TKF.

Remark 8.5 As an emerging variant of the prestigious KF, the TKF is renowned for its powerful competence in handling a specific type of measurement nonlinearity, i.e. the censored observation. With the introduction of new definitions (for measurement expectation, residual as well as variance), the TKF is capable of formalizing a completely recursive computation prototype of state estimates and the associated EECs. In addition to censored observations, many practical systems are confronted with the parameter uncertainties, which can be modeled by multiplicative noises (also called state-dependent noises). For example, in wireless sensor networks where active microwave sensors are utilized to collect the data on the terrain backscatter, the sampled measurements are often corrupted by a special type of noise (i.e. the speckle noise), which can be seen as a multiplicative noise. Another practical example can be found in the equalizer design problem for communication channels of wireless sensor networks where the channel identification errors are normally modeled as multiplicative noises. Different from the additive noise, the second-order statistics of the multiplicative noise are usually unknown, as they depend on the state of the system, which leads to additional difficulties. If inadequately treated, the multiplicative noises will undoubtedly provoke biased outputs and thus impair the performance of the designed TKF. Note that the situation could become even worse in the multi-sensor case due to the accumulation of local estimation biases/errors at the fusion center. As such, there is a practical need to establish a holistic multi-sensor Tobit Kalman fusion estimation framework to handle both the censored observations and the parametric uncertainties.

8.3 BOUNDEDNESS AND MONOTONICITY

In this section, we are set to establish a holistic characterization of the estimation performance of the FTKFE algorithm. Given the censoring probabilities, we first carry out a rigorous mathematical analysis on the monotonicity of the fused EEC with respect to the censoring threshold. Then, we draw attention to the boundedness of the fused EEC.

By virtue of Theorem 8.1, $P_{\tilde{x}_{i,k+1}}$ is rearranged as

$$P_{\tilde{x}_{i,k+1}} = A_k P_{\tilde{x}_{i,k}} A_k^T + F_k P_{x_k} F_k^T + Q_{i,k} - A_k P_{\tilde{x}_{i,k}} C_{i,k}^T \bar{\gamma}_{i,k}$$

$$\times \left[\bar{\gamma}_{i,k} C_{i,k} P_{\tilde{x}_{i,k}} C_{i,k}^T \bar{\gamma}_{i,k} + \bar{\gamma}_{i,k} G_{i,k} P_{x_k} G_{i,k}^T \bar{\gamma}_{i,k} + \mathcal{R}_{i,k} \right]^{-1} \tag{8.37}$$

$$\times \bar{\gamma}_{i,k} C_{i,k} P_{\tilde{x}_{i,k}} A_k^T.$$

Denote $\mathbb{S}_+^{n_x}$ as the set of n_x by n_x positive semidefinite matrices. For any $X_i \in \mathbb{S}_+^{n_x}$ we define matrix functions $g_k(\bullet)$ and $h_k(\cdot) : \mathbb{S}_+^{n_x} \to \mathbb{S}_+^{n_x}$ as follows:

$$g_k\left(\bar{\gamma}_{i,k}, \mathcal{I}_i, X_{i,k}\right)$$

$$\triangleq A_k X_{i,k} A_k^T + F_k P_{x_k} F_k^T + Q_{i,k} - A_k X_{i,k} C_{i,k}^T \bar{\gamma}_{i,k} \tag{8.38}$$

$$\times \left[\bar{\gamma}_{i,k} C_{i,k} P_{\tilde{x}_{i,k}} C_{i,k}^T \bar{\gamma}_{i,k} + \bar{\gamma}_{i,k} G_{i,k} P_{x_k} G_{i,k}^T \bar{\gamma}_{i,k} + \mathcal{R}_{i,k} \right]^{-1} \bar{\gamma}_{i,k} C_{i,k} X_{i,k} A_k^T,$$

$$h_k\left(\bar{\gamma}_{i,k}, \mathcal{I}_i, \mathcal{K}_{i,k}, X_{i,k}\right)$$

$$\triangleq \left(1 - \bar{\gamma}_{i,k}^2\right)\left(A_k X_{i,k} A_k^T + F_k P_{x_k} F_k^T + Q_{i,k}\right)$$

$$+ \bar{\gamma}_{i,k}^2 \left(\mathcal{F}_{i,k} X_{i,k} \mathcal{F}_{i,k}^T + F_k P_{x_k} F_k^T + Q_{i,k}\right) \tag{8.39}$$

$$+ \bar{\gamma}_{i,k}^2 \mathcal{K}_{i,k} \left[\bar{\gamma}_{i,k} G_{i,k} P_{x_k} G_{i,k}^T \bar{\gamma}_{i,k} + \mathcal{R}_{i,k} \right] \mathcal{K}_{i,k}^T,$$

where

$$\mathcal{F}_{i,k} \triangleq A_k + \mathcal{K}_{i,k} \bar{\gamma}_{i,k} C_{i,k},$$

$$\mathfrak{F}_{i,k} \triangleq A_k + \mathfrak{K}_{i,k} \bar{\gamma}_{i,k} C_{i,k},$$

$$\mathfrak{K}_{i,k} \triangleq -A_k X_{i,k} C_{i,k}^T \bar{\gamma}_{i,k} \left[\bar{\gamma}_{i,k} C_{i,k} X_{i,k} C_{i,k}^T \bar{\gamma}_{i,k} + \bar{\gamma}_{i,k} G_{i,k} P_{x_k} G_{i,k}^T \bar{\gamma}_{i,k} + \mathcal{R}_{i,k} \right]^{-1},$$

and $\mathcal{K}_{i,k}$ is defined to be a matrix that minimizes $h_k\left(\bar{\gamma}_{i,k}, \mathcal{I}_i, \mathcal{K}_{i,k}, X_{i,k}\right)$.

To facilitate later analysis, the following lemma is presented to uncover some useful properties in relation to $g_k\left(\bar{\gamma}_{i,k}, \mathcal{I}_i, X_{i,k}\right)$ and $h_k\left(\bar{\gamma}_{i,k}, \mathcal{I}_i, \mathcal{K}_{i,k}, X_{i,k}\right)$.

Lemma 8.2 (1) Following (8.38)–(8.39), we have

$$g_k\left(\bar{\gamma}_{i,k}, \mathcal{I}_i, X_{i,k}\right) = \min_{\mathcal{K}_{i,k}} h_k\left(\bar{\gamma}_{i,k}, \mathcal{I}_i, \mathcal{K}_{i,k}, X_{i,k}\right)$$

$$= h_k\left(\bar{\gamma}_{i,k}, \mathcal{I}_i, \mathfrak{K}_{i,k}, X_{i,k}\right),$$

indicating that $\mathfrak{K}_{i,k}$ is the optimal gain parameter that minimizes the matrix function $h_k\left(\bar{\gamma}_i, \mathcal{I}_i, \mathcal{K}_{i,k}, X_{i,k}\right)$ and the minimum value is $g_k\left(\bar{\gamma}_{i,k}, \mathcal{I}_i, X_{i,k}\right)$.

(2) If $\mathcal{I}_i \leq \mathcal{I}_i'$, then

$$h_k\left(\bar{\gamma}_{i,k}, \mathcal{I}_i, \mathcal{K}_{i,k}, X_{i,k}\right) \leq h_k\left(\bar{\gamma}_{i,k}, \mathcal{I}_i', \mathcal{K}_{i,k}, X_{i,k}\right),$$

indicating that a higher censoring threshold (i.e. less informative measurements) leads to a higher EEC.

(3) If $\overline{\gamma}_{i,k} \leq \overline{\gamma}'_{i,k}$, then

$$h_k\left(\overline{\gamma}_{i,k}, \mathcal{I}_i, \mathcal{K}_{i,k}, X_{i,k}\right) \geq h_k\left(\overline{\gamma}'_{i,k}, \mathcal{I}_i, \mathcal{K}_{i,k}, X_{i,k}\right),$$

indicating that a smaller censoring probability threshold (i.e. a less informative measurement) leads to a higher EEC.

(4) If $X_{i,k} \leq X'_{i,k}$, then

$$h_k\left(\overline{\gamma}_{i,k}, \mathcal{I}_i, \mathcal{K}_{i,k}, X_{i,k}\right) \leq h_k(\overline{\gamma}_{i,k}, \mathcal{I}_i, \mathcal{K}_{i,k}, X'_{i,k}),$$

indicating that the matrix function $h_k\left(\overline{\gamma}_{i,k}, \mathcal{I}_i, \mathcal{K}_{i,k}, X_{i,k}\right)$ is monotonically non-decreasing with respect to $X_{i,k}$.

Proof The proof of this lemma is divided into four parts. In the first part, we pay attention to expressions

$$\mathcal{K}_{i,k} \triangleq -A_k X_{i,k} C_{i,k}^T \overline{\gamma}_{i,k} \left[\overline{\gamma}_{i,k} C_{i,k} X_{i,k} C_{i,k}^T \overline{\gamma}_{i,k} + \overline{\gamma}_{i,k} G_{i,k} P_{x_k} G_{i,k}^T \overline{\gamma}_{i,k} + \mathcal{R}_{i,k} \right]^{-1},$$

$$\mathfrak{F}_{i,k} \triangleq A_k + \mathcal{K}_{i,k} \overline{\gamma}_{i,k} C_{i,k},$$

and obtain

$$\mathfrak{F}_{i,k} X_{i,k} C_{i,k}^T \overline{\gamma}_{i,k} + \mathcal{K}_{i,k} \left[\overline{\gamma}_{i,k} G_{i,k} P_{x_k} G_{i,k}^T \overline{\gamma}_{i,k} + \mathcal{R}_{i,k} \right]$$
$$= A_k X_{i,k} C_{i,k}^T \overline{\gamma}_{i,k} + \mathcal{K}_{i,k} \left[\overline{\gamma}_{i,k} C_{i,k} X_{i,k} C_{i,k}^T \overline{\gamma}_{i,k} + \overline{\gamma}_{i,k} G_{i,k} P_{x_k} G_{i,k}^T \overline{\gamma}_{i,k} + \mathcal{R}_{i,k} \right] \qquad (8.40)$$
$$= 0.$$

Based on (8.40) and the expressions of $\mathcal{K}_{i,k}$ and $\mathfrak{F}_{i,k}$, (8.38) is rearranged into

$$g_k\left(\overline{\gamma}_{i,k}, \mathcal{I}_i, X_{i,k}\right)$$
$$= \left(1 - \overline{\gamma}_{i,k}^2\right)\left(A_k X_{i,k} A_k^T + F_k P_{x_k} F_k^T + Q_{i,k}\right)$$
$$+ \overline{\gamma}_{i,k}^2 \left(\mathfrak{F}_{i,k} X_{i,k} A_k^T + F_k P_{x_k} F_k^T + Q_{i,k}\right) \qquad (8.41)$$
$$+ \overline{\gamma}_{i,k}^2 \left(\mathfrak{F}_{i,k} X_{i,k} C_{i,k}^T \overline{\gamma}_{i,k} \mathcal{K}_{i,k} \left[\overline{\gamma}_{i,k} G_{i,k} P_{x_k} G_{i,k}^T \overline{\gamma}_{i,k} + \mathcal{R}_{i,k} \right]\right) \mathcal{K}_{i,k}^T.$$

With a little more effort, we attain

$$g_k\left(\overline{\gamma}_{i,k}, \mathcal{I}_i, X_{i,k}\right) = \left(1 - \overline{\gamma}_{i,k}^2\right)\left(A_k X_{i,k} A_k^T + F_k P_{x_k} F_k^T + Q_{i,k}\right)$$
$$+ \overline{\gamma}_{i,k}^2 \left(\mathfrak{F}_{i,k} X_{i,k} \mathfrak{F}_{i,k}^T + F_k P_{x_k} F_k^T + Q_{i,}\right)$$
$$+ \overline{\gamma}_{i,k}^2 \mathfrak{F}_{i,k} \left[\overline{\gamma}_{i,k} G_{i,k} P_{x_k} G_{i,k}^T \overline{\gamma}_{i,k} + \mathcal{R}_{i,k} \right] \mathcal{K}_{i,k}^T \qquad (8.42)$$
$$= h_k\left(\overline{\gamma}_{i,k}, \mathcal{I}_i, \mathcal{K}_{i,k}, X_{i,k}\right)$$

Noting $\mathcal{F}_{i,k} \triangleq A_k + \mathcal{K}_{i,k}\overline{\gamma}_{i,k}C_{i,k}$, (8.39) is rearranged into

$$
\begin{aligned}
h_k &\left(\overline{\gamma}_{i,k}, \mathcal{I}_i, \mathcal{K}_{i,k}, X_{i,k}\right) \\
&= \left(1 - \overline{\gamma}_{i,k}^2\right)\left(A_k X_{i,k} A_k^T + F_k P_{x_k} F_k^T + Q_{i,k}\right) \\
&\quad + \overline{\gamma}_{i,k}^2 \mathcal{K}_{i,k}\left[\overline{\gamma}_{i,k}^2 G_{i,k} P_{x_k} G_{i,k}^T + \mathcal{R}_{i,k}\right]\mathcal{K}_{i,k}^T \\
&\quad + \overline{\gamma}_{i,k}^2 \left(\left(A_k + \mathcal{K}_{i,k}\overline{\gamma}_{i,k}C_{i,k}\right)X_{i,k}\left(A_{i,k} + \mathcal{K}_{i,k}\overline{\gamma}_{i,k}C_{i,k}\right)^T + F_k P_{x_k} F_k^T + Q_{i,k}\right).
\end{aligned}
\tag{8.43}
$$

Due to the fact that the sum of the second and third terms on the right-hand side of (8.43) is quadratic and convex with respect to $\mathcal{K}_{i,k}$, the optimal $\mathcal{K}_{i,k}$ that minimizes $h_k\left(\overline{\gamma}_{i,k}, \mathcal{I}_i, \mathcal{K}_{i,k}, X_{i,k}\right)$ can be acquired by solving

$$
\frac{\partial h_k\left(\overline{\gamma}_{i,k}, \mathcal{I}_i, \mathcal{K}_{i,k}, X_{i,k}\right)}{\partial \mathcal{K}_{i,k}} = 0,
$$

and this yields

$$
\begin{aligned}
\mathcal{K}_{i,k} &= -A_k X_{i,k} C_{i,k}^T \overline{\gamma}_{i,k}\left[\overline{\gamma}_i C_{i,k} X_{i,k} C_{i,k}^T \overline{\gamma}_{i,k} + \overline{\gamma}_{i,k} G_{i,k} P_{x_k} G_{i,k}^T \overline{\gamma}_{i,k} + \mathcal{R}_{i,k}\right]^{-1} \\
&= \hat{\mathcal{K}}_{i,k}.
\end{aligned}
$$

As such, we have

$$
\begin{aligned}
\min_{\mathcal{K}_{i,k}} h_k\left(\overline{\gamma}_{i,k}, \mathcal{I}_i, \mathcal{K}_{i,k}, X_{i,k}\right) &= h_k\left(\overline{\gamma}_{i,k}, \mathcal{I}_i, \hat{\mathcal{K}}_{i,k}, X_{i,k}\right) \\
&= g_k(\overline{\gamma}_{i,k}, \mathcal{I}_i, X_{i,k}).
\end{aligned}
$$

In the second part of the proof, by supposing $\mathcal{I}_i \leq \mathcal{I}_i'$, one observes from (8.3)–(8.5) and (8.10) that $\overline{\gamma}_{i,k} \geq \overline{\gamma}_{i,k}'$. Moving forward from (8.38), we have

$$
\begin{aligned}
g_k &\left(\overline{\gamma}_{i,k}, \mathcal{I}_i, X_{i,k}\right) \\
&\leq A_k X_{i,k} A_k^T + F_k P_{x_k} F_k^T + Q_{i,k} - A_k X_{i,k} C_{i,k}^T \overline{\gamma}_{i,k}' \\
&\quad \times \left[\overline{\gamma}_{i,k}' C_{i,k} X_{i,k} C_{i,k}^T \overline{\gamma}_{i,k}' + \overline{\gamma}_{i,k}' G_{i,k} P_{x_k} G_{i,k}^T \overline{\gamma}_{i,k}' + \mathcal{R}_{i,k}\right]^{-1}\overline{\gamma}_{i,k}' C_{i,k} X_{i,k} A_k^T \\
&= g_k\left(\overline{\gamma}_{i,k}', \mathcal{I}_i', X_{i,k}\right),
\end{aligned}
$$

which implies that $g_k\left(\overline{\gamma}_{i,k}, \mathcal{I}_i, X_{i,k}\right)$ is monotonically non-decreasing in regard to $\mathcal{I}_i$.

In the third part of the proof, it follows evidently from the proof of Lemma 8.2 (2) that $g_k\left(\overline{\gamma}_{i,k}, \mathcal{I}_i, X_{i,k}\right)$ is monotonically non-increasing in regard to $\overline{\gamma}_{i,k}$.

Finally, in the fourth part of the proof, we let

$$
\hat{\mathcal{K}}_{i,k}' \triangleq -A_k X_{i,k}' C_{i,k}^T \overline{\gamma}_{i,k}\left[\overline{\gamma}_{i,k} C_{i,k} X_{i,k}' C_{i,k}^T \overline{\gamma}_{i,k} + \overline{\gamma}_{i,k} G_{i,k} P_{x_k} G_{i,k}^T \overline{\gamma}_{i,k} + \mathcal{R}_{i,k}\right]^{-1}.
$$

For any $X_{i,k}$, $X'_{i,k} \in \mathbb{S}$, suppose that $X_{i,k} \leq X'_{i,k}$. Parallel to the proof of Lemma 8.2 (1), we can verify that $\mathfrak{K}'_{i,k}$ is the optimal gain matrix that minimizes $h_k\left(\bar{\gamma}_{i,k}, \mathcal{I}_i, \mathcal{K}_{i,k}, X'_{i,k}\right)$. Accordingly, we have

$$g_k\left(\bar{\gamma}_{i,k}, \mathcal{I}_i, X_{i,k}\right) = h_k\left(\bar{\gamma}_{i,k}, \mathcal{I}_i, \mathfrak{K}_{i,k}, X_{i,k}\right)$$
$$\leq h_k\left(\bar{\gamma}_{i,k}, \mathcal{I}_i, \mathfrak{K}'_{i,k}, X_{i,k}\right)$$
$$\leq h_k(\bar{\gamma}_{i,k}, \mathcal{I}_i, \mathfrak{K}'_{i,k}, X'_{i,k}),$$

which further leads to

$$g_k\left(\bar{\gamma}_{i,k}, \mathcal{I}_i, X_{i,k}\right) \leq g_k(\bar{\gamma}_{i,k}, \mathcal{I}_i, X'_{i,k}), \tag{8.44}$$

indicating that $g_k\left(\bar{\gamma}_{i,k}, \mathcal{I}_i, X_{i,k}\right)$ is monotonically non-decreasing with respect to $X_{i,k}$.

With slight notation abuse, we use $P_{\tilde{x}_{i,k}^-}$ and $P'_{\tilde{x}_{i,k}^-}$, respectively, to stand for the local EECs in (8.20) with censoring thresholds $\mathcal{I}_i$ and $\mathcal{I}'_i$. Let $P_{\tilde{x}_k^-}$ and $P'_{\tilde{x}_k^-}$ be the fused EEC (8.28) with censoring threshold sets $\left\{\mathcal{I}_1, \mathcal{I}_2, \ldots, \mathcal{I}_p\right\}$ and $\left\{\mathcal{I}'_1, \mathcal{I}'_2, \ldots, \mathcal{I}'_p\right\}$, respectively, and denote $\mathcal{I} \triangleq \sum_{i=1}^p \mathcal{I}_i$ and $\mathcal{I}' \triangleq \sum_{i=1}^p \mathcal{I}'_i$. Then, the monotonicity issue is addressed as follows.

Theorem 8.2 Let the ith LTKF in Theorem 8.1 be realized with scalar thresholds $\mathcal{I}_i$ and $\mathcal{I}'_i$, respectively. If $\mathcal{I}_i \leq \mathcal{I}'_i$, then $P_{\tilde{x}_{i,k}^-} \leq P'_{\tilde{x}_{i,k}^-}$ holds for all $k \geq 0$, i.e. $P_{\tilde{x}_{i,k}^-}$ is monotonically non-decreasing with respect to $\mathcal{I}_i$. Besides, let the p LTKFs be realized with two sets of thresholds $\left\{\mathcal{I}_1, \mathcal{I}_2, \ldots, \mathcal{I}_p\right\}$ and $\left\{\mathcal{I}'_1, \mathcal{I}'_2, \ldots, \mathcal{I}'_p\right\}$, respectively. If $\mathcal{I} \leq \mathcal{I}'$, then $P_{\tilde{x}_k^-} \leq P'_{\tilde{x}_k^-}$ holds for all $k > 0$, i.e. $P_{\tilde{x}_{i,k}^-}$ is monotonically non-decreasing with respect to $\mathcal{I}_i$.

Proof We prove the monotonicity property via mathematical induction. For $k = 0$, the result $P_{\tilde{x}_{i,0}^-} \leq P'_{\tilde{x}_{i,0}^-}$ certainly holds since the initial conditions are set as $P_{\tilde{x}_{i,0}^-} \triangleq \epsilon_i^{-1} P_0$ and $P'_{\tilde{x}_{i,0}^-} \triangleq \epsilon_i^{-1} P_0$.

For k > 0, suppose that $P_{\tilde{x}_{i,k}^-} \leq P'_{\tilde{x}_{i,k}^-}$. A direct comparison between (8.37) and (8.38) tells

$$P_{\tilde{x}_{i,k+1}^-} = g_k\left(\bar{\gamma}_i, \mathcal{I}_i, P_{\tilde{x}_{i,k}^-}\right),$$

and the reminiscence of $\mathcal{I}_i \leq \mathcal{I}'_i$ produces $\bar{\gamma}_i \geq \bar{\gamma}'_i$. Moving forward to time k + 1, we have $P_{\tilde{x}_{i,k+1}^-} = g_k\left(\bar{\gamma}_i, \mathcal{I}_i, P_{\tilde{x}_{i,k}^-}\right) \leq g_k\left(\bar{\gamma}'_i, \mathcal{I}'_i, P'_{\tilde{x}_{i,k}^-}\right)$, which further leads to

$$P_{\tilde{x}_{i,k+1}^-} \leq P'_{\tilde{x}_{i,k+1}^-}, \tag{8.45}$$

where the inequalities hold from Lemma 8.2. The first part of Theorem 8.3 is thus proved.

We now go ahead to the proof of the second part. It follows evidently from $\mathcal{I} \leq \mathcal{I}'$ that there admits at least one censoring threshold pair $\left(\mathcal{I}_i, \mathcal{I}_i'\right)$ $(i=1,2,\ldots,p)$ such that $\mathcal{I}_i \leq \mathcal{I}_i'$, and therefore $\overline{\gamma}_i \geq \overline{\gamma}_i'$.

For k = 0, the conclusion $P_{\tilde{x}_0} \leq P_{\tilde{x}_0}'$ holds noticeably from (8.28). For k > 0, supposing $P_{\tilde{x}_k} \leq P_{\tilde{x}_k}'$, there exists at least one matrix pair $\left(P_{\tilde{x}_{i,k}}, P_{\tilde{x}_{i,k}}'\right)$ $(i=1,2,\ldots,p)$ such that $P_{\tilde{x}_{i,k}^-} \leq P_{\tilde{x}_{i,k}^-}'$. Consequently,

$$P_{\tilde{x}_{i,k+1}^-} = g_k\left(\overline{\gamma}_i, \mathcal{I}_i, P_{\tilde{x}_{i,k}^-}\right) \leq g_k\left(\overline{\gamma}_i, \mathcal{I}_i, P_{\tilde{x}_{i,k}^-}'\right) \leq g_k\left(\overline{\gamma}_i', \mathcal{I}_i', P_{\tilde{x}_{i,k}^-}'\right),$$

which further leads to

$$P_{\tilde{x}_{i,k+1}^-} \leq P_{\tilde{x}_{i,k+1}^-}'. \tag{8.46}$$

Noting the semi-positive definiteness of $P_{\tilde{x}_{i,k+1}^-}$ and $P_{\tilde{x}_{i,k+1}^-}'$ in conjunction with (8.46), we confirm that

$$P_{\tilde{x}_{i,k+1}^-}^{-1} \geq \left(P_{\tilde{x}_{i,k+1}^-}'\right)^{-1}$$

holds for at least one pair $\left(P_{\tilde{x}_{i,k+1}^-}, P_{\tilde{x}_{i,k+1}^-}'\right)$, which implies

$$\sum_{i=1}^{p} P_{\tilde{x}_{i,k+1}^-}^{-1} \geq \sum_{i=1}^{p}\left(P_{\tilde{x}_{i,k+1}^-}'\right)^{-1}.$$

As a result, we have the confirmation that

$$P_{\tilde{x}_k} = \left(\sum_{i=1}^{p} P_{\tilde{x}_{i,k}^-}^{-1}\right)^{-1} \leq \left(\sum_{i=1}^{p}\left(P_{\tilde{x}_{i,k}^-}'\right)^{-1}\right)^{-1} = P_{\tilde{x}_k}'.$$

In addition to the monotonicity, the boundedness is also a key performance measure for the established algorithm. As such, we are motivated to study the boundedness property of the local and fused algorithms. Since the evolution of $P_{\tilde{x}_{i,k}}$ is determined by P_{x_k}, we first explore the lower and upper bounds on P_{x_k}:

Assumption 8.2 There exist positive real numbers $\overline{a}, \overline{f}, \overline{q}, \overline{c}_i, \underline{a}, \underline{f}, \underline{q}, \underline{g}_i, \underline{r}$ and χ such that

$$\underline{\gamma}_i \leq \overline{\gamma}_{i,k}, G_{i,k}G_{i,k}^T \geq \underline{g}_i I, \hat{x}_k \hat{x}_k^T \leq \chi I,$$

$$\overline{\varphi}_{i,k} \leq \overline{\varphi}_i, \underline{a} \leq A_k A_k^T \leq \overline{a}I, \underline{q}I \leq Q_k \leq \overline{q}I,$$

$$\underline{r}I \leq \mathcal{R}_{i,k}, C_{i,k}C_{i,k}^T \leq \overline{c}_i I, \underline{f}I \leq F_k F_k^T \leq \overline{f}I,$$

are true for all $k \geq 0$.

Lemma 8.3 Given a constant $e > 0$ and Assumption 8.2, there exist lower and upper bounds $\underline{\rho}_k$ and $\overline{\rho}_k$ such that the state covariance P_{x_k} satisfies

$$\underline{\rho}_k I \leq P_{x_k} \leq \overline{\rho}_k I, \tag{8.47}$$

where

$$\underline{\rho}_k = \underline{a}^k \underline{\rho}_0 + \underline{q} + (k-1)\underline{a}\,\underline{q}, \tag{8.48}$$

$$\overline{\rho}_k = (1+e)\overline{p}_{i,k} + (1+e^{-1})\chi, \tag{8.49}$$

$\underline{\rho}_0 = \lambda_{\min}\left(P_{x_0}\right)$ *is the minimum eigenvalue of* P_{x_0}*, and* $\overline{p}_{i,k}$ *is to be determined later.*

Proof On one hand, it follows evidently from (8.23) and Assumption 8.2 that

$$
\begin{aligned}
P_{x_k} &= A_{k-1}P_{x_{k-1}}A_{k-1}^T + F_{k-1}P_{x_{k-1}}F_{k-1}^T + Q_{k-1} \\
&\geq A_{k-1}P_{x_{k-1}}A_{k-1}^T + Q_{k-1} \\
&\geq A_{k-1}(A_{k-2}P_{x_{k-2}}A_{k-2}^T + Q_{k-2})A_{k-1}^T + Q_{k-1} \\
&\quad\vdots \\
&\geq A_{k-1}A_{k-2}\cdots A_0 P_{x_0}A_0^T \cdots A_{k-2}^T A_{k-1}^T + Q_{k-1} \\
&\quad + A_{k-1}Q_{k-2}A_{k-1}^T + A_{k-2}Q_{k-3}A_{k-2}^T + \cdots + A_1 Q_0 A_1^T \\
&\geq [\underline{a}^k \underline{\rho}_0 + \underline{q} + (k-1)\underline{a}\,\underline{q}]I \\
&= \underline{\rho}_k I,
\end{aligned}
$$

where $\underline{\rho}_k$ is defined by (8.48).

On the other hand, for any constant $e > 0$ and vectors x and y, taking plain use of the famous inequality

$$xy^T + yx^T \leq exx^T + e^{-1}yy^T,$$

we have

$$
\begin{aligned}
P_{x_k} &\triangleq \mathbb{E}\left\{\left(\hat{x}_{i,k} + \tilde{x}_{i,k}\right)\left(\hat{x}_{i,k} + \tilde{x}_{i,k}\right)^T\right\} \\
&= \mathbb{E}\left\{\tilde{x}_{i,k}\tilde{x}_{i,k}^T + \hat{x}_{i,k}\hat{x}_{i,k}^T + \tilde{x}_{i,k}\hat{x}_{i,k}^T + \hat{x}_{i,k}\tilde{x}_{i,k}^T\right\} \\
&\leq (1+e)P_{\tilde{x}_{i,k}} + (1+e^{-1})\hat{x}_{i,k}\hat{x}_{i,k}^T \\
&\leq \left[(1+e)\overline{p}_{i,k} + (1+e^{-1})\chi\right]I \\
&= \overline{\rho}_k I.
\end{aligned}
$$

where $\overline{\rho}_k$ is defined by (8.49).

Combining these two inequalities together results in (8.47).

On the basis of Lemma 3.12, the lower and upper bounds on the local EEC $P_{\tilde{x}_{i,k}}$ are provided by the following two theorems.

Theorem 8.3 Under Assumption 8.2, there exists a lower bound $\underline{p}_k$ such that the local EEC $P_{\tilde{x}_{i,k}}$ satisfies

$$P_{\tilde{x}_{i,k}} \geq \underline{p}_{i,k} I, \tag{8.50}$$

where

$$\underline{p}_{i,k} = \left[\frac{1}{(\underline{f}\underline{\rho}_k + \epsilon_i^{-1}\underline{q})} + \frac{\overline{c}_i}{\underline{\gamma}_i \underline{g}_i \underline{\rho}_k + \underline{r}} \right]^{-1}. \tag{8.51}$$

Proof Denoting

$$\tilde{R}_{i,k} \triangleq \overline{\gamma}_{i,k} G_{i,k} P_{x_k} G_{i,k}^T \overline{\gamma}_{i,k} + \mathcal{R}_{i,k},$$

and referring to (8.20) and (8.22), we have

$$\begin{aligned}
P_{\tilde{x}_{i,k}} &= P_{\tilde{x}_{i,k}^-} - P_{\tilde{x}_k^-} C_{i,k}^T \overline{\gamma}_{i,k} \left[\overline{\gamma}_{i,k} C_{i,k} P_{\tilde{x}_k^-} C_{i,k}^T \overline{\gamma}_{i,k} + \tilde{R}_{i,k} \right]^{-1} \overline{\gamma}_{i,k} C_{i,k} P_{\tilde{x}_k^-} \\
&= \left[P_{\tilde{x}_{i,k}^-}^{-1} + C_{i,k}^T \overline{\gamma}_{i,k} \tilde{R}_{i,k}^{-1} \overline{\gamma}_{i,k} C_{i,k} \right]^{-1}.
\end{aligned} \tag{8.52}$$

Keeping Assumption 8.2 and (8.47) in mind, the lower bounds on $\tilde{R}_{i,k}$ and $P_{\tilde{x}_{i,k}^-}$ can be obtained as

$$\begin{aligned}
\tilde{R}_{i,k} &= \overline{\gamma}_{i,k} G_{i,k} P_{x_k} G_{i,k}^T \overline{\gamma}_{i,k} + \mathcal{R}_{i,k} \\
&\geq [\underline{\gamma}_i \underline{g}_i \underline{\rho}_k + \underline{r}]I,
\end{aligned} \tag{8.53}$$

$$\begin{aligned}
P_{\tilde{x}_{i,k}^-} &= A_{k-1} P_{\tilde{x}_{i,k-1}} A_{k-1}^T + F_{k-1} P_{x_{k-1}} F_{k-1}^T + Q_{i,k-1} \\
&\geq F_{k-1} P_{x_{k-1}} F_{k-1}^T + Q_{i,k-1} \\
&\geq \left(\underline{f}\underline{\rho}_k + \epsilon_i^{-1}\underline{q} \right) I.
\end{aligned} \tag{8.54}$$

Putting (8.53)–(8.54) into (8.52), we have

$$\begin{aligned}
P_{\tilde{x}_{i,k}} &= \left[P_{\tilde{x}_{i,k}^-}^{-1} + C_{i,k}^T \overline{\gamma}_{i,k} \tilde{R}_{i,k}^{-1} \overline{\gamma}_{i,k} C_{i,k} \right]^{-1}, \\
&\geq \left[\frac{1}{\left(\underline{f}\underline{\rho}_k + \epsilon_i^{-1}\underline{q}\right)} + \frac{\overline{c}_i}{\underline{\gamma}_i \underline{g}_i \underline{\rho}_k + \underline{r}} \right]^{-1} I \\
&\triangleq \underline{p}_{i,k} I,
\end{aligned}$$

where $\underline{p}_k$ is defined by (8.51).

  Theorem 8.4 Under Assumption 8.2, there exists an upper bound $\overline{p}_{i,k}$ such that the local EEC $P_{\tilde{x}_{i,k}}$ satisfies

$$P_{\tilde{x}_{i,k}} \leq \overline{p}_{i,k} I, \tag{8.55}$$

for all $k \geq 0$, where for $k > 0$,

$$\overline{p}_{i,k} = \overline{p}_{i,0}\left[\overline{a}+(1+e)\overline{f}\right]^{k} + \left[\left(1+e^{-1}\right)\overline{f}\chi + \epsilon_{i}^{-1}\overline{q}\right]\sum_{j=0}^{k-1}\left[\overline{a}+(1+e)\overline{f}\right]^{j}, \quad (8.56)$$

and $\overline{p}_{i,0} = \lambda_{\max}\left(P_{\tilde{x}_{i,0}}\right)$ is the maximum eigenvalue of $P_{\tilde{x}_{i,0}}$.

Proof Let us prove our result by mathematical induction, which consists of the initial and inductive steps.

Initial Step. For $k = 0$, one immediately has

$$P_{\tilde{x}_{i,0}} \leq \overline{p}_{i,0}I = \lambda_{\max}\left(P_{\tilde{x}_{i,0}}\right)I,$$

where $P_{\tilde{x}_{i,0}}$ is defined by $\overline{p}_{i,0} \triangleq \lambda_{\max}\left(P_{\tilde{x}_{i,0}}\right)$.

Inductive Step. The initial step shows $P_{\tilde{x}_{i,t}} \leq \overline{p}_{i,t}I$ is true for $t = 0$. Next, given $P_{\tilde{x}_{i,t}} \leq \overline{p}_{i,t}I$ is true for $t = k$, we need to show $P_{\tilde{x}_{i,t}} \leq \overline{p}_{i,t}I$ is also true for $t = k + 1$. (8.52) tells

$$P_{\tilde{x}_{i,k}}^{-1} = P_{\tilde{x}_{i,k}^{-}}^{-1} + C_{i,k}^{T}\overline{\gamma}_{i,k}\tilde{R}_{i,k}^{-1}\overline{\gamma}_{i,k}C_{i,k},$$

which gives rise to

$$P_{\tilde{x}_{i,k+1}} \leq P_{\tilde{x}_{i,k+1}^{-}} = A_{k}P_{\tilde{x}_{k}}A_{k}^{T} + F_{k}P_{x_{k}}F_{k}^{T} + Q_{i,k}.$$

Consequently, we have

$$P_{\tilde{x}_{i,k+1}} \leq \overline{a}\overline{p}_{i,k}I + \overline{f}\left[(1+e)\overline{p}_{i,k}I + \left(1+e^{-1}\right)\chi I\right] + \epsilon_{i}^{-1}\overline{q}I$$

$$= \overline{p}_{i,0}\left[\overline{a}+(1+e)\overline{f}\right]^{k+1}I + \left[\left(1+e^{-1}\right)\overline{f}\chi I + \epsilon_{i}^{-1}\overline{q}I\right]\sum_{j=0}^{k}\left[\overline{a}+(1+e)\overline{f}\right]^{j}$$

$$= \overline{p}_{i,k+1}I,$$

from Assumption 8.2, Lemma 8.3 and the assumption that $P_{\tilde{x}_{i,t}} \leq \overline{p}_{i,t}I$ is true for $t = k$. As a result, we conclude that $P_{\tilde{x}_{i,t}} \leq \overline{p}_{i,t}I$ is true for $t = k + 1$.

Combining the two steps together, we draw the conclusion that $P_{\tilde{x}_{i,k}} \leq \overline{p}_{i,k}I$ is true for all $k \geq 0$ where $\overline{p}_{i,0} = \lambda_{\max}\left(P_{\tilde{x}_{i,0}}\right)$ $\overline{p}_{i,k}$ and $\overline{p}_{i,k}$ $(k > 0)$ is given by (8.56).

Based on the federated fusion rule and Theorems 8.3–8.4, there exist a lower bound $\underline{p}_{k}$ and an upper bound $\overline{p}_{k}$ such that the fused EEC $P_{\tilde{x}_{k}}$ satisfies

$$\underline{p}_{k}I \leq P_{\tilde{x}_{k}} \leq \overline{p}_{k}I, \quad (8.57)$$

where

$$\underline{p}_{k} = \left(\sum_{i=1}^{p}\underline{p}_{i,k}^{-1}\right)^{-1}, \quad (8.58)$$

$$\overline{p}_{k} = \left(\sum_{i=1}^{p}\overline{p}_{i,k}^{-1}\right)^{-1}. \quad (8.59)$$

Remark 8.6 It is apparent from the foregoing analysis that the monotonicity and boundedness properties of our FTKFE comprise two steps: (1) the requirement of monotonicity and boundedness in the local case; and (2) the extension from the local case to the fused case. On one hand, Lemma 8.2 and Theorems 8.3.–8.4 should be carefully verified to guarantee the monotonicity and boundedness properties at the local level. On the other hand, the federated fusion rule should be utilized so as to generalize such properties from the local case to the fused case. The involved complexity is obviously increased by the concurrent presence of the MSS, censored observations and multiplicative noises.

Remark 8.7 Different from the conventional TKF, the design routine of our FTKFE (as shown in the previous section) is closely dependent on the MSS and multiplicative noises. As such, it is natural to envision that impacts from the MSS, censored observations and multiplicative noises should all be reflected in the procedure of the performance assessment. Clearly, such impacts are characterized by the monotonicity property in Lemma 8.2 and Theorem 8.2 and the boundedness result in Theorems 8.3–8.4. More specifically, $\overline{\gamma}_{i,k}$ and $\mathcal{I}_i$ in Lemma 8.2 certainly quantify the effect from the censored observation, $F_k P_{x_k} F_k^T$ and $\overline{\gamma}_i G_{i,k} P_{x_k} G_{i,k}^T \overline{\gamma}_i$ explicitly describe the effect from the multiplicative noise effect, and variables and matrices with subscript i together with (8.58)–(8.59) noticeably sketch the MSS-induced effect.

Remark 8.8 It is noteworthy that the introduction of the MSS and multiplicative noises brings on board substantial challenges to both filter design and performance analysis as compared to the conventional TKF for the following reasons. (1) Because of the MSS, we turn to devise a set of LTKFs whose outputs are integrated to output the global OSE via resorting to the federated fusion rule. Nevertheless, the federated fusion estimation provokes an intricate relationship between the local and global EECs, demanding significant effort in carrying out the monotonicity analysis as elucidated in the proof of Theorem 8.3. (2) Due to the multiplicative noises, we turn to seek the specifically tailored regression model in Lemma 8.1 and the recursive structure of P_{x_k} in (8.25). Nonetheless, the appearance of term P_{x_k} gives rise to extra complexity in conducting the performance analysis, since the boundedness of P_{x_k} is required to ensure the boundedness of $P_{\tilde{x}_k}$ (see Lemma 8.3). (3) The MSS along with the multiplicative noises apparently lead to distinctive design and analysis procedures of our FTKFE as manifested in Lemmas 8.1–8.3 and Theorems 8.1–8.4.

8.4 ILLUSTRATIVE EXAMPLES

In this section, the oscillator example and target tracking example are employed to justify the feasibility of our FTKFE. Denote the root mean-squared error (RMSE) of x_k^j (the jth dimension of the state) as $\text{RMSE}j = \sqrt{(1/M) \sum_{i=1}^{n} X_i^2 \sum_{i=1}^{M} \left(x_k^j - \hat{x}_k^{j(i)} \right)^2}$, and denote the average root mean-squared error (ARMSE) of x_k^j as ARMSE $j \triangleq \dfrac{1}{N} \sum_{k=1}^{N} \sqrt{\dfrac{1}{M} \sum_{i=1}^{M} \left(x_k^j - \hat{x}_k^{j(i)} \right)^2}$, where M is the number of Monte Carlo trials and N is the number of time steps in each trial.

8.4.1 OSCILLATOR EXAMPLE

Let system (8.1)–(8.3) have the following parameters:

$$Q_k = \operatorname{diag}\{0.0025, 0.0025\},\ R_{1,k} = 1,\ \epsilon_1 = \epsilon_2 = 0.5,$$

$$F_k = 0.1I_2,\ G_{1,k} = \begin{bmatrix} 0.1 & 0 \end{bmatrix},\ G_{2,k} = \begin{bmatrix} 0 & 0.1 \end{bmatrix},\ R_{2,k} = 1,$$

$$P_0 = I_2,\ \bar{x}_0 = \begin{bmatrix} 5 & 0 \end{bmatrix}^T,\ \mathcal{I}_1 = \mathcal{I}_2 = 0,\ w = 0.052\pi,\ e = 0.5,$$

$$A_k = \begin{bmatrix} \cos(w) & -\sin(w) \\ \sin(w) & \cos(w) \end{bmatrix},\ C_{1,k} = \begin{bmatrix} 1 & 0 \end{bmatrix},\ C_{2,k} = \begin{bmatrix} 0 & 1 \end{bmatrix}.$$

The example concerns the estimation of ballistic roll rates of the noisy dynamic model with uncertain magnetometer data. Here, the dynamic model is contaminated by parametric uncertainties and ambient disturbances entering into the system via α_k and ω_k, respectively, and the magnetometer data is inclined to censored observations, ambient disturbances $\upsilon_{i,k}$ and parametric uncertainties $\beta_{i,k}$. In the follow-up simulations, we first make a performance comparison between the federated Tobit Kalman filter (FTKF) (which barely tackles the MSS and censored observations) and FTKFE (which simultaneously tackles the MSS, censored observations along with parametric uncertainties) to show the superiority of our FTKFE. Then, traces of fused EECs

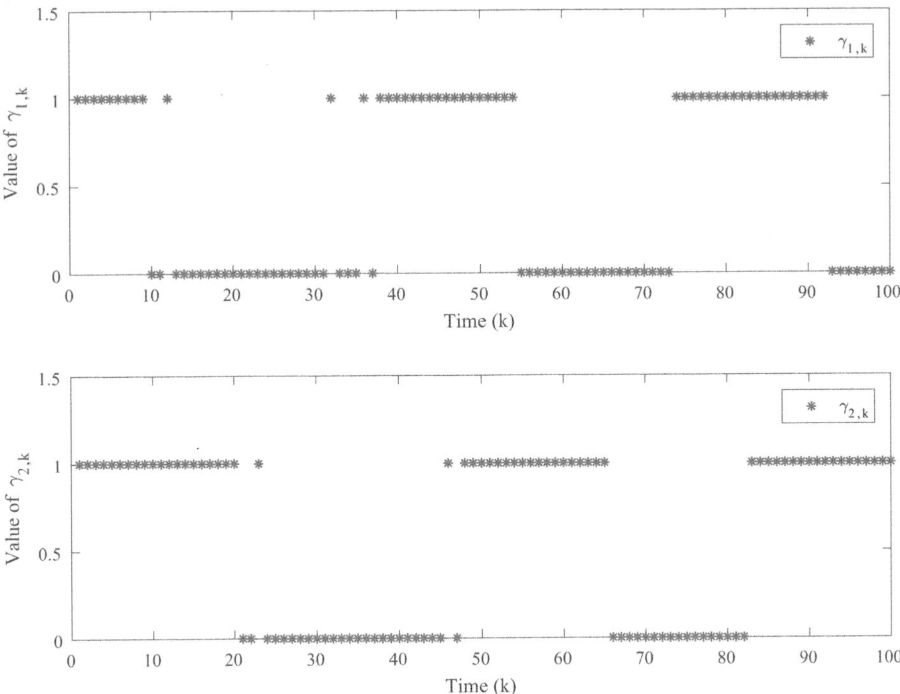

FIGURE 8.1 Values of $\gamma_{1,k}$ and $\gamma_{2,k}$.

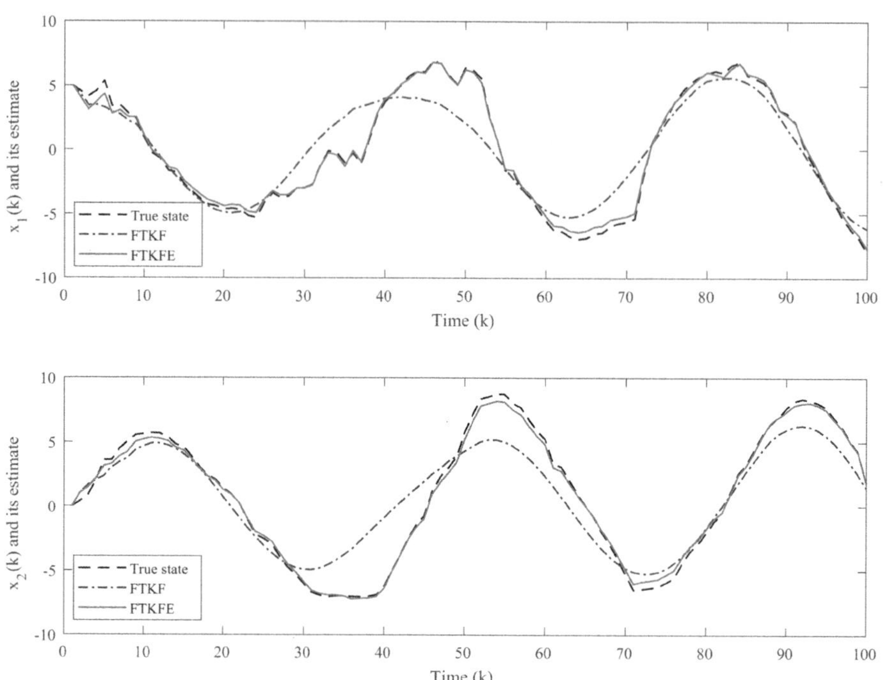

FIGURE 8.2   States and estimates.

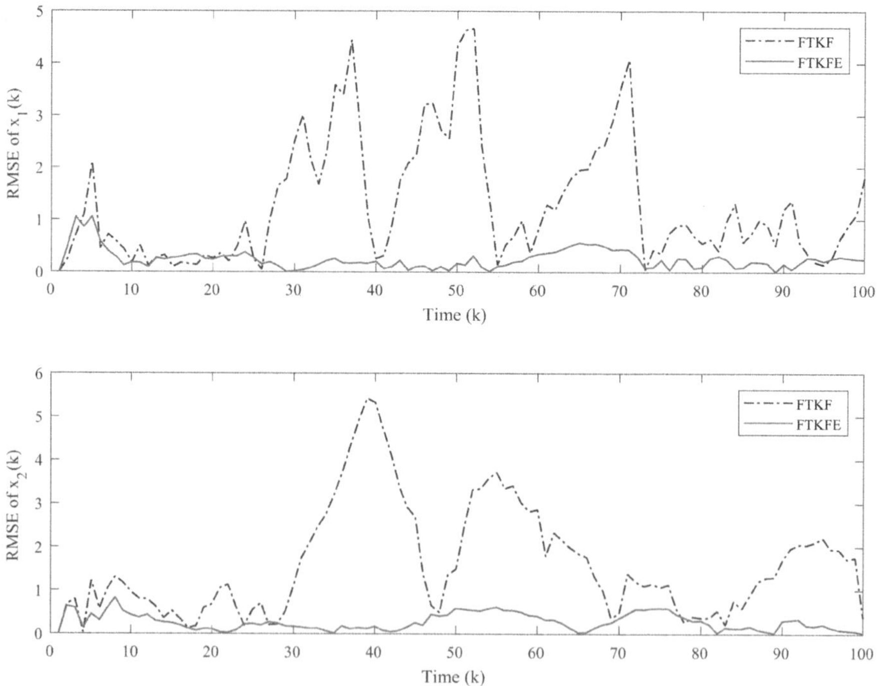

FIGURE 8.3   Comparison of RMSE1 and RMSE2.

under different censoring thresholds are demonstrated to explain the monotonicity property. At last, the variation trend of the EEC along the time axis is illustrated to verify the boundedness behavior.

Figure 8.1 illustrates the values of $\gamma_{1,k}$ and $\gamma_{2,k}$ used in the simulation, Figure 8.2 depicts the true values and estimates of x_k^1 and x_k^2 generated by the two algorithms, and Figure 8.3 plots the RMSE comparison results. Since it is analytically tough to calculate the actual RMSE values versus time, the presented RMSE values are experimentally acquired through 1000 independent Monte Carlo trials. It is confirmed from Figure 8.2 that in the case of censored observations and multiplicative noises, our FTKFE succeeds in tracking states x_k^1 and x_k^2, while the FTKF has explicit deviations from states x_k^1 and x_k^2. Besides, it is given by Figure 8.3 that the RMSE curve of our FTKFE always resides lower than that of the FTKF. The reason is that in our FTKFE, the multiplicative noises that appear in both state and observation equations are carefully addressed by establishing a modified multi-sensor Tobit Kalman fusion estimation scheme, whilst they are not tackled in the FTKF.

Next, to further elucidate the monotonicity and boundedness properties of our FTKFE, we denote the censoring thresholds of the two sensors as $\{\mathcal{I}_1,\mathcal{I}_2\}$, $\{\mathcal{I}_1',\mathcal{I}_2'\}$, $\{\mathcal{I}_1'',\mathcal{I}_2''\}$, and let the state transition matrix be $0.9A_k$, $\mathcal{I}_1 = \mathcal{I}_2 = -2$, $\mathcal{I}_1' = \mathcal{I}_2' = 0$, $\mathcal{I}_1'' = \mathcal{I}_2'' = 2$. Then, we have $\mathcal{I} = -4$, $\mathcal{I}' = 0$ and $\mathcal{I}'' = 4$. Meanwhile, it can be effort-

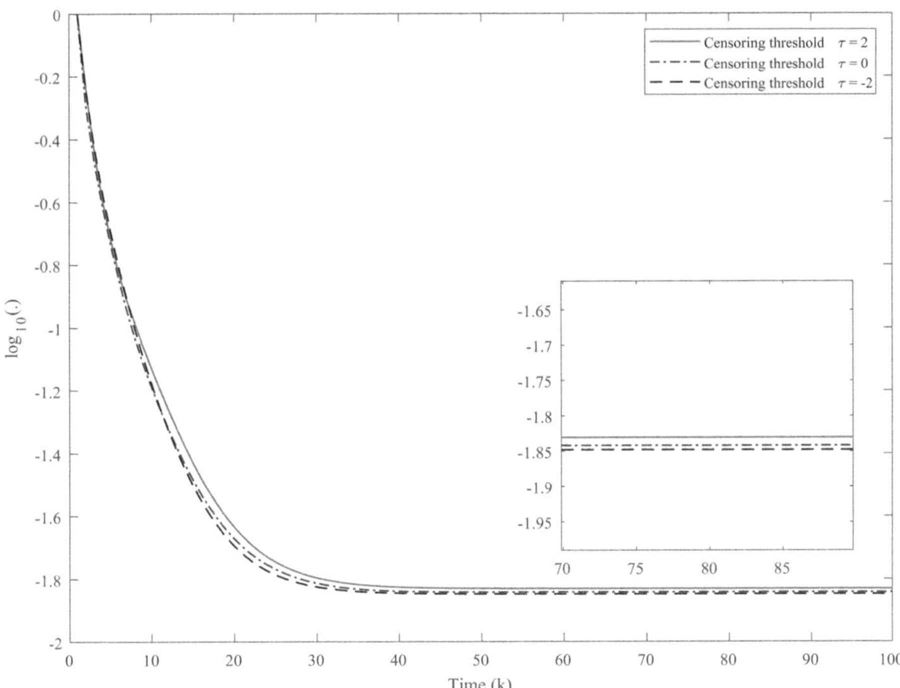

FIGURE 8.4 $\log_{10}\left(\mathrm{tr}\{P_{\tilde{x}_k}\}\right)$ under different censoring thresholds.

lessly verified that $\bar{\varphi}_1 = \bar{\varphi}_2 = 0.5$, $\underline{a} = 0.8$, $\bar{a} = 0.81$, $\underline{q} = 0.002$, $\bar{q} = 0.0025$, $\underline{\gamma} = 0.1$, $\underline{g}_1 = \underline{g}_2 = 0.005$, $\chi = 100$, $\underline{r}_1 = \underline{r}_2 = 0.5$, $\bar{c}_1 = \bar{c}_2 = 0.02$, $\underline{f} = 0.005$ and $\bar{f} = 0.01$.

Accordingly, Figure 8.4 plots the variation trends of trace curves $\log_{10}\left(\mathrm{tr}\left(P_{\tilde{x}_k}\right)\right)$ on the condition of three sets of censoring thresholds. It is confirmed from Figure 8.4 that trace curves $\log_{10}\left(\mathrm{tr}\left(P_{\tilde{x}_k}\right)\right)$ with larger censoring thresholds always reside higher than those with smaller censoring thresholds. This observation evidently verifies the correctness of the monotonicity result claimed in Theorem 8.2. Additionally, it is observed from Figure 8.5 that the trace curve $\log_{10}\left(\mathrm{tr}\left(P_{\tilde{x}_k}\right)\right)$ always resides between trace curves $\log_{10}\left(\mathrm{tr}\left(\underline{p}_k I\right)\right)$ and $\log_{10}\left(\mathrm{tr}\left(\bar{p}_k I\right)\right)$, which justifies the boundedness statement in (8.57) that $\underline{p}_k$ and $\bar{p}_k$ can be, respectively, treated as reasonable upper and lower bounds on $P_{\tilde{x}_k}$.

8.4.2 Target Tracking Example

Consider an example of distributive target tracking with three sensors. According to Newton's force principle, the movement of the target can be modeled as a constant acceleration nominal system corrupted by modeling errors and ambient disturbances

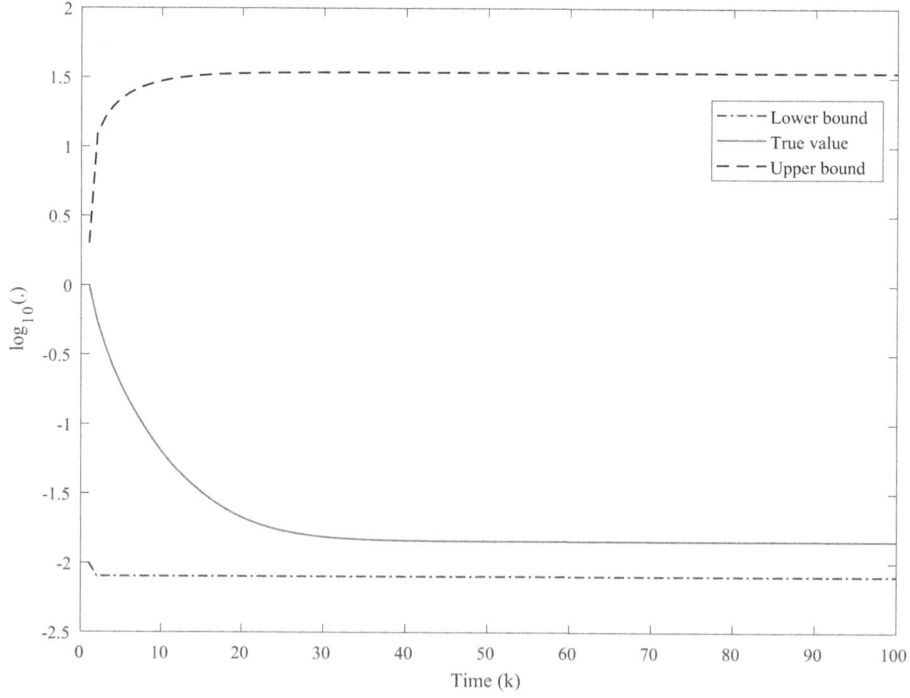

FIGURE 8.5   Traces of $P_{\tilde{x}_k}$ and its bounds.

entering the system via multiplicative noise and process noise, respectively. The three sensors sample the target position, velocity and acceleration, respectively, with the same sampling period $T = 0.1s$. The measurement sampling is subject to censoring phenomena and parametric uncertainties. This situation can be modeled by (8.1)–(8.4) with the following parameters.

$$C_{1,k} = \begin{bmatrix} 1 & 0 & 0 \end{bmatrix}, C_{2,k} = \begin{bmatrix} 0 & 1 & 0 \end{bmatrix}, C_{3,k} = \begin{bmatrix} 0 & 0 & 1 \end{bmatrix},$$

$$G_{1,k} = 0.1C_{1,k}, G_{2,k} = 0.1C_{2,k}, G_{3,k} = 0.1C_{3,k}, \mathcal{I}_2 = 50,$$

$$R_{1,k} = R_{2,k} = R_{3,k} = 1, Q_k = 0.0025I_3, P_0 = I_3, \mathcal{I}_3 = 0.5,$$

$$A_k = \begin{bmatrix} 1 & T & 0.5T^2 \\ 0 & 1 & T \\ 0 & 0 & 1 \end{bmatrix}, \bar{x}_0 = \begin{bmatrix} 10 \\ 0 \\ 0 \end{bmatrix}, F_k = 0.1I_3, \mathcal{I}_1 = 500.$$

Figure 8.6 plots the true values and estimates of the target position, velocity and acceleration, and Figure 8.7 sketches the RMSE curves of the FTKF and FTKFE after 1000 independent Monte Carlo trials. It is spotted from Figures 8.6–8.7 that in comparison with the FTKF, our FTKFE can always provide better tracking accuracy of the target position, velocity and acceleration, and moreover, the RMSE curves

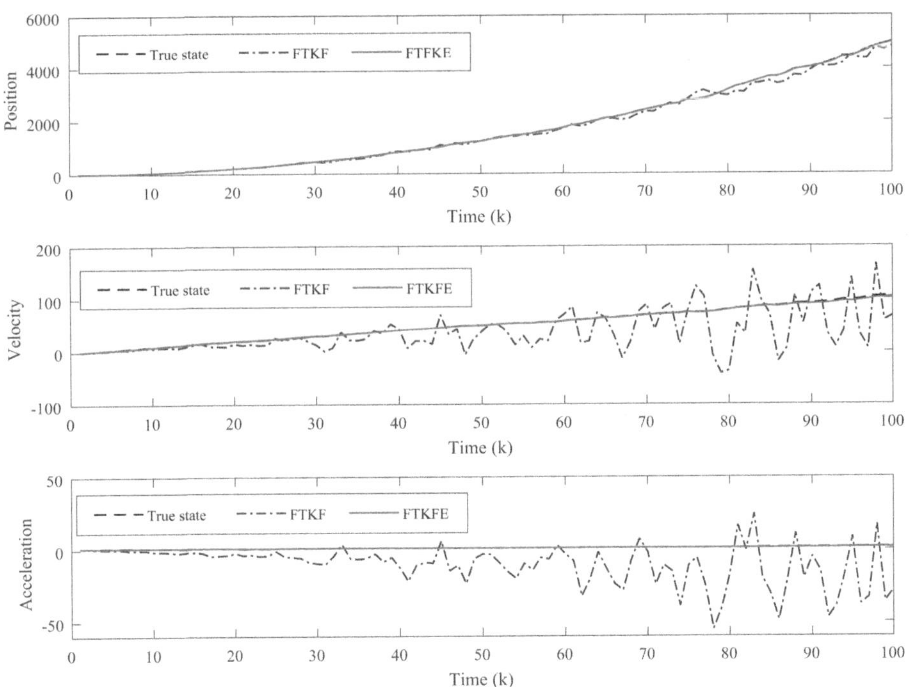

FIGURE 8.6 Position, velocity and acceleration estimation.

with respect to the target position, velocity and acceleration produced by our FTKFE always locate lower than those produced by the FTKF. These demonstration results obviously showcase the effectiveness and applicability of the proposed FTKFE in addressing the distributive target tracking problem subject to the measurement censoring and parametric uncertainties.

For different multiplicative noise covariances $\Sigma_{\beta_{i,k}}$ and censoring thresholds $[\mathcal{I}_1, \mathcal{I}_2, \mathcal{I}_3]$, the associated ARMSEs of our developed filtering fusion algorithm based on 1000 Monte Carlo trials (with each trial comprised of 100 time steps) are demonstrated in Tables 8.2–8.3. Looking at the two tables, we can draw the following conclusions: (1) as the noise covariance $\Sigma_{\beta_{i,k}}$ decreases, the filtering accuracy in regard to the target position, velocity and acceleration improves and (2) as the censoring thresholds $[\mathcal{I}_1, \mathcal{I}_2, \mathcal{I}_3]$ descend, the filtering accuracy with respect to the target position, velocity and acceleration increases. This is due to the fact that smaller multiplicative noise covariances and lower censoring thresholds will undoubtedly lead to performance improvement of the developed filtering fusion algorithm.

Remark 8.9 According to [182, 183], the posterior Cramer-Rao lower bound (PCRLB) is defined as the inverse of the Fisher information matrix (FIM) for a random vector and provides a lower bound on the mean squared error (MSE) of any estimator of the random parameter. If we want to obtain the PCRLB of our proposed

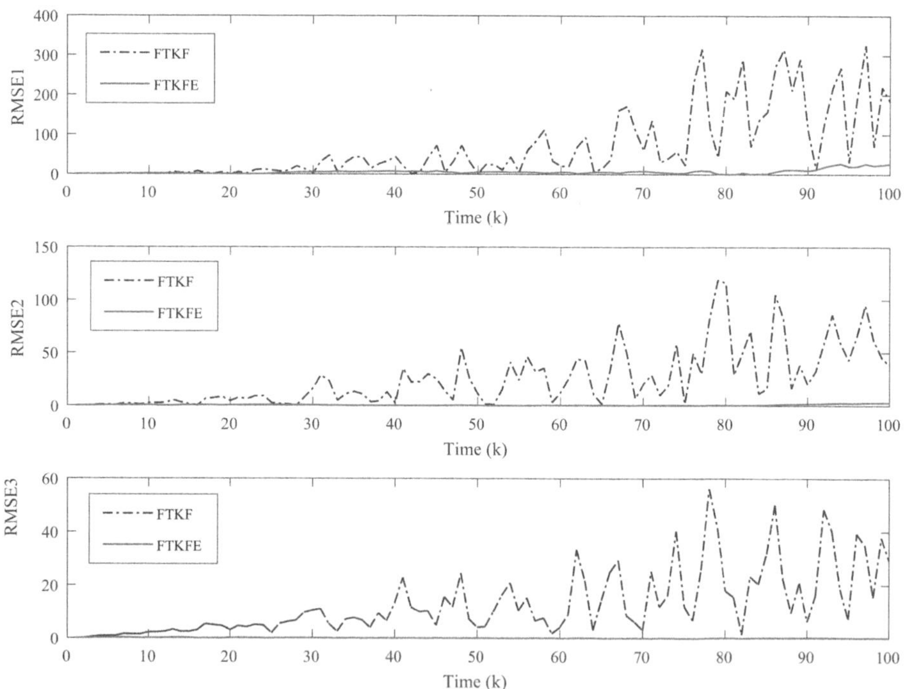

FIGURE 8.7 Comparison of RMSE1, RMSE2 and RMSE3.

TABLE 8.2
ARMSEs for Different Noise Covariances $R_{m,k}$

$\Sigma_{\beta_{i,k}}$	ARMSE1	ARMSE2	ARMSE3
9	17.7522	12.6909	1.8304
4	12.0548	9.2543	1.5251
1	8.1738	5.3240	1.0912

TABLE 8.3
ARMSEs for Different Thresholds $\left[\mathcal{I}_1,\mathcal{I}_2,\mathcal{I}_3\right]$

$\left[\mathcal{I}_1,\mathcal{I}_2,\mathcal{I}_3\right]$	ARMSE1	ARMSE2	ARMSE3
[500, 50, 0.5]	8.1738	5.3240	1.0912
[300, 30, 0.3]	4.7521	2.5625	0.6995
[100, 10, 0.1]	1.5020	0.9035	0.2015

filter, we would need to calculate the FIM by taking the expectation of the partial derivation of the joint PDF (of the measurements and states up to the current time) with respect to the system states. Unfortunately, it is extremely difficult (if not impossible) to find an analytic expression of the joint PDF as well as the PCRLB because of the nonlinearity embedded in the measurement censoring and multiplicative noises. As such, in this chapter, we turn to adopt approaches of using the orthogonality principle and calculating the conditional expectation in the filter design. Note that such approaches have been commonly used in the design of Kalman-type filters in the literature. Then, we assess the performance of our designed filter through (1) studying the monotonicity of the fused error covariance with respect to the censoring threshold; and (2) examining the boundedness property of the fused error covariance in the case of bounded system parameters and noise covariances. One of our future research topics is to further explore the possibility of utilizing the PCRLB to assess the performance of our designed fusion estimator.

Remark 8.10 Note that the FTKF algorithm is based on approaches of the federated Kalman filtering fusion and conventional Tobit Kalman filtering, while the proposed FTKFE algorithm is based on approaches of the federated Kalman filtering fusion and modified Tobit Kalman filtering with parametric uncertainties. In comparison with the conventional Tobit Kalman filtering, the modified Tobit Kalman filtering with parametric uncertainties introduces a suite of new terms including P_{x_k}, $\hat{\zeta}_{i,k}^-$ and $P_{\bar{z}_k^-}$, the computation of which mainly consists of matrix multiplications. As a result, both the conventional and modified Tobit Kalman filtering have the same time complexity of $\mathcal{O}\left(n_x^3\right)$. Additionally, one knows from the computation formula of the federated fusion that the federated Kalman filtering fusion has a time complexity of

$\mathcal{O}\left(pn_x^3\right)$ where n_x is the dimension of system state and p is the total number of sensors in the system. As such, we draw the conclusion that both the FTKF and FTKFE have the same time complexity of $\mathcal{O}\left(\left(p+1\right)n_x^3\right)$, but our proposed FTKFE has better computational accuracy and filtering performance, as shown in the simulation study.

8.5 SUMMARY

In this chapter, we have addressed the multi-sensor Tobit Kalman fusion estimation problem in the presence of censored observations and multiplicative noises. For each sensor, in order to accommodate the multiplicative noise effect, the conventional Tobit regression model has been rectified, and a unified framework has been constructed via the proposed LTKF. Thanks to the federated fusion rule, local estimates generated by all the LTKFs have been incorporated to obtain the FTKFE. Further assessment has been conducted by analyzing the monotonicity of the fused EEC with respect to the censoring thresholds and by discussing the boundedness of the fused EEC. The monotonicity and convergence results ride on not only the multi-sensor nature but also the censoring probability along with the multiplicative noise feature. Finally, the applicability of the FTKFE has been validated by a simulation experiment.

9 Protocol-Based Fusion Estimator Design for State-Saturated Systems with Dead-Zone-Like Censoring under Deception Attacks

In practical applications of the multi-sensor fusion estimation (MSFE), network components are often required to operate on, or transmit information via shared communication mediums. The ever-increasing usage of shared networks inevitably renders more chances of being maliciously attacked, which would then lead to hijacked/falsified/manipulated control commands and measurement outputs. As such, recurring research attention has recently been drawn towards secure filtering/control questions against diverse security threats, with typical examples including deception attacks, replay attacks and denial-of-service attacks. Amongst them, the deception attack stands out as the most hazardous one because, in this case, the adversaries are capable of arbitrarily injecting malicious data so as to impair or even destabilize the target systems.

In addition to cyber-attacks, information communication within a shared network often faces network-induced challenges (e.g. time delays and packet dropouts) due to limited communication capacity. In this regard, it is of momentous importance to leverage communication protocols with a view toward regulating data transmission. Among various communication protocols that have been adopted, the weighted try-once-discard protocol (WTODP) has proven to be particularly efficient because of its dynamic behavior for the urgency-dependent scheduling that is related to the significance of diverse missions. Unfortunately, so far, there have been very few results available on the protocol-based MSFE subject to deception attacks, leaving an open problem for us to research.

To date, most MSFE-related references have been based on an implicit assumption that the states of the target plants are not subjected to any constraints. This assumption is, however, likely to be violated in the case of state saturations resulting from physical limitations of devices. In fact, the state saturation is a typical system nonlinearity that occurs frequently in a great number of engineering systems whose state variables are forced to stay in a predefined bounded set. State saturations, if

DOI: 10.1201/9781003461623-9

inappropriately handled, might degrade (or even invalidate) the performance of the fusion estimation algorithms developed under the saturation-free conditions. As a result, some initial attempts have been made on state-saturated fusion estimation problems, but the corresponding results on the MSFE problems for nonlinear systems have been scarce, not to mention the case in which both the cyber-attack and the protocol scheduling are simultaneously taken into account.

The primary objective of this chapter is to design a protocol-based fusion estimator for state-saturated systems with dead-zone-like censoring phenomena under deception attacks. This appears to be a non-trivial task for the following challenges. (1) How to propose a new regression model that reflects the joint effects of the WTODP, dead-zone-like censoring and deception attacks? (2) How to handle the tight couplings of the state saturations, the constrained innovations, the covariance bounds as well as the filter gains? (3) How to prove the boundedness of the error covariance so as to evaluate the performance of the proposed fusion estimation framework? In the chapter, we are going to handle these identified challenges one by one and hope to come up with a desired protocol-based fusion estimator.

9.1 PROBLEM FORMULATION

Consider a state-saturated system described by the following model:

$$x_{k+1} = \sigma\left(A_k x_k\right) + \omega_k, \tag{9.1}$$

$$z_{m,k} = C_{m,k} x_k + \upsilon_{m,k}, \, m = 1, 2, \ldots, p \tag{9.2}$$

where $x_k \in \mathbb{R}^n$ and $z_{m,k} \in \mathbb{R}$ are, respectively, the state and observation vectors; p is the number of sensors; A_k and $C_{m,k}$ are known matrices; ω_k and $\upsilon_{m,k}$ are zero-mean white Gaussian noises with covariances Q_k and $R_{m,k}$, respectively. Here, $\sigma(\cdot):$ $\mathbb{R}^n \mapsto \mathbb{R}^n$ is the saturation function defined by

$$\sigma(t) \triangleq \begin{bmatrix} \sigma_1(t_1) & \sigma_2(t_2) & \cdots & \sigma_n(t_n) \end{bmatrix} \tag{9.3}$$

where, for $s = 1, 2, \ldots, n$,

$$\sigma_s(t_s) \triangleq sign(t_s) \min\left\{t_s^{\max}, |t_s|\right\}, \tag{9.4}$$

and $t_s^{\max}$ is the known saturation level of t_s which is the sth entry of t.

In the present research, the sensor measurements $z_{m,k} \in \mathbb{R}$ $(m = 1, 2, \ldots, p)$ are sent to a set of remote estimators via a shared communication network of limited bandwidth. To suppress data collisions from happening for better efficiency of resource utilization, the WTODP is deployed to regulate the data communication in the channel linking the sensor and the estimator. Under the WTODP scheduling, the communication priority of any measurement $z_{m,k}$ is determined according to the difference between the current sensor measure and the last stored sensor measure. At time k, the sensor having the highest priority is granted to update its measurement, and if multiple sensors are assigned with the highest priority, a randomly selected sensor is permitted to carry out the measurement update.

Define $\hbar_k \in \{1, 2, \ldots, p\}$ as the sensor acquiring the network access at time k and $\Gamma_{m,\hbar_k} \triangleq \delta(\hbar_k - m)$ $(m = 1, 2, \ldots, p)$ as the update coefficient regulating the scheduling behavior of sensor m with $\Omega_{m,k}$ being the update weight. According to the WTODP and the zero-order holder strategy, at time k, the *actual* measurement (from sensor m) arriving at the remote estimator is

$$\bar{z}_{m,k} = \sum_{j=0}^{p-1} \Gamma_{m,\hbar_{k-j}} z_{m,k-j}, \tag{9.5}$$

where $\hbar_k = \arg\max_{1 \le m \le p} \left(z_{m,k} - \bar{z}_{m,k-1}\right)^T \Omega_{m,k} \left(z_{m,k} - \bar{z}_{m,k-1}\right)$ with "arg" standing for "argument" and indicating that the value of m maximizes $\left(z_{m,k} - \bar{z}_{m,k-1}\right)^T \Omega_{m,k} \left(z_{m,k} - \bar{z}_{m,k-1}\right)$ for $k - j \le 0$, $\hbar_{k-j} \triangleq j$ and $z_{m,k-j} \triangleq z_{m,0}$.

As a matter of fact, owing to the opening-up feature of the network environment, data transmission within the network is quite vulnerable to cyber-attacks launched by adversaries. In other words, $\bar{z}_{m,k}$ in (9.5) may be intentionally falsified by the so-called randomly occurring deception attacks during its transmission over the network, and this attacking behavior can often be modeled by

$$\bar{y}_{m,k} = \bar{z}_{m,k} + \gamma_{m,k} u_{m,k}, \tag{9.6}$$

where $\bar{y}_{m,k}$ are measurements undergoing contamination, and $u_{m,k} \in \mathbb{R}$ are deception attack signals of the form:

$$u_{m,k} = -\bar{z}_{m,k} + \xi_{m,k}, \tag{9.7}$$

with $\xi_{m,k} \in \mathbb{R}$ being zero-mean white Gaussian noises with covariances $\Xi_{m,k}$. Notice that $u_{m,k}$ might not always take effect owing to intricate defense measures and network conditions. From the defenders' viewpoint, $u_{m,k}$ are likely to appear in a random fashion and, thus, Bernoulli-distributed variables $\gamma_{m,k}$ are utilized to characterize the randomly occurring attacking behavior with probability distributions

$$\begin{cases} \mathrm{Prob}\{\gamma_{m,k} = 1\} = \bar{\gamma}_{m,k}, \\ \mathrm{Prob}\{\gamma_{m,k} = 0\} = 1 - \bar{\gamma}_{m,k}, \end{cases} \tag{9.8}$$

where $\bar{\gamma}_{m,k} \in [0, 1]$ are known constants.

At the input terminal of each local estimator, an additional detection device is equipped with the aim to check whether $\bar{y}_{m,k}$ need to be censored or not. Accordingly, the Tobit model with two-side censoring is

$$y_{m,k} = \begin{cases} \tau_m^l, & \bar{y}_{m,k} \le \tau_m^l, \\ \bar{y}_{m,k}, & \tau_m^l < \bar{y}_{m,k} < \tau_m^r, \\ \tau_m^r, & \bar{y}_{m,k} \ge \tau_m^r, \end{cases} \tag{9.9}$$

where $y_{m,k}$ are censored observations, and τ_m^l and τ_m^r are, respectively, the left- and right-censoring thresholds with regard to $y_{m,k}$.

Bearing in mind model (9.9), we define the following Bernoulli random variables $\beta_{m,k}^l$ and $\beta_{m,k}^r$ to regulate the dead-zone-like censoring phenomena of $\bar{y}_{m,k}$.

$$\beta_{m,k}^{l} = \begin{cases} 1, & \overline{y}_{m,k} \leq \tau_m^l, \\ 0, & \overline{y}_{m,k} > \tau_m^l, \end{cases} \tag{9.10}$$

$$\beta_{m,k}^{r} = \begin{cases} 1, & \overline{y}_{m,k} \geq \tau_m^r, \\ 0, & \overline{y}_{m,k} < \tau_m^r \end{cases} \tag{9.11}$$

with the following probability distributions:

$$\begin{cases} \mathrm{Prob}\{\beta_{m,k}^{l} = 1\} = \overline{\beta}_{m,k}^{l}, \\ \mathrm{Prob}\{\beta_{m,k}^{r} = 1\} = \overline{\beta}_{m,k}^{r}, \\ \mathrm{Prob}\{\beta_{m,k}^{l} = 0\} = 1 - \overline{\beta}_{m,k}^{l}, \\ \mathrm{Prob}\{\beta_{m,k}^{r} = 0\} = 1 - \overline{\beta}_{m,k}^{r}. \end{cases} \tag{9.12}$$

where $\beta_{m,k}^{l}$ and $\beta_{m,k}^{l}$ are known non-negative constants which are uncorrelated with the noise ω_k. Based on (9.10)–(9.11), $y_{m,k}$ in (9.9) can be rewritten as follows:

$$y_{m,k} = \left(1 - \beta_{m,k}^{l} - \beta_{m,k}^{r}\right)\overline{y}_{m,k} + \beta_{m,k}^{l}\tau_m^l + \beta_{m,k}^{r}\tau_m^r. \tag{9.13}$$

Assumption 9.1 The initial state x_0 has the mean $\overline{x}_0$ and covariances $P_{\tilde{x}_0}$. The random variables x_0, $\xi_{m,k}$, $\gamma_{m,k}$, ω_k and $\upsilon_{m,k}$ are mutually independent.

Let $\hat{x}_{m,k} \triangleq \mathbb{E}\{x_k \mid y_{m,1:k}\}$, $\hat{x}_{m,k}^- \triangleq \mathbb{E}\{x_k \mid y_{m,1:k-1}\}$, be, respectively, the estimate and one-step prediction of x_k based on the information from sensor m where $y_{m,1:k} \triangleq \{y_{m,1}, y_{m,2}, \dots, y_{m,k}\}$ is the measurement sequence of sensor m till time k. Denote $\tilde{x}_{m,k} \triangleq x_k - \hat{x}_{m,k}$, $\tilde{x}_{m,k}^- \triangleq x_k - \hat{x}_{m,k}^-$, $P_{\tilde{x}_{m,k}} \triangleq \mathbb{E}\{\tilde{x}_{m,k}\tilde{x}_{m,k}^T\}$, $P_{\tilde{x}_{m,k}^-} \triangleq \mathbb{E}\{\tilde{x}_{m,k}^-(\tilde{x}_{m,k}^-)^T\}$, respectively, as the local filtering error, prediction error, FEC and prediction error covariance.

In this chapter, we are interested in devising a fusion estimator of the following structure based on the dynamical system (9.1) and the measurement model (9.13).

$$\begin{cases} \hat{x}_k = \sum_{m=1}^{p} W_{m,k}\hat{x}_{m,k}, \\ \hat{x}_{m,k}^- = \sigma\left(A_{k-1}\hat{x}_{m,k-1}\right), \\ \hat{x}_{m,k} = \hat{x}_{m,k}^- + K_{m,k}\left(y_{m,k} - \hat{y}_{m,k}^-\right), \end{cases} \tag{9.14}$$

where $\hat{x}_k$ is the fused estimate attained under the federated fusion rule, and the matrix weights $W_{m,k}$ and filter gains $K_{m,k}$ are to be determined later.

Remark 9.1 The dead-zone-like censoring modeled by (9.9), which is widely encountered in systems equipped with commercial yet cheap off-the-shelf sensors, has its distinctive feature of censored Gaussian (rather than pure Gaussian) noise distribution at/near the censored region, and this circumvents the direct employment of the conventional Kalman filter. In addition to dead-zone-like censoring, state saturations (stemming possibly from physical constraints on system states) may also arise in system modeling, and such kind of saturations, if inadequately treated, will undoubtedly provoke deteriorated (or even invalidated) performance of the designed fusion estimator. Note that the situation could become even worse in the case of deception attacks as the false data injected by adversaries unavoidably leads to false measurement signals at the input terminal of the filter. As such, there is a practical need to establish a holistic protocol-based fusion estimation framework to handle the dead-zone-like censoring, deception attacks and state saturations.

Denote

$$\hat{\eta}_{m,k}^- \triangleq \left(1-\overline{\gamma}_{m,k}\right)\Gamma_{m,\hbar_k}C_{m,k}\hat{x}_{m,k}^- + \left(1-\overline{\gamma}_{m,k}\right)\sum_{j=1}^{p-1}\Gamma_{m,\hbar_{k-j}}C_{m,k-j}\hat{x}_{m,k-j},$$

$$\mathcal{R}_{m,k} \triangleq \overline{\gamma}_{m,k}\left(1-\overline{\gamma}_{m,k}\right)\left(\sum_{j=0}^{p-1}\Gamma_{m,\hbar_{k-j}}^2 R_{m,k-j} + \Xi_{m,k}\right).$$

As can be seen from (9.13), the random variables $\beta_{m,k}^l$ and $\beta_{m,k}^r$ are put forward to describe the dead-zone-like censoring of $\overline{y}_{m,k}$. More specifically, if $\overline{y}_{m,k}$ has censoring, i.e. $\beta_{m,k}^l = \beta_{m,k}^r = 0$, one has $y_{m,k} = \overline{y}_{m,k}$, which means that the value of the measurement equals that of the latent measurement. If $\overline{y}_{m,k}$ has left censoring, i.e. $\beta_{m,k}^l = 1$ and $\beta_{m,k}^r = 0$, one has $y_{m,k} = \tau_m^l$, which implies that the threshold τ_m^l is assigned to the sensor measurement. If $\overline{y}_{m,k}$ has right censoring, i.e. $\beta_{m,k}^l = 0$ and $\beta_{m,k}^r = 1$, one has $y_{m,k} = \tau_m^r$, which indicates that the threshold τ_m^r is assigned to the sensor measurement. Note that the censoring probabilities $\overline{\beta}_{m,k}^l$ and $\overline{\beta}_{m,k}^r$ are assumed to be known *a priori* either via statistical experiments or through the following approximations.

$$\begin{cases} \overline{\beta}_{m,k}^l \approx \Phi\left(\dfrac{\tau_m^l - \hat{\eta}_{m,k}^-}{\sqrt{\mathcal{R}_{m,k}}}\right), \\[4mm] \overline{\beta}_{m,k}^r \approx \Phi\left(\dfrac{\hat{\eta}_{m,k}^- - \tau_m^r}{\sqrt{\mathcal{R}_{m,k}}}\right), \end{cases} \tag{9.15}$$

Remark 9.2 The presented local filter structure (9.14) accounts for the effects from the dead-zone-like censoring, state saturations, deception attacks and WTODP. Specifically, the state-saturated filter form is employed to better reflect the saturation phenomenon of the interested plant, and the measurement $y_{m,k}$ embraces the scheduling rule of the WTODP, the occurring fashion of the dead-zone-like censoring and the injection manner of the deception attack.

The objective of this chapter is to devise the protocol-based fusion estimator (9.14) such that (1) upper bounds on both local and fused FECs are guaranteed; (2) appropriate filter parameters $W_{m,k}$ and $K_{m,k}$ are determined through minimizing the attained upper bounds; and (3) the fusion estimation performance is ensured through assessing the boundedness property of the minimized upper bound on the fused FEC.

9.2 MAIN RESULTS

In this section, we will first derive the upper bounds on both local and fused FECs through solving certain matrix difference equations then design the filter parameters to minimize the attained upper bounds and finally investigate the boundedness of the upper bound on the fused FEC.

Lemma 9.1 [28] For $\forall a, b \in \mathbb{R}$, there exists a real number $c \in [0,1]$ such that $\sigma(a) - \sigma(b) = c(a-b)$ where $\sigma(\cdot)$ is defined by (9.4).

Lemma 9.2 [184] Let a matrix function $\varpi_k(X): \mathbb{R}^{n \times n} \to \mathbb{R}^{n \times n}$ be given such that $\varpi_k(X) \le \varpi_k(Y)$ holds for any matrices X and Y satisfying $X = X^T \ge 0$, $Y = Y^T \ge 0$ and $X \le Y$. Then, given the initial condition $X_0 = Y_0$, there exist solutions X_k and Y_k to difference equations $X_{k+1} = \varpi_k(X_k)$ and $Y_{k+1} \le \varpi_k(Y_k)$ such that $X_k \le Y_k$ holds for all $k \ge 0$.

Denote

$$\eta_{m,k} \triangleq \left(1 - \gamma_{m,k}\right) \sum_{j=0}^{p-1} \Gamma_{m,h_{k-j}} C_{m,k-j} x_{k-j},$$

$$\beta_{m,k} \triangleq 1 - \beta_{m,k}^l - \beta_{m,k}^r, \vartheta_{m,k}^l \triangleq \frac{\tau_m^l - \eta_{m,k}}{\mathcal{R}_{m,k}}, \vartheta_{m,k}^r \triangleq \frac{\tau_m^r - \eta_{m,k}}{\mathcal{R}_{m,k}},$$

$$P_{\hat{y}_{m,k}} \triangleq \mathbb{E}\left\{\left(y_{m,k} - \hat{y}_{m,k}\right)\left(y_{m,k} - \hat{y}_{m,k}\right)^T\right\}, \bar{y}_{m,k} \triangleq \eta_{m,k} + v_{m,k},$$

$$v_{m,k} \triangleq \left(1 - \gamma_{m,k}\right) \sum_{j=0}^{p-1} \Gamma_{m,h_{k-j}} \upsilon_{m,k-j} + \gamma_{m,k} \xi_{m,k}, \hat{y}_{m,k} \triangleq \mathbb{E}\left\{y_{m,k} \mid \eta_{m,k}\right\}.$$

In the following, we formulate a generalized Tobit regression model to cater for the joint effects from the dead-zone-like censoring, the state saturations, the deception attacks and the adopted WTODP.

Lemma 9.3 The expectation and the variance of y_k conditioned on $\eta_{m,k}$ are given by

$$\hat{y}_{m,k} = \beta_{m,k}^l \tau_m^l + \beta_{m,k}^r \tau_m^r + \beta_{m,k}\left[\eta_{m,k} - \sqrt{\mathcal{R}_{m,k}} \lambda\left(\vartheta_{m,k}^r, \vartheta_{m,k}^l\right)\right], \tag{9.16}$$

$$P_{\hat{y}_{m,k}} = \mathcal{R}_{m,k}\left[1 + \varphi\left(\vartheta_{m,k}^r, \vartheta_{m,k}^l\right)\right], \tag{9.17}$$

where

$$\lambda\left(\vartheta_{m,k}^r, \vartheta_{m,k}^l\right) = \frac{\phi\left(\vartheta_{m,k}^r\right) - \phi\left(\vartheta_{m,k}^l\right)}{\Phi\left(\vartheta_{m,k}^r\right) - \Phi\left(\vartheta_{m,k}^l\right)}, \tag{9.18}$$

$$\varphi\left(\vartheta_{m,k}^r, \vartheta_{m,k}^l\right) = \frac{\vartheta_{m,k}^l \phi\left(\vartheta_{m,k}^l\right) - \vartheta_{m,k}^r \phi\left(\vartheta_{m,k}^r\right)}{\Phi\left(\vartheta_{m,k}^r\right) - \Phi\left(\vartheta_{m,k}^l\right)} - \lambda^2\left(\vartheta_{m,k}^r, \vartheta_{m,k}^l\right). \tag{9.19}$$

Here, $\phi(\cdot)$ and $\Phi(\cdot)$ are, respectively, the probability density functions (PDFs) and CDFs of Gaussian random variables $\vartheta_{m,k} \in \left\{ \vartheta_{m,k}^r, \vartheta_{m,k}^l \right\}$,

$$\phi\left(\vartheta_{m,k}\right) = \frac{1}{\sqrt{2\pi}} e^{-\frac{\vartheta_{m,k}^2}{2}}, \tag{9.20}$$

$$\Phi\left(\vartheta_{m,k}\right) = \int_{-\infty}^{\tau} \frac{1}{\sqrt{2\pi}} e^{-\frac{\vartheta_{m,k}^2}{2}} d_{\vartheta_{m,k}}, \tag{9.21}$$

and $\tau_m \in \left\{ \tau_m^r, \tau_m^l \right\}$.

Proof The definition of $v_{m,k}$ tells us that $v_{m,k}$ are zero-mean Gaussian noises with covariances $\mathcal{R}_{m,k}$. Keeping (9.13) and $\overline{y}_{m,k} \triangleq \eta_{m,k} + v_{m,k}$ in mind, the PDFs of $y_{m,k}$ are

$$\begin{aligned}
f\left(y_{m,k} \mid \eta_{m,k}\right) = {} & \frac{1}{\sqrt{\mathcal{R}_{m,k}}} \phi\left(\frac{y_{m,k} - \eta_{m,k}}{\sqrt{\mathcal{R}_{m,k}}}\right) \alpha\left(y_{m,k} - \tau_m^l\right) \alpha\left({}_m\tau^r - y_{m,k}\right) \\
& + \delta\left(y_{m,k} - \tau_m^l\right) \Phi\left(\frac{y_{m,k} - \eta_{m,k}}{\sqrt{\mathcal{R}_{m,k}}}\right) \\
& + \delta\left(y_{m,k} - \tau_m^r\right) \left(1 - \Phi\left(\frac{y_{m,k} - \eta_{m,k}}{\sqrt{\mathcal{R}_{m,k}}}\right)\right),
\end{aligned} \tag{9.22}$$

where $\alpha\left(y_{m,k} - \tau_m^l\right)$ and $\alpha\left(\tau_m^r - y_{m,k}\right)$ are unit step functions, and $\phi\left(\dfrac{y_{m,k} - \eta_{m,k}}{\sqrt{\mathcal{R}_{m,k}}}\right)$ and $\Phi\left(\dfrac{y_{m,k} - \eta_{m,k}}{\sqrt{\mathcal{R}_{m,k}}}\right)$ are calculated by (9.20)–(9.21).

It follows from (9.10)–(9.11) that (9.22) can be reformulated as

$$\begin{aligned}
f\left(y_{m,k} \mid \eta_{m,k}\right) = {} & \beta_{m,k} \frac{1}{\sqrt{\mathcal{R}_{m,k}}} \frac{\phi\left(\dfrac{y_{m,k} - \eta_{m,k}}{\sqrt{\mathcal{R}_{m,k}}}\right)}{\Phi\left(\vartheta_{m,k}^r\right) - \Phi\left(\vartheta_{m,k}^l\right)} \\
& + \beta_{m,k}^l \frac{\Phi\left(\dfrac{y_{m,k} - \eta_{m,k}}{\sqrt{\mathcal{R}_{m,k}}}\right)}{\Phi\left(\vartheta_{m,k}^r\right)} + \beta_{m,k}^r \frac{\left(1 - \Phi\left(\dfrac{y_{m,k} - \eta_{m,k}}{\sqrt{\mathcal{R}_{m,k}}}\right)\right)}{1 - \Phi\left(\vartheta_{m,k}^r\right)},
\end{aligned} \tag{9.23}$$

based on which we obviously have (9.16), where $\lambda\left(\vartheta_{m,k}^r, \vartheta_{m,k}^l\right)$ is computed by (9.18).

Noticing (9.16), one has

$$\mathbb{E}\left\{y_{m,k}|\eta_{m,k},\beta_{m,k}^{l}=1,\beta_{m,k}^{r}=0\right\}=\tau_{m}^{l}, \tag{9.24}$$

$$\mathbb{E}\left\{y_{m,k}|\eta_{m,k},\beta_{m,k}^{l}=0,\beta_{m,k}^{r}=1\right\}=\tau_{m}^{r}, \tag{9.25}$$

$$\mathbb{E}\left\{y_{m,k}|\eta_{m,k},\beta_{m,k}^{l}=0,\beta_{m,k}^{r}=0\right\}=\eta_{m,k}-\sqrt{\mathcal{R}_{m,k}}\,\lambda\left(\vartheta_{m,k}^{r},\vartheta_{m,k}^{l}\right). \tag{9.26}$$

Moving forward, one obtains

$$var\left\{y_{m,k}|\eta_{m,k},\beta_{m,k}^{l}=1,\beta_{m,k}^{r}=0\right\}=0, \tag{9.27}$$

$$var\left\{y_{m,k}|\eta_{m,k},\beta_{m,k}^{l}=0,\beta_{m,k}^{r}=1\right\}=0, \tag{9.28}$$

$$\begin{aligned}
&var\left\{y_{m,k}|\eta_{m,k},\beta_{m,k}^{l}=0,\beta_{m,k}^{r}=0\right\}\\
&=\mathbb{E}\left\{y_{m,k}^{2}|\eta_{m,k},\beta_{m,k}^{l}=0,\beta_{m,k}^{r}=0\right\}-\left(\mathbb{E}\left\{y_{m,k}|\eta_{m,k}\beta_{m,k}^{l}=0,\beta_{m,k}^{r}=0\right\}\right)^{2}.
\end{aligned} \tag{9.29}$$

The notice of (9.26) along with (9.29) tells

$$\begin{aligned}
&\mathbb{E}\left\{y_{m,k}^{2}|\eta_{m,k},\beta_{m,k}^{l}=0,\beta_{m,k}^{r}=0\right\}\\
&=\eta_{m,k}^{2}+\mathcal{R}_{m,k}-\sqrt{\mathcal{R}_{m,k}}\,\eta_{m,k}\lambda\left(\vartheta_{m,k}^{r},\vartheta_{m,k}^{l}\right)\\
&\quad+\sqrt{\mathcal{R}_{k}}\,\frac{\left(\tau_{m}^{l}\phi\left(\vartheta_{m,k}^{l}\right)-\tau_{m}^{r}\phi\left(\vartheta_{m,k}^{r}\right)\right)}{\Phi\left(\vartheta_{m,k}^{r}\right)-\Phi\left(\vartheta_{m,k}^{l}\right)}.
\end{aligned} \tag{9.30}$$

Inserting (9.16) and (9.30) into (9.29) generates

$$var\left\{y_{m,k}|\eta_{m,k},\beta_{m,k}^{l}=0,\beta_{m,k}^{r}=0\right\}=\mathcal{R}_{m,k}\left[1+\varphi\left(\vartheta_{m,k}^{r},\vartheta_{m,k}^{l}\right)\right], \tag{9.31}$$

where $\varphi\left(\vartheta_{m,k}^{r},\vartheta_{m,k}^{l}\right)$ is provided by (9.19). Combining (9.27)–(9.28) and (9.31) together, one has (9.17).

Remark 9.3 The proposed generalized regression model given by Lemma 9.3 contains expectations and variances of $y_{m,k}$ conditional on $\eta_{m,k}$ and censoring sequences $\beta_{m,1:k}^{l}$ and $\beta_{m,1:k}^{r}$. Note that the regression model in the standard TKF deals with the left-censoring issue with unknown $\beta_{1:k}^{l}$. Different from the standard TKF, in this chapter, the model (9.16)–(9.17) involves variables (e.g. $\beta_{m,k}^{r}$ and $\vartheta_{m,k}^{r}$) induced by both left and right censoring owing to the dead-zone-like censoring issue. Also, the left-censoring probabilities $\bar{\beta}_{m,k}^{l}$ is now replaced by the true censoring variables $\beta_{m,k}^{l}$ and $\beta_{m,k}^{r}$ as a result from the acknowledgement of $\beta_{m,1:k}^{l}$ and $\beta_{m,1:k}^{r}$. Furthermore, because of the consideration of the deception attacks and WTODP, $C_{k}x_{k}$ (the product of the measurement coefficient C_{k} and state x_{k}) and R_{k} (the measurement noise covariance) are now replaced by $\eta_{m,k}$ and $\mathcal{R}_{m,k}$, respectively. Note that these new features would bring in additional difficulties to the subsequent algorithm development.

Denote

$$\tilde{v}_{m,k} \triangleq \beta_{m,k} v_{m,k} + \beta_{m,k} \sqrt{\overline{\mathcal{R}}_{m,k}} \lambda\left(\hat{\vartheta}^r_{m,k}, \hat{\vartheta}^l_{m,k}\right),$$

$$\hat{\vartheta}^l_{m,k} \triangleq \frac{\tau^l_m - \hat{\eta}^-_{m,k}}{\overline{\mathcal{R}}_{m,k}}, \hat{\vartheta}^r_{m,k} \triangleq \frac{\tau^r_m - \hat{\eta}^-_{m,k}}{\overline{\mathcal{R}}_{m,k}}, t \triangleq \sum_{s=1}^{n} \left(t^{\max}_s\right)^2.$$

With the newly derived Tobit regression model in hand, we then move forward to present the calculation formula of the local FEC $P_{\tilde{x}_{m,k}}$.

Lemma 9.4 The local FEC of estimator (9.14) is given by

$$
\begin{aligned}
P_{\tilde{x}_{m,k}} &= \left(I - K_{m,k}\beta_{m,k}\left(1 - \overline{\gamma}_{m,k}\right)\Gamma_{m,h_k} C_{m,k}\right) P_{\tilde{x}^-_{m,k}} \\
&\quad \times \left(I - K_{m,k}\beta_{m,k}\left(1 - \overline{\gamma}_{m,k}\right)\Gamma_{m,h_k} C_{m,k}\right)^T \\
&\quad + K_{m,k} P_{\tilde{v}_{m,k}} K^T_{m,k} + \overline{\gamma}_{m,k}\left(1 - \overline{\gamma}_{m,k}\right)\beta^2_{m,k} K_{m,k} \\
&\quad \times \sum_{j=1}^{p-1} \Gamma_{m,h_{k-j}} C_{m,k-j} P_{x_{k-j}} C^T_{m,k-j} \Gamma^T_{m,h_{k-j}} K^T_{m,k} \\
&\quad + \beta^2_{m,k} bar\gamma_{m,k}\left(1 - \overline{\gamma}_{m,k}\right) K_{m,k} \sum_{j=1}^{p-1} \Gamma_{m,h_{k-j}} C_{m,k-j} \\
&\quad \times P_{\tilde{x}_{m,k-j}} C^T_{m,k-j} \Gamma^T_{m,h_{k-j}} K^T_{m,k} + \Pi_{m,k} + \Pi^T_{m,k},
\end{aligned}
\tag{9.32}
$$

where

$$P_{\tilde{v}_{m,k}} = \mathcal{R}_{m,k}\left[1 + \varphi\left(\hat{\vartheta}^r_{m,k}, \hat{\vartheta}^l_{m,k}\right)\right], \tag{9.33}$$

$$\Pi_{m,k} = \mathbb{E}\left\{\left(I - K_{m,k}\beta_{m,k}\left(1 - \overline{\gamma}_{m,k}\right)\right)\tilde{x}^-_{m,k}\tilde{v}^T_{m,k} K^T_{m,k}\right\}, \tag{9.34}$$

$$P_{x_k} = P_{\tilde{x}_{m,k}} + \mathbb{E}\left\{\hat{x}_{m,k}\tilde{x}^T_{m,k}\right\} + \mathbb{E}\left\{\hat{x}_{m,k}\tilde{x}^T_{m,k}\right\} + \hat{x}_{m,k}\tilde{x}^T_{m,k}, \tag{9.35}$$

$$P_{\tilde{x}^-_{m,k+1}} = \mathbb{E}\left\{c_{m,k} A_k \tilde{x}_{m,k}\tilde{x}^T_{m,k} A^T_k c_{m,k}\right\} + Q_{m,k}, \tag{9.36}$$

Here, $Q_{m,k}$ is the redistributed process noise covariance to be determined later, and $c_{m,k} \in [0,1]$ is a real number.

Proof Defining the expectation of measurement $y_{m,k}$ conditioned on the measurement sequence $y_{1:m,k-1}$ by $\hat{y}^-_{m,k} \triangleq \mathbb{E}\{y_{m,k} \mid y_{1:m,k-1}\}$ and noting that (9.16) exhibits the expectation of measurement $y_{m,k}$ conditioned on the pseudo-state $\eta_{m,k}$, we have

$$\hat{y}^-_{m,k} \approx \beta^l_{m,k}\tau^l_m + \beta^r_{m,k}\tau^r_m + \beta_{m,k}\left[\hat{\eta}^-_{m,k} - \sqrt{\overline{\mathcal{R}}_{m,k}} \lambda\left(\hat{\vartheta}^r_{m,k}, \hat{\vartheta}^l_{m,k}\right)\right], \tag{9.37}$$

where the approximation holds as the pseudo-state $\eta_{m,k}$ can be approximated by $\hat{\eta}_{m,k}^{-} \triangleq \mathbb{E}\{\eta_{m,k} \mid y_{1:m,k-1}$ given the available measurement information $y_{1:m,k-1}$.

Remembering the definitions of $\bar{y}_{m,k}$ and $\tilde{v}_{m,k}$, and subtracting (9.37) from (9.6), one obtains

$$\tilde{y}_{m,k}^{-} = \beta_{m,k}\tilde{\eta}_{m,k}^{-} + \tilde{v}_{m,k},\tag{9.38}$$

where

$$\begin{aligned}
\tilde{\eta}_{m,k}^{-} &\triangleq \eta_{m,k} - \hat{\eta}_{m,k}^{-} \\
&= \left(\bar{\gamma}_{m,k} - \gamma_{m,k}\right)\Gamma_{m,\hbar_k}C_{m,k}x_k + \left(1 - \bar{\gamma}_{m,k}\right)\Gamma_{m,\hbar_k}C_{m,k}\tilde{x}_{m,k}^{-} \\
&\quad + \left(\bar{\gamma}_{m,k} - \gamma_{m,k}\right)\sum_{j=1}^{p-1}\Gamma_{m,\hbar_{k-j}}C_{m,k-j}x_{k-j} \\
&\quad + \left(1 - \bar{\gamma}_{m,k}\right)\sum_{j=1}^{p-1}\Gamma_{m,\hbar_{k-j}}C_{m,k-j}\tilde{x}_{m,k-j}^{-}.
\end{aligned}\tag{9.39}$$

Based on dynamic equation (9.1) and filter (9.14), the local FEC is derived as

$$\tilde{x}_{m,k+1} = \tilde{x}_{m,k+1}^{-} - K_{k+1}\left(y_{m,k+1} - \hat{y}_{m,k+1}^{-}\right).\tag{9.40}$$

The error dynamics in (9.40) can be further expanded as

$$\begin{aligned}
\tilde{x}_{m,k} &= \tilde{x}_{m,k}^{-} - K_{m,k}\beta_{m,k}\tilde{\eta}_{m,k}^{-} - K_{m,k}\tilde{v}_{m,k} \\
&= \left(I - K_{m,k}\beta_{m,k}\left(1 - \bar{\gamma}_{m,k}\right)\Gamma_{m,\hbar_k}C_{m,k}\right)\tilde{x}_{m,k}^{-} - K_{m,k}\tilde{v}_{m,k} \\
&\quad - K_{m,k}\beta_{m,k}\left(\bar{\gamma}_{m,k} - \gamma_{m,k}\right)\sum_{j=0}^{p-1}\Gamma_{m,\hbar_{k-j}}C_{m,k-j}x_{k-j} \\
&\quad - K_{m,k}\beta_{m,k}\left(1 - \bar{\gamma}_{m,k}\right)\sum_{j=1}^{p-1}\Gamma_{m,\hbar_{k-j}}C_{m,k-j}\tilde{x}_{m,k-j},
\end{aligned}\tag{9.41}$$

which gives rise to (9.32) where $P_{x_k} \triangleq \mathbb{E}\{x_k x_k^{T}\}$, $P_{\tilde{v}_{m,k}} \triangleq \mathbb{E}\{\tilde{v}_{m,k}\tilde{v}_{m,k}^{T}\}$ and

$$\Pi_{m,k} = \mathbb{E}\left\{\left(I - K_{m,k}\beta_{m,k}\left(1 - \bar{\gamma}_{m,k}\right)\right)\tilde{x}_{m,k}^{-}\tilde{v}_{m,k}^{T}K_{m,k}^{T}\right\}.\tag{9.42}$$

Paying attention to dynamic equation (9.1), filter (9.14) and Lemma 9.3, we have

$$\begin{aligned}
\tilde{x}_{m,k+1}^{-} &= \sigma\left(A_k x_k\right) + \omega_k - \sigma\left(A_k \hat{x}_k\right) \\
&= c_{m,k}A_k\tilde{x}_{m,k} + \omega_k,
\end{aligned}\tag{9.43}$$

where $c_{m,k} \in [0,1]$ are real numbers. As a result, terms P_{x_k} and $P_{\tilde{x}_{m,k+1}^-}$ on the right-hand side of (9.42) are calculated by (9.35) and (9.36).

Additionally, it follows from the definition of $\tilde{v}_{m,k}$ that

$$
\begin{aligned}
\tilde{v}_{m,k} &\approx \beta_{m,k}\left(\eta_{m,k}+v_{m,k}\right)+\beta_{m,k}^l \tau_m^l +\beta_{km,}^r \tau_m^r -\beta_{m,k}^l \tau_m^l \\
&\quad -\beta_k^r \tau^r -\beta_k\left[\eta_{m,k}-\sqrt{\bar{\mathcal{R}}_{m,k}}\,\lambda\left(\vartheta_{m,k}^r,\vartheta_{m,k}^l\right)\right] \\
&= y_{m,k}-\mathbb{E}\left\{y_{m,k}\mid\eta_{m,k}\right\},
\end{aligned}
\tag{9.44}
$$

which gives rise to

$$
P_{\tilde{v}_{m,k}} \triangleq \mathbb{E}\left\{\tilde{v}_{m,k}\tilde{v}_{m,k}^T\right\} = P_{y_{m,k}},
\tag{9.45}
$$

which is approximated by (9.33).

It is worth noting that the local FEC presented in Lemma 9.4 is actually stochastic because of its dependence on random left- and right-censoring variables $\beta_{m,k}^l$ and $\beta_{m,k}^r$, and this indicates that the local FEC, filter gain as well as state estimate effectively become functions of $\beta_{m,k}^l$ and $\beta_{m,k}^r$. Besides, due to the existence of the state saturation and the cross-terms in (9.32), it is technically difficult to find the desired filter parameters by directly minimizing the local FEC (9.32). As a result, an alternative way to continue the design procedure of filter (9.14) is to (1) explore the statistical property of $\mathcal{P}_{\tilde{x}_k}^{\min}$ via the evaluation index $\mathbb{E}\left\{P_{\tilde{x}_k}\right\}$; (2) find the upper bound on such an evaluation index; and (3) minimize the resultant upper bound so as to obtain the desired filter parameters. For the reason of notation brevity, let us define $M_{m,k} \triangleq \mathbb{E}\left\{P_{\tilde{x}_{m,k}}\right\}$ and $\bar{\beta}_{m,k} \triangleq 1-\bar{\beta}_{m,k}^l -\bar{\beta}_{m,k}^r$.

Theorem 9.1 Let scalars $a_{m,k}>0$ and $f_{m,k}>0$ and initial values $M_{m,0}^u>0$ be given. Assume that there exist matrix sequences $M_{m,k}^u$ satisfying the following difference equation:

$$
\begin{aligned}
M_{m,k}^u =&\left(1+a_{m,k}\right)\left(I-K_{m,k}\bar{\beta}_{m,k}\left(1-\bar{\gamma}_{m,k}\right)\Gamma_{m,h_k}C_{m,k}\right) \\
&\times N_{m,k}^u\left(I-K_{m,k}\bar{\beta}_{m,k}\left(1-\bar{\gamma}_{m,k}\right)\Gamma_{m,h_k}C_{m,k}\right)^T \\
&+\bar{\gamma}_{m,k}\left(1-\bar{\gamma}_{m,k}\right)\bar{\beta}_{m,k}^2 K_{m,k}\sum_{j=0}^{p-1}\Gamma_{m,h_{k-j}}C_{m,k-j} \\
&\times S_{m,k-j}^u C_{m,k-j}^T\Gamma_{m,h_{k-j}}^T K_{m,k}^T +\bar{\beta}_{m,k}^2\bar{\gamma}_{m,k}\left(1-\bar{\gamma}_{m,k}\right) \\
&\times K_{m,k}\sum_{j=1}^{p-1}\Gamma_{m,h_{k-j}}C_{m,k-j}M_{m,k-j}^u C_{m,k-j}^T\Gamma_{m,h_{k-j}}^T \\
&\times K_{m,k}^T +\left(1+a_{m,k}^{-1}\right)K_{m,k}P_{\tilde{v}_{m,k}}K_{m,k}^T,
\end{aligned}
\tag{9.46}
$$

where

$$S_{m,k}^{u} = \min\left\{4tI + Q_{m,k-1}, (1+f_{m,k})N_{m,k}^{u} + (1+f_{m,k}^{-1})\hat{x}_{m,k}^{-}(\hat{x}_{m,k}^{-})^{T}\right\}, \tag{9.47}$$

$$N_{m,k}^{u} = \min\left\{4t, tr\left(A_{k-1}M_{m,k-1}^{u}A_{k-1}^{T}\right)\right\}I + Q_{m,k-1}. \tag{9.48}$$

Then, $M_{m,k}^{u}$ is the upper bound on $M_{m,k}$, i.e. $M_{m,k} \leq M_{m,k}^{u}$ hold for all $k \geq 0$. Furthermore, $M_{m,k}^{u}$ is minimized by the following filter gain:

$$\begin{aligned}
K_{m,k} &= \left(1+a_{m,k}\right)N_{m,k}^{u}C_{m,k}^{T}\Gamma_{m,\hbar_{k}}\left(1-\bar{\gamma}_{m,k}\right)\bar{\beta}_{m,k} \\
&\quad \times\left[\left(1+a_{m,k}^{-1}\right)P_{\tilde{v}_{m,k}} + \bar{\beta}_{m,k}^{2}\bar{\gamma}_{m,k}\left(1-\bar{\gamma}_{m,k}\right)\right. \\
&\quad \times\sum_{j=1}^{p-1}\Gamma_{m,\hbar_{k-j}}C_{m,k-j}M_{m,k-j}^{u}C_{m,k-j}^{T}\Gamma_{m,\hbar_{k-j}} \\
&\quad +\left(1-\bar{\gamma}_{m,k}\right)\bar{\gamma}_{m,k}\bar{\beta}_{m,k}^{2}\sum_{j=0}^{p-1}\Gamma_{m,\hbar_{k-j}}C_{m,k-j} \\
&\quad \times S_{m,k-j}^{u}C_{m,k-j}^{T}\Gamma_{m,\hbar_{k-j}} +\left(1+a_{m,k}\right)\bar{\beta}_{m,k}^{2} \\
&\quad \left. \times\bar{\gamma}_{m,k}\left(1-\bar{\gamma}_{m,k}\right)\Gamma_{m,\hbar_{k}}C_{m,k}N_{m,k}^{u}C_{m,k}^{T}\Gamma_{m,\hbar_{k}}\right]^{-1},
\end{aligned} \tag{9.49}$$

and the minimized upper bound $\bar{M}_{m,k}^{\min}$ is calculated by

$$M_{m,k}^{u,\min} = \left(1+a_{m,k}\right)\left(I - K_{m,k}\bar{\beta}_{m,k}\left(1-\bar{\gamma}_{m,k}\right)\Gamma_{m,\hbar_{k}}C_{m,k}\right)N_{m,k}^{u,\min}. \tag{9.50}$$

Proof On one hand, recalling $x_{k} = \hat{x}_{m,k}^{-} + \tilde{x}_{m,k}^{-}$ and Lemma 9.1, the state covariance $P_{x_{k}}$ is formulated as

$$\begin{aligned}
P_{\hat{x}_{k}} &= P_{\tilde{x}_{m,k}^{-}} +\left\{\hat{x}_{m,k}^{-}\left(\tilde{x}_{m,k}^{-}\right)^{T}\right\} + \mathbb{E}^{T}\left\{\hat{x}_{m,k}^{-}\left(\tilde{x}_{m,k}^{-}\right)^{T}\right\} + \hat{x}_{m,k}^{-}\left(\hat{x}_{m,k}^{-}\right)^{T} \\
&\leq \left(1+f_{m,k}\right)P_{\tilde{x}_{m,k}^{-}} +\left(1+f_{m,k}^{-1}\right)\hat{x}_{m,k}^{-}\left(\hat{x}_{m,k}^{-}\right)^{T},
\end{aligned} \tag{9.51}$$

where $f_{m,k} > 0$ are given real numbers.

On the other hand, recalling (9.1), $P_{x_{k}}$ is also formulated as

$$\begin{aligned}
P_{x_{k}} &= \mathbb{E}\left\{\sigma\left(A_{k-1}x_{k-1}\right)\sigma^{T}\left(A_{k-1}x_{k-1}\right)\right\} + Q_{k-1} \\
&\leq \mathbb{E}\left\{tr\left(\sigma\left(A_{k-1}x_{k-1}\right)\sigma^{T}\left(A_{k-1}x_{k-1}\right)\right)I\right\} + Q_{k-1} \\
&\leq tI + Q_{k-1}.
\end{aligned} \tag{9.52}$$

and thus we have

$$P_{\hat{x}_{k}} \leq \min\left\{4tI + Q_{k-1}, \left(1+f_{m,k}\right)P_{\tilde{x}_{m,k}^{-}} +\left(1+f_{m,k}^{-1}\right)\hat{x}_{m,k}^{-}\left(\hat{x}_{m,k}^{-}\right)^{T}\right\}. \tag{9.53}$$

Taking expectation on both sides of (9.53) leads to

$$\mathbb{E}P_{\hat{x}_k} \leq \min\left\{4tI + Q_{k-1}, \left(1+f_{m,k}\right)\mathbb{E}\left\{P_{\tilde{x}_{m,k}^-}\right\} + \left(1+f_{m,k}^{-1}\right)\hat{x}_{m,k}^-\left(\hat{x}_{m,k}^-\right)^T\right\}.$$
$$\triangleq S_{m,k}. \tag{9.54}$$

By means of the matrix operation and trace property, we have

$$\mathbb{E}\left\{c_{m,k}A_k\tilde{x}_{m,k}\tilde{x}_{m,k}^T A_k^T c_{m,k}\right\} \leq \mathbb{E}\left\{\left\|c_{m,k}A_k\tilde{x}_{m,k}\right\|^2\right\}I$$
$$\leq \mathbb{E}\left\{\left\|A_k\tilde{x}_{m,k}\right\|^2\right\}I \tag{9.55}$$
$$= tr\left(A_k P_{\tilde{x}_{m,k}} A_k^T\right)I.$$

Inserting (9.55) into (9.44) and taking expectation on both sides of the resultant inequality results in

$$\mathbb{E}\left\{P_{\tilde{x}_{m,k+1}^-}\right\} \leq tr\left(A_k M_{m,k} A_k^T\right)I + Q_k. \tag{9.56}$$

In addition to (9.44), we also have

$$P_{\tilde{x}_{m,k+1}^-}$$
$$\leq \mathbb{E}\left\{tr\left(\left(\sigma\left(A_k x_k\right)-\sigma\left(A_k\hat{x}_k\right)\right)\left(\sigma\left(A_k x_k\right)-\sigma\left(A_k\hat{x}_k\right)\right)^T\right)I\right\} + Q_k$$
$$\leq \mathbb{E}\left\{tr\left(\begin{bmatrix}2t_1^{\max} & 2t_2^{\max} & \cdots & 2t_n^{\max}\end{bmatrix}^T \begin{bmatrix}2t_1^{\max} & 2t_2^{\max} & \cdots & 2t_n^{\max}\end{bmatrix}\right)I\right\} + Q_k \tag{9.57}$$
$$\leq 4tI + Q_k,$$

which leads to

$$\mathbb{E}\left\{P_{\tilde{x}_{m,k}^-}\right\} \leq \min\left\{4tI, tr\left(A_{k-1}M_{m,k-1}A_{k-1}^T\right)I\right\} + Q_{k-1}$$
$$\triangleq N_{m,k}. \tag{9.58}$$

Moving forward, for any real numbers $a_{m,k} > 0$, we have

$$\Pi_{m,k} + \Pi_{m,k}^T$$
$$\leq a_{m,k}\mathbb{E}\left\{\left(I - K_{m,k}\beta_{m,k}\left(1-\bar{\gamma}_{m,k}\right)\Gamma_{m,h_k}C_{m,k}\right)\tilde{x}_{m,k}^-\right.$$
$$\left.\times\left(\tilde{x}_{m,k}^-\right)^T\left(I - K_{m,k}\beta_{m,k}\left(1-\bar{\gamma}_{m,k}\right)\right)^T\right\}$$
$$+ a_{m,k}^{-1}\mathbb{E}\left\{K_{m,k}\tilde{v}_{m,k}\tilde{v}_{m,k}^T K_{m,k}^T\right\} \tag{9.59}$$
$$= a_{m,k}\left(I - K_{m,k}\beta_{m,k}\left(1-\bar{\gamma}_{m,k}\right)\Gamma_{m,h_k}C_{m,k}\right)P_{\tilde{x}_{m,k}^-}$$
$$\times\left(I - K_{m,k}\beta_{m,k}\left(1-\bar{\gamma}_{m,k}\right)\Gamma_{m,h_k}C_{m,k}\right)^T + a_{m,k}^{-1}K_{m,k}P_{\tilde{v}_{m,k}}K_{m,k}^T.$$

Taking expectation on both sides of (9.59) yields

$$
\begin{aligned}
\mathbb{E}\left\{\Pi_{m,k}\right\}+\mathbb{E}\left\{\Pi_{m,k}^{T}\right\} \leq\ & a_{m,k}\left(I-K_{m,k}\beta_{m,k}\left(1-\bar{\gamma}_{m,k}\right)\Gamma_{m,h_{k}}C_{m,k}\right)\mathbb{E}\left\{P_{\tilde{x}_{m,k}^{-}}\right\} \\
& \times\left(I-K_{m,k}\beta_{m,k}\left(1-\bar{\gamma}_{m,k}\right)\Gamma_{m,h_{k}}C_{m,k}\right)^{T} \\
& +a_{m,k}^{-1}K_{m,k}P_{\tilde{v}_{m,k}}K_{m,k}^{T}.
\end{aligned}
\tag{9.60}
$$

Putting (9.54), (9.58) and (9.60) into (9.32) leads to

$$
\begin{aligned}
M_{m,k} \leq\ & \left(1+a_{m,k}\right)\left(I-K_{m,k}\bar{\beta}_{m,k}\left(1-\bar{\gamma}_{m,k}\right)\Gamma_{m,h_{k}}C_{m,k}\right) \\
& \times N_{m,k}\left(I-K_{m,k}\bar{\beta}_{m,k}\left(1-\bar{\gamma}_{m,k}\right)\Gamma_{m,h_{k}}C_{m,k}\right)^{T} \\
& +\bar{\gamma}_{m,k}\left(1-\bar{\gamma}_{m,k}\right)\bar{\beta}_{m,k}^{2}K_{m,k}\sum_{j=0}^{p-1}\Gamma_{m,h_{k-j}}C_{m,k-j} \\
& \times S_{m,k-j}C_{m,k-j}^{T}\Gamma_{m,h_{k-j}}^{T}K_{m,k}^{T}+\bar{\beta}_{m,k}^{2}\bar{\gamma}_{m,k}\left(1-\bar{\gamma}_{m,k}\right) \\
& \times K_{m,k}\sum_{j=1}^{p-1}\Gamma_{m,h_{k-j}}C_{m,k-j}M_{m,k-j}C_{m,k-j}^{T}\Gamma_{m,h_{k-j}}^{T} \\
& \times K_{m,k}^{T}+\left(1+a_{m,k}^{-1}\right)K_{m,k}P_{\tilde{v}_{m,k}}K_{m,k}^{T}.
\end{aligned}
\tag{9.61}
$$

Inspired by (9.61), let us define the following matrix function $\varpi_{k}\left(M_{m,k}^{u}\right):\mathbb{R}^{n\times n}\to\mathbb{R}^{n\times n}$ by

$$
\begin{aligned}
M_{m,k+1}^{u} \triangleq\ & \varpi_{k}\left(M_{m,k}^{u}\right) \\
=\ & \left(1+a_{m,k+1}\right)\left(I-K_{m,k+1}\bar{\beta}_{m,k+1}\left(1-\bar{\gamma}_{m,k+1}\right)\right. \\
& \times\Gamma_{m,h_{k+1}}C_{m,k+1}\Big)N_{m,k+1}^{u} \\
& \times\left(I-K_{m,k+1}\bar{\beta}_{m,k+1}\left(1-\bar{\gamma}_{m,k+1}\right)\Gamma_{m,h_{k+1}}C_{m,k+1}\right)^{T} \\
& +\bar{\gamma}_{m,k+1}\left(1-\bar{\gamma}_{m,k+1}\right)\bar{\beta}_{m,k+1}^{2}K_{m,k+1}\sum_{j=0}^{p-1}\Gamma_{m,h_{k+1-j}} \\
& \times C_{m,k+1-j}S_{m,k-j}C_{m,k+1-j}^{T}\Gamma_{m,h_{k+1-j}}^{T}K_{m,k+1}^{T} \\
& +\bar{\beta}_{m,k+1}^{2}\bar{\gamma}_{m,k+1}\left(1-\bar{\gamma}_{m,k+1}\right)K_{m,k+1}\sum_{j=1}^{p-1}\Gamma_{m,h_{k+1-j}} \\
& \times C_{m,k+1-j}M_{m,k+1-j}^{u}C_{m,k+1-j}^{T}\Gamma_{m,h_{k+1-j}}^{T}K_{m,k+1}^{T} \\
& +\left(1+a_{m,k+1}^{-1}\right)K_{m,k+1}P_{\tilde{v}_{m,k+1}}K_{m,k+1}^{T},
\end{aligned}
\tag{9.62}
$$

where $S_{m,k}^{u}$ and $N_{m,k}^{u}$ are defined by (9.47)–(9.48).

Keeping in mind (9.62), one verifies that function $\varpi_k\left(M_{m,k}^u\right)$ satisfies conditions given by Lemma 9.2. Consequently, given initial conditions $M_{m,0}^u = M_{m,0}$, it follows plainly from (9.59)–(9.62) and Lemma 9.2 that there exists a solution $M_{m,k}^u$ to the equation $M_{m,k+1}^u = \varpi_k\left(M_{m,k}^u\right)$ such that for all $k \geq 0$, $M_{m,k} \leq M_{m,k}^u$ hold, i.e. $M_{m,k}^u$ is an upper bound on $M_{m,k}$.

At last, setting the partial derivative of $tr\left(M_{m,k}^u\right)$ with respect to $K_{m,k}$ to be zero results in (9.49). Substituting (9.49) into (9.46), we have the minimized upper bound $\bar{M}_{m,k}^{\min}$ in (9.50).

With the minimized upper bounds $M_{m,k}^{u,\min}$ in hand, according to the federated fusion rule, the weight matrix $W_{m,k}$ can now be determined by

$$W_{m,k} = M_k^{u,\min}\left(M_{m,k}^{u,\min}\right)^{-1}, \tag{9.63}$$

where $M_k^{u,\min}$ is the minimized fused upper bound given by

$$M_k^{u,\min} = \left(\sum_{m=1}^{p}\left(M_{m,k}^{u,\min}\right)^{-1}\right)^{-1}. \tag{9.64}$$

Up till now, we have accomplished the design procedure of fusion estimator (9.14). Particularly, it is worth pointing out that abiding by the information-sharing principle of the federated fusion rule, at time $k-1$, the local estimate, upper bound and noise covariance should be initialized as

$$\begin{cases} \hat{x}_{m,k-1} \triangleq \hat{x}_{k-1}, \\ M_{m,k-1}^{u,\min} \triangleq \epsilon_m^{-1} M_{k-1}^{u,\min}, \\ Q_{m,k-1} \triangleq \epsilon_m^{-1} Q_k, \end{cases} \tag{9.65}$$

where $\sum_{m=1}^{p} \epsilon_m \triangleq 1$.

Remark 9.4 With the exact acknowledgement of the censoring probability given by (9.15), the censoring information hidden in $\bar{\beta}_{m,k}^l$, $\bar{\beta}_{m,k}^r$ as well as $\bar{\beta}_{m,k} \triangleq 1 - \bar{\beta}_{m,k}^l - \bar{\beta}_{m,k}^l$ can be incorporated in the design of the fusion estimator manifested in Theorems 9.1. Taking into account deception attacks and state saturations, our protocol-based filter gains $K_{m,k}$ are calculated through minimizing upper bounds $M_{m,k}^u$ (on local FECs). Note that the computation of $K_{m,k}$ is dependent on $M_{m,k-j}^u$ $(j=1,2,\ldots,p-1)$, which can be recursively attained in terms of (9.46).

Given this proposed fusion paradigm consisting of (9.14), Lemma 9.3 and Theorem 9.1, we next assess its fusion performance via exploring the boundedness property of $M_{m,k}^{u,\min}$.

Assumption 9.2 There exists a real number $\bar{q} > 0$ such that $Q_k \leq \bar{q}I$ is true for all $k \geq 0$.

Theorem 9.2 Given scalars $a_{m,k} > 0$, Assumptions 9.1–9.2, filter (9.14) and gain (9.49), there exists an upper bound $\bar{M}_k$ such that

$$M_k^{u,\min} \leq \bar{M}_k, \tag{9.66}$$

holds for all $k \geq 0$, where

$$\bar{M}_k = \left(4t + \bar{q}\right)\left(\sum_{m=1}^{p}\left(1 + a_{m,k}\right)^{-1}\right)^{-1}. \tag{9.67}$$

Proof Expanding out the terms on the right side of (9.46), $M^u_{m,k}$ becomes

$$\begin{aligned}
M^u_{m,k} &= \left(1 + a_{m,k}\right)N^u_{m,k} - \left(1 + a_{m,k}\right)K_{m,k}\bar{\beta}_{m,k}\left(1 - \bar{\gamma}_{m,k}\right)\Gamma_{m,h_k}C_{m,k}N^u_{m,k} \\
&\quad - \left(1 + a_m\right)N^u_{m,k}C^T_{m,k}\Gamma_{m,h_k}\left(1 - \bar{\gamma}_{m,k}\right)\beta_{m,k}K^T_{m,k} + K_{m,k}U_{m,k}K^T_{m,k},
\end{aligned} \tag{9.68}$$

where

$$\begin{aligned}
U_{m,k} &= \left(1 + a_{m,k}\right)\bar{\beta}_{m,k}\left(1 - \bar{\gamma}_{m,k}\right)\Gamma_{m,h_k}C_{m,k}N^u_{m,k}C^T_{m,k}\Gamma_{m,h_k}\left(1 - \bar{\gamma}_{m,k}\right)\bar{\beta}_{m,k} \\
&\quad + \bar{\gamma}_{m,k}\left(1 - \bar{\gamma}_{m,k}\right)\bar{\beta}^2_{m,k}\sum_{j=0}^{p-1}\Gamma_{m,h_{k-j}}C_{m,k-j}S^u_{m,k-j}C^T_{m,k-j}\Gamma^T_{m,h_{k-j}} \\
&\quad + \bar{\beta}^2_{m,k}\bar{\gamma}_{m,k}\left(1 - \bar{\gamma}_{m,k}\right)\sum_{j=1}^{p-1}\Gamma_{m,h_{k-j}}C_{m,k-j}M^u_{m,k-j}C^T_{m,k-j}\Gamma^T_{m,h_{k-j}} \\
&\quad + \left(1 + a^{-1}_{m,k}\right)P_{\bar{v}_{m,k}}.
\end{aligned} \tag{9.69}$$

Multiplying both sides of (9.49) by $U_{m,k}K^T_{m,k}$ leads to

$$K_{m,k}U_{m,k}K^T_{m,k} = \left(1 + a_{m,k}\right)N^u_{m,k}C^T_{m,k}\Gamma_{m,h_k}\left(1 - \bar{\gamma}_{m,k}\right)\bar{\beta}_{m,k}K^T_{m,k}. \tag{9.70}$$

Substituting (9.70) into (9.68) and noting $K_{m,k}$ in (9.49) can be rewritten as $K_{m,k} = \left(1 + a_{m,k}\right)N^u_{m,k}C^T_{m,k}\Gamma_{m,h_k}\left(1 - \bar{\gamma}_{m,k}\right)\bar{\beta}_{m,k}U^{-1}_{m,k}$, we have

$$\begin{aligned}
M^{u,\min}_{m,k} &= \left(1 + a_{m,k}\right)N^u_{m,k} - \left[\left(1 + a_{m,k}\right)N^u_{m,k}C^T_{m,k}\Gamma_{m,h_k}\left(1 - \bar{\gamma}_{m,k}\right)\bar{\beta}_{m,k}\right]U^{-1}_{m,k} \\
&\quad \times \left[\left(1 + a_{m,k}\right)N^u_{m,k}C^T_{m,k}\Gamma_{m,h_k}\left(1 - \bar{\gamma}_{m,k}\right)\bar{\beta}_{m,k}\right]^T \\
&\leq \left(1 + a_{m,k}\right)N^u_{m,k} \\
&\leq \left(1 + a_{m,k}\right)\left(4t + \bar{q}\right)I,
\end{aligned} \tag{9.71}$$

and therefore, we confirm

$$M^{u,\min}_k \leq \left(4t + \bar{q}\right)\left(\sum_{m=1}^{p}\left(1 + a_{m,k}\right)^{-1}\right)^{-1} \triangleq \bar{M}_k, \tag{9.72}$$

which proves that $M^{u,\min}_k$ is upper bounded by $\bar{M}_k$.

Theorem 9.2 ensures the boundedness of the minimized fused upper bound $M_k^{u,\min}$ under Assumption 9.2. Accounting for possible energy constraints in physical systems, such an assumption is reasonable, and an appropriate positive number $\bar{q}$ satisfying $Q_k \leq \bar{q}I$ can certainly be found.

Remark 9.5 So far, we have accomplished the design task of our protocol-based fusion estimator for state-saturated systems under deception attacks and dead-zone-like censoring. In comparison to the available filtering literature on state-saturated systems, our primary results own the following distinctive merits: (1) the addressed problem is new, as multiple engineering-oriented phenomena (e.g. state saturation, WTODP, dead-zone-like censoring and deceptions attack) are comprehensively considered; (2) the devised fusion paradigm is new, as the state-saturated estimator is purposely built by means of intensive stochastic analysis; and (3) the performance assessment is new in the sense of guaranteeing the boundedness of the minimized upper bound on the fused FEC. In the subsequent section, the usefulness of our developed fusion paradigm will be testified via an illustrative example.

9.3 AN ILLUSTRATIVE EXAMPLE

Consider dynamic system (9.1) with parameters:

$$A_k = \begin{bmatrix} 0.88 & 0.22 \\ 0.22 + 0.11\sin(k) & 0.66 \end{bmatrix}, Q_k = \begin{bmatrix} 0.1 & 0 \\ 0 & 0.3 \end{bmatrix},$$

and measurement model (9.2) with two sensors. For each sensor, we set measurement coefficients as $C_{1,k} = \begin{bmatrix} 1+0.05\cos(k) & 0.7 \end{bmatrix}$ and $C_{2,k} = \begin{bmatrix} 1 & 0.7+0.035\cos(k) \end{bmatrix}$, noise covariances as $R_{1,k} = R_{2,k} = 1$, weight matrices as $\Omega_{1,k} = 0.2$ and $\Omega_{2,k} = 0.3$, attacking probabilities as $\bar{\gamma}_{1,k} = 0.3$ and $\bar{\gamma}_{2,k} = 0.2$, attacking parameters as $\Xi_{1,k} = \Xi_{2,k} = 0.1$ and censoring thresholds as $\tau_1^l = \tau_2^l = -5$ and $\tau_1^r = \tau_2^r = 5$. Additionally, let saturation levels be $t_1^{\max} = 2$ and $t_2^{\max} = 3$, real numbers be $a_{1,k} = 0.1$, $a_{2,k} = 0.2$, $f_{1,k} = 0.1$ and $f_{2,k} = 0.2$ and initial values be $x_0 = \begin{bmatrix} 2 & 1 \end{bmatrix}^T$ and $M_{1,0}^u = M_{2,0}^u = 0.1I_2$.

Denote the mean squared errors (MSEs) of x_k^1 and x_k^2 (the first and second dimensions of the state), respectively, as $\text{MSE}j = (1/M)\sum_{i=1}^{M}\left(x_k^{j(i)} - \hat{x}_k^{j(i)}\right)^2$ where $j = 1,2,\ldots,n$ and $M = 1000$ is the number of Monte Carlo trials. In our follow-up simulations, we first make a performance comparison between the fused protocol-based Tobit Kalman filter (FPTKF) and the local protocol-based Tobit Kalman filter (LPTKF) to elaborate the performance superiority of our FPTKF over its local counterparts, i.e. LPTKF1 and LPTKF2, which build themselves on measurements from sensor 1 and sensor 2, respectively. Then, traces of $M_k^{u,\min}$ and its upper bound $\bar{M}_k$ are demonstrated to verify the correctness of our boundedness result showcased in Theorem 9.2.

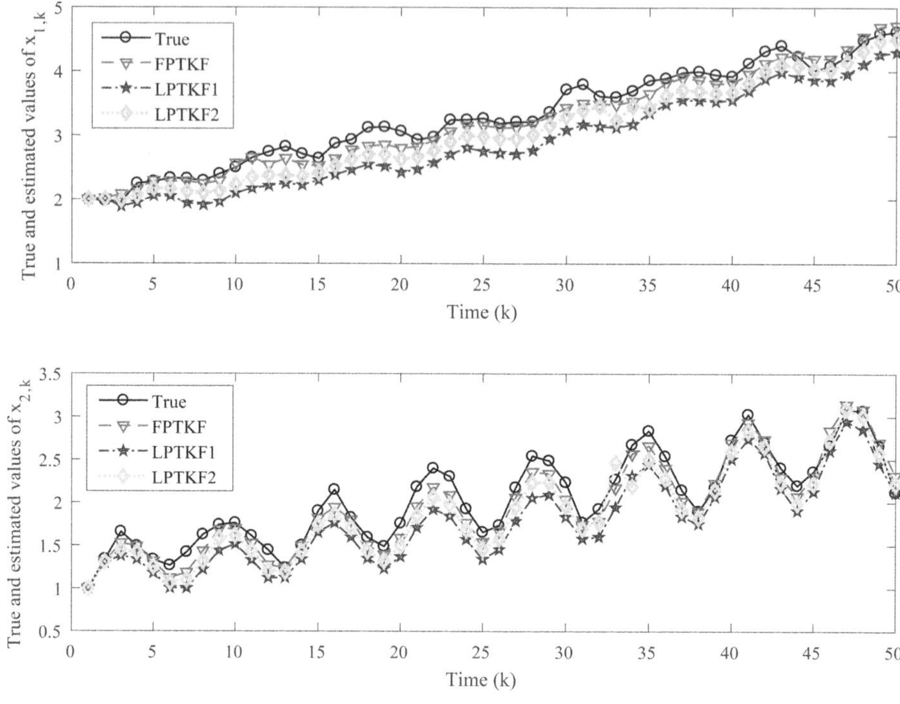

FIGURE 9.1 True and estimated values.

After implementing the LPTKF1, LPTKF2 and FPTKF, Figure 9.1 depicts the true and estimated values of system states, and Figure 9.2 plots the associate values of log(MSE1) and log(MSE2). It is evidently observed from Figures 9.1–9.2 that the tracking performance of the FPTKF outperforms that of the LPTKF1 and LPTKF2. This is because (1) in the FPTKF, all available sensor information is properly integrated to accomplish the task of state estimation and (2) in both LPTKF1 and LPTKF2, estimation results are obtained via making use of partial sensor observations.

Finally, the relationship between the trace of the mean error covariance $M_k^{u,\min}$ and the trace of its upper bound $\bar{M}_k$ (computed by our FPTKF) are sketched in Figure 9.3. It is evidently observed from Figure 9.3 that the curve of $\log_{10}\left(\mathrm{tr}\left(M_k^{u,\min}\right)\right)$ always resides lower than that of $\log_{10}\left(\mathrm{tr}\left(\bar{M}_k\right)\right)$, which justifies the boundedness statement shown in Theorem 9.2 that $\bar{M}_k$ can be treated as a proper upper bound on $M_k^{u,\min}$.

9.4 SUMMARY

In this chapter, we have addressed the multi-sensor Tobit Kalman fusion estimation problem for state-saturated systems with dead-zone-like censoring and deception attacks under the WTODP. For each sensor, in order to curb the effects from both

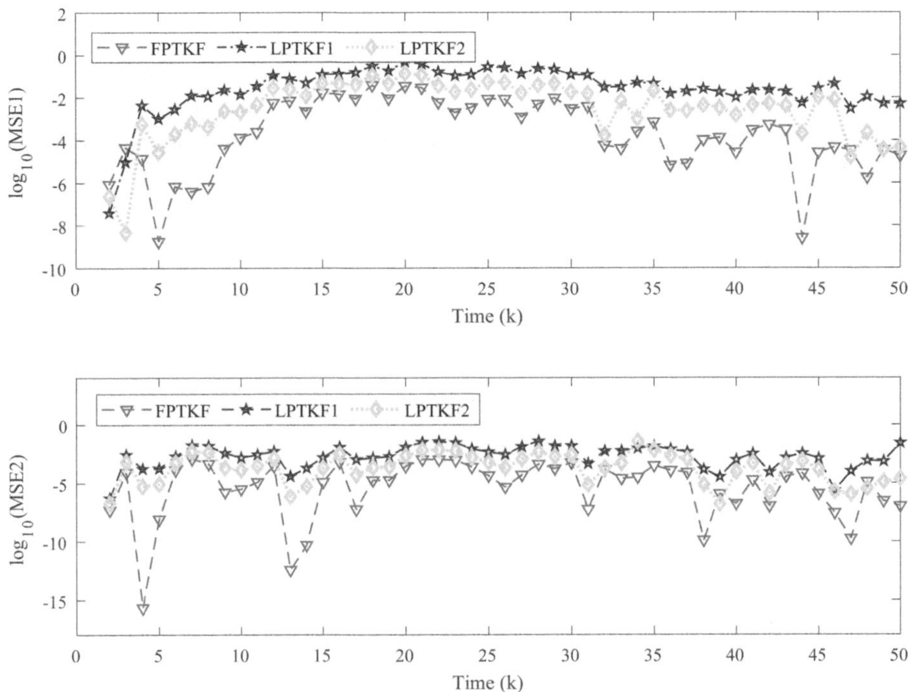

FIGURE 9.2   Comparison in $\log_{10}$ (MSE1) and $\log_{10}$ (MSE2).

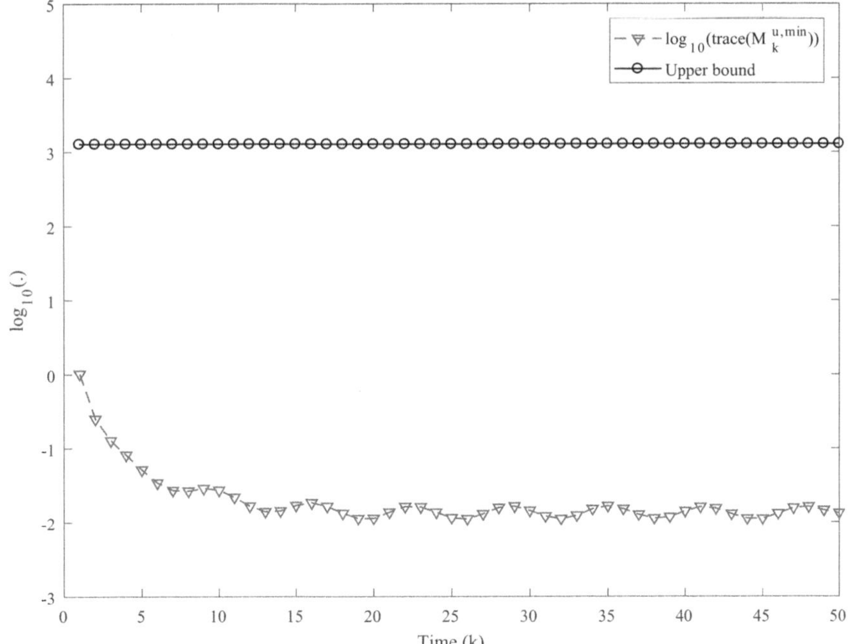

FIGURE 9.3 $\log_{10}\left(\mathrm{tr}\left(M_k^{u,\min}\right)\right)$ and its upper bound.

the deception attacks and WTODP, the conventional Tobit regression model has been rectified, and a unified local estimation framework has been constructed via the proposed LPTKF. Thanks to the federated fusion rule, local estimates generated by all the LPTKFs have been incorporated to obtain the FPTKF by properly parameterizing the filter parameters. Further assessment has been conducted by analyzing the boundedness of the fused FEC, and finally, the applicability of the FPTKF has been validated by a simulation experiment.

10 Variance-Constrained Filtering Fusion for Nonlinear Cyber-Physical Systems with the Denial-of-Service Attacks and Stochastic Communication Protocol

Cyber-physical systems (CPSs) encompass a wide collection of physical and computer infrastructures/components that are connected and intertwined with each other in a cooperative way. Typical examples of CPSs include smart grids, water plants, robotics systems, tracking systems etc. The real-world implementation of CPSs often requires system components to operate on a shared communication network, which, unfortunately, results in a higher chance of malicious cyber-attacks towards control commands and system outputs. In order to safely and efficiently carry out the desired cyber-physical process, increasing research efforts have recently been made on the secure filtering/control problems for CPSs under deception attacks, replay attacks and denial-of-service (DoS) attacks [185–192].

It should be pointed out that most CPS-related literature has been concerned with filtering/control problems for *linear* systems only despite the fact that practical CPSs often exhibit inherently nonlinear characteristics [83, 188, 193, 194]. Clearly, the underlying nonlinearities with the CPSs would inevitably bring in substantial challenges to the development of the filtering/control algorithms [195–199]. Note that the cooperative control problem has recently been studied in [26] for nonlinear CPSs subject to multiple DoS attacks. Although the filtering problem for nonlinear CPSs has drawn some preliminary attention, the corresponding results have not, and the filtering error covariance is of a major concern, and this constitutes one of our motivations to look into variance-constrained filter design for a class of nonlinear CPSs.

In this chapter, we endeavor to develop a new variance-constrained fusion estimator for a general class of nonlinear CPSs that encompass the complexities brought

DOI: 10.1201/9781003461623-10

from system nonlinearity, SCP scheduling and DoS attacks. To be more specific, a set of variance-constrained local filters is devised by first accommodating the addressed complexities via a dedicated introduced filter structure and then minimizing certain upper bounds on local filtering error covariances. By means of the matrix inequality and stochastic analysis technique, a variance-constrained fusion estimator is put forward through fully integrating the estimates from all local filters and rigorously studying the boundedness of the fused error covariance.

10.1 PROBLEM FORMULATION

Consider the following nonlinear system:

$$\begin{cases} x_{s+1} = f(x_s) + \omega_s, \\ z_{m,s} = h_m(x_s) + v_{m,s}, \ m = 1, 2, \ldots, p, \end{cases} \tag{10.1}$$

where $x_s \in \mathbb{R}^n$ and $z_{m,s} \in \mathbb{R}$ are, respectively, the state vector and the measurement vector of the mth sensor; p is the total number of the sensors; $f(x_s): \mathbb{R}^n \to \mathbb{R}^n$ and $h_m(x_s): \mathbb{R} \to \mathbb{R}$ are known nonlinear functions; and $\omega_k \in \mathbb{R}^{n_x}$ and $v_{m,s} \in \mathbb{R}$ are zero-mean Gaussian white noises with covariances Q_k and $R_{m,s}$, respectively.

Define a sequence of uncorrelated random variables $\{\alpha_k\}_{k \geq 0} \in \{1, 2, \ldots, p\}$ with the following probability distributions that regulate the scheduling behavior of the sensors:

$$\mathrm{Prob}\{\alpha_s = j\} = \bar{\alpha}_j, j = 1, 2, \ldots, p, \tag{10.2}$$

where $\bar{\alpha}_j \left(\sum_{j=1}^{p} \bar{\alpha}_j = 1 \right)$ are the known occurrence probabilities under which sensor j is permitted to access the network channel. Furthermore, define the update coefficient of the SCP by

$$\Phi_{m,\alpha_s} \triangleq \delta(\alpha_s - m),$$

where $\delta(\cdot)$ is a Kronecker delta function. According to the zero-input strategy, the *actual* measurements (of sensor m) arriving at the mth remote estimator are

$$\bar{z}_{m,s} = \Phi_{m,\alpha_s} z_{m,s}. \tag{10.3}$$

Due to the typical opening-up characteristics of the considered networked environment, the data communication might be prone to cyber-attacks initiated by malicious adversaries. In this chapter, the sensor measurements $\bar{z}_{m,k}$ in (10.3) are subject to intentional falsification from the so-called randomly occurring DoS attacks. Let us define a sequence of Bernoulli-distributed variables $\beta_{m,k}$ to describe the DoS behavior with the following distributions:

$$\begin{cases} \mathrm{Prob}\{\beta_{m,s} = 1\} = \bar{\beta}_m, \\ \mathrm{Prob}\{\beta_{m,s} = 0\} = 1 - \bar{\beta}_m \end{cases} \tag{10.4}$$

where $\bar{\beta}_m$ are the success rates of launched DoS attacks. Accordingly, the DoS behavior can be modeled by

$$y_{m,s} = \left(1 - \beta_{m,s}\right)\bar{z}_{m,s},\tag{10.5}$$

where $y_{m,s}$ are the falsified sensor measurements.

Assumption 10.1 The initial state x_0 has the mean $\hat{x}_0$ and covariance $P_{\tilde{x}_0}$. Also, x_0, ω_s, $v_{m,s}$, α_s and $\beta_{m,s}$ are mutually independent.

Assumption 10.2 $f(0) = 0$, $h_m(0) = 0$, and there exist non-negative constants a, c_m $(m = 1, 2, \ldots, p)$ and matrices A, C_m such that the following pseudo-Lipschitz conditions:

$$\left\| f\left(x_s + \Delta_{x_s}\right) - f\left(x_s\right) - A\Delta_{x_s} \right\| \le a \left\| \Delta_{x_s} \right\|,$$

$$\left\| h_m\left(x_s + \Delta_{x_s}\right) - h_m\left(x_s\right) - C_m\Delta_{x_s} \right\| \le c_m \left\| \Delta_{x_s} \right\|$$

hold for all $s \ge 0$ where $\|\cdot\|$ is the Euclidean vector norm.

Remark 10.1 Assumption 10.2 reflects the "distance" between the originally non-linear model (1) and the "nominal" linear model, whose system parameters are A and C_m. Such nonlinear descriptions are actually similar to the Lipschitz conditions on the nonlinear functions $f(\cdot)$ and $h_m(\cdot)$. In applications, the linearization technique may be exploited to quantify the maximum possible derivations from the nominal model. One of the future research topics is to describe the nonlinearities in a more general way.

Define the fused state estimation, the local state estimation, and the local state prediction, respectively, by

$$\hat{x}_s \triangleq \mathbb{E}\left\{x_s \mid y_{1:s}\right\}, \hat{x}_{m,s} \triangleq \mathbb{E}\left\{x_s \mid y_{m,1:s}\right\}, \hat{x}_{m,s}^- \triangleq \mathbb{E}\left\{x_s \mid y_{m,1:s-1}\right\}$$

where

$$y_{m,1:s} \triangleq \left\{y_{m,1}, y_{m,2}, \ldots, y_{m,s}\right\}, y_{1:s} \triangleq \left\{y_{1,1:s}, y_{2,1:s}, \ldots, y_{p,1:s}\right\}.$$

Let

$$\tilde{x}_{m,s} \triangleq x_s - \hat{x}_{m,s}, \tilde{x}_{m,s}^- \triangleq x_s - \hat{x}_{m,s}^-,$$

$$P_{\tilde{x}_{m,s}} \triangleq \mathbb{E}\left\{\tilde{x}_{m,s}\tilde{x}_{m,s}^T \mid y_{m,1:s}\right\}, P_{\tilde{x}_{m,s}^-} \triangleq \mathbb{E}\left\{\tilde{x}_{m,s}^-\left(\tilde{x}_{m,s}^-\right)^T \mid y_{m,1:s-1}\right\}$$

be, respectively, the local state estimation error, prediction error, estimation error covariance and prediction error covariance. Denote

$$P_{\tilde{x}_s} \triangleq \mathbb{E}\left\{\left(x_s - \hat{x}_s\right)\left(x_s - \hat{x}_s\right)^T \mid y_{1:s}\right\},$$

as the fused error covariance.

In this chapter, we are devoted to constructing the following variance-constrained nonlinear fusion estimator:

$$
\begin{cases}
\hat{x}_{m,s}^{-} = f\left(\hat{x}_{m,s-1}\right), \\
\hat{x}_{m,s} = \hat{x}_{m,s}^{-} + K_{m,s}\left(y_{m,s} - \hat{y}_{m,s}^{-}\right), \\
\hat{x}_{s} = \sum_{m=1}^{p} W_{m,s}\hat{x}_{m,s}
\end{cases}
\tag{10.6}
$$

such that there exist upper bounds $\mathcal{P}_{\tilde{x}_{m,s}}$ and $\mathcal{P}_{\tilde{x}_{s}}$ satisfying

$$
P_{\tilde{x}_{m,s}} \leq \mathcal{P}_{\tilde{x}_{m,s}},
\tag{10.7}
$$

$$
P_{\tilde{x}_{s}} \leq \mathcal{P}_{\tilde{x}_{s}}
\tag{10.8}
$$

where $\hat{x}_{s}$ and $P_{\tilde{x}_{s}}$ are the fused estimate and error covariance under the federated fusion criterion; $K_{m,s}$ and $W_{m,s}$ are, respectively, the local estimator gain and weight matrix to be designed later; and $\hat{y}_{m,s}^{-}$ is the prediction given by

$$
\hat{y}_{m,s}^{-} = \left(1 - \bar{\beta}_{m}\right)\Phi_{m,\alpha_{s}}h_{m}\left(\hat{x}_{s}^{-}\right).
\tag{10.9}
$$

Remark 10.2 Due to the presence of the SCP scheduling method, the structure of estimator (10.6) is largely affected by the update coefficient $\Phi_{m,\alpha_{s}}$. The treatment of this coefficient in determining estimator gains $K_{m,s}$ and weight matrices $W_{m,s}$ constitutes the first challenge of this chapter. However, it is observed from estimator (10.6) that the gain matrices $K_{m,s}$ should be recursively computed based on the error covariances $P_{\tilde{x}_{s}}$. Accounting for the time-varying and nonlinear nature of the system, it would be difficult to ensure the convergence of $P_{\tilde{x}_{s}}$, and therefore, we turn to pursue certain upper bounds on the error covariances (under the variance constraints) so as to guarantee the non-divergence of the designed algorithm, and this constitutes the second challenge of this chapter. Additionally, since the measurement $y_{m,s}$ integrates the information from both system nonlinearities and DoS attacks, such information should be fully reflected in both the parameter determination and the boundedness investigation, which constitutes the third challenge of this chapter.

Pertaining to these challenges, our tasks in this chapter are to (1) devise the variance-constrained nonlinear fusion estimator (10.6) by appropriately determining parameters $K_{m,s}$ and $W_{m,s}$; (2) assess fusion estimation performance through studying the boundedness property of obtained covariance bounds; and (3) study the effects from the system nonlinearities, SCP scheduling and DoS attacks with parameter determination and a boundedness investigation.

10.2 MAIN RESULTS

Lemma 10.1 [184] Let the matrix function $w_k(X): \mathbb{R}^{n \times n} \to \mathbb{R}^{n \times n}$ be given such that

$$w_s(X) \le w_s(Y),$$

holds for any matrices X and Y satisfying

$$X \le Y,$$

$$X = X^T \ge 0$$

$$Y = Y^T \ge 0.$$

Then, given the initial condition $X_0 = Y_0$, there exist solutions X_k and Y_k to difference equations

$$X_{s+1} = w_k(X_s),$$

$$Y_{s+1} \ge w_s(Y_s)$$

such that

$$X_s \le Y_s$$

holds for all $k \ge 0$.

TABLE 10.1
Pseudocode: Variance-Constrained Filtering Fusion

Algorithm: variance-constrained filtering fusion

Input: $\hat{x}_0$, $\mathcal{P}_{\tilde{x}_0}^{\min}$, $y_{m,1:s}$

Output: $\hat{x}_s$, $\mathcal{P}_{\tilde{x}_s}^{\min}$.

1: let $\hat{x}_{m,0} = \hat{x}_0$, $\mathcal{P}_{\tilde{x}_{m,0}}^{\min} = \epsilon_m^{-1} \mathcal{P}_{\tilde{x}_0}$, $Q_{m,0} = \epsilon_m^{-1} Q_0$.

2: **for** $s = 1 : N$ **do**

3:　　calculate local prediction $\hat{x}_{m,s}^-$ by (10.6);

4:　　calculate filter gain $K_{m,s}$ by (10.52)–(10.20);

5:　　calculate minimal bound $\mathcal{P}_{\tilde{x}_{m,s}}^{\min}$ by (10.21)–(10.29);

6:　　calculate the local estimate $\hat{x}_{m,s}$ by (10.6);

7　　calculate fused bound $\mathcal{P}_{\tilde{x}_s}^{\min}$ by (10.44);

7　　calculate weight matrix $W_{m,s}$ by (10.41);

7　　calculate fused estimate $\hat{x}_s$ by (10.6);

8:　　reallocate $\hat{x}_{m,s}$, $\mathcal{P}_{\tilde{\xi}_{m,s}}^{\min}$ and $Q_{m,s}$ by (10.53);

9: **end for**

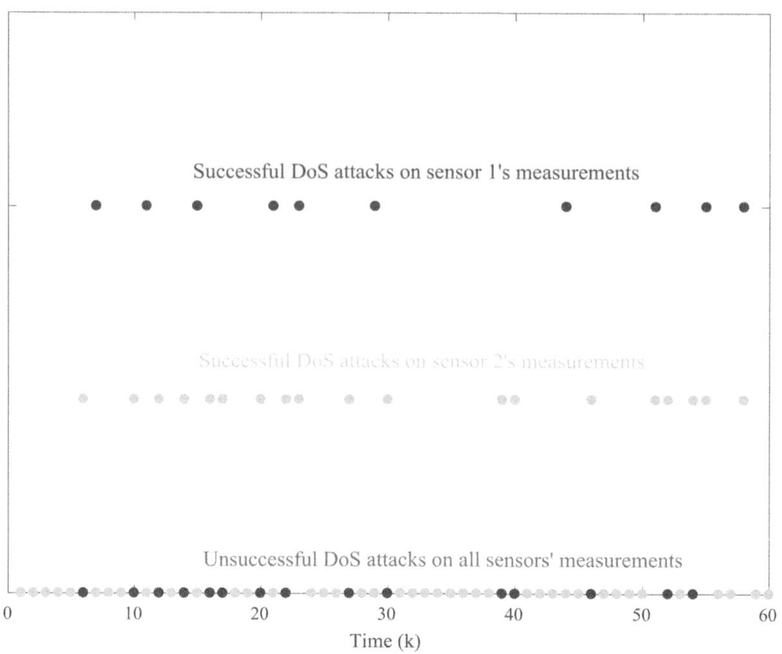

FIGURE 10.1 DoS attacks on two sensors with success rates $\bar{\beta}_1 = 0.2$ and $\bar{\beta}_2 = 0.3$. The black/gray dots on or above the horizontal axis indicate unsuccessful or successful DoS attacks on sensor 1/2 measurements. For instance, at time k, if a black dot is located on or above the horizontal axis, it indicates that the measurement of sensor 1 is free from or under DoS attacks.

Let constants $a_m \geq 0$, $c_m \geq 0$, and matrices A, C_m satisfy the conditions in Assumption 10.2. Define

$$\bar{\gamma}_m \triangleq \mathbb{E}\{\gamma_{m,s}\}, \tag{10.10}$$

$$\gamma_{m,s} \triangleq \left(1 - \beta_{m,s}\right)\Phi_{m,\alpha_s}, \tag{10.11}$$

$$\tilde{\gamma}_m \triangleq \mathbb{E}\left\{\left(\gamma_{m,s} - \bar{\gamma}_m\right)^2\right\}, \tag{10.12}$$

$$\xi_{m,s} \triangleq f(x_s) - f(\hat{x}_{m,s}) - A\tilde{x}_{m,s}, \tag{10.13}$$

$$\zeta_{m,s} \triangleq h(x_s) - h(\hat{x}_{m,s}^-) - C_m\tilde{x}_{m,s}^-. \tag{10.14}$$

Furthermore, for any constants $d_{i,m} > 0$ $\left(i = 1, 2\ldots, 5, m = 1, 2\ldots, p\right)$, let us define

$$e_{1,m} \triangleq 1 - d_{1,m}^{-1} - d_{2,m}^{-1}, e_{2,m} \triangleq \tilde{\gamma}_m\left(1 + d_{3,m} + d_{4,m}\right),$$

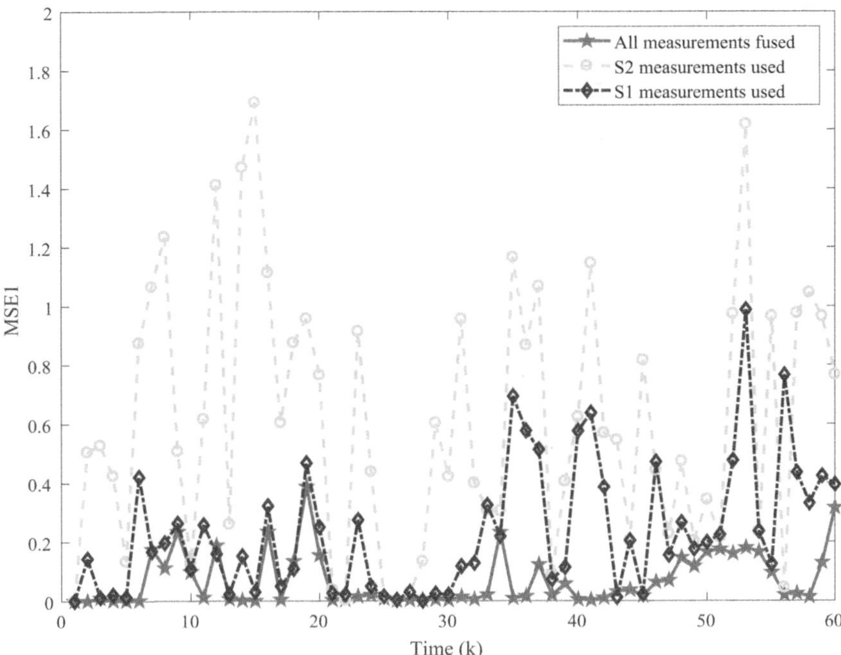

FIGURE 10.2 MSE1 comparison.

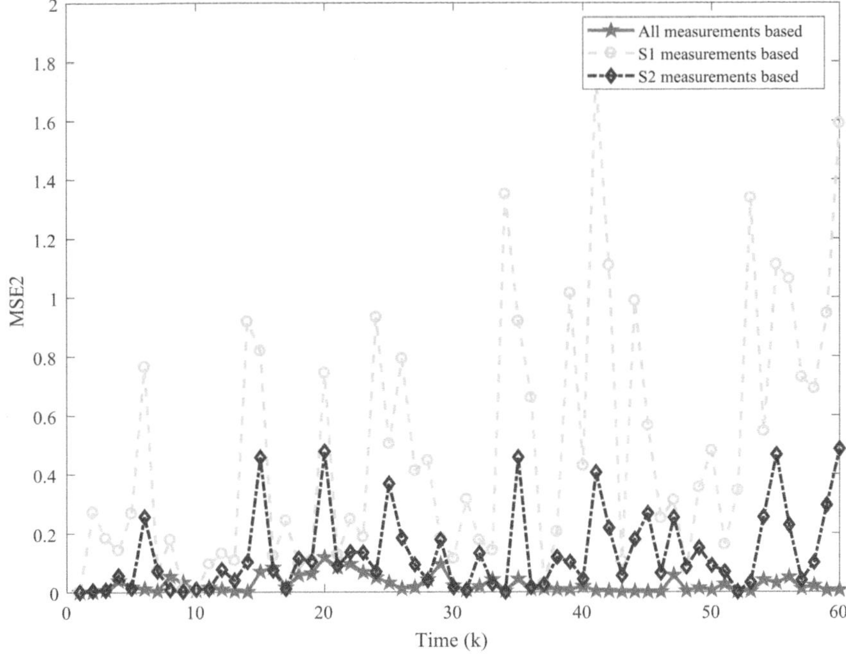

FIGURE 10.3 MSE2 comparison.

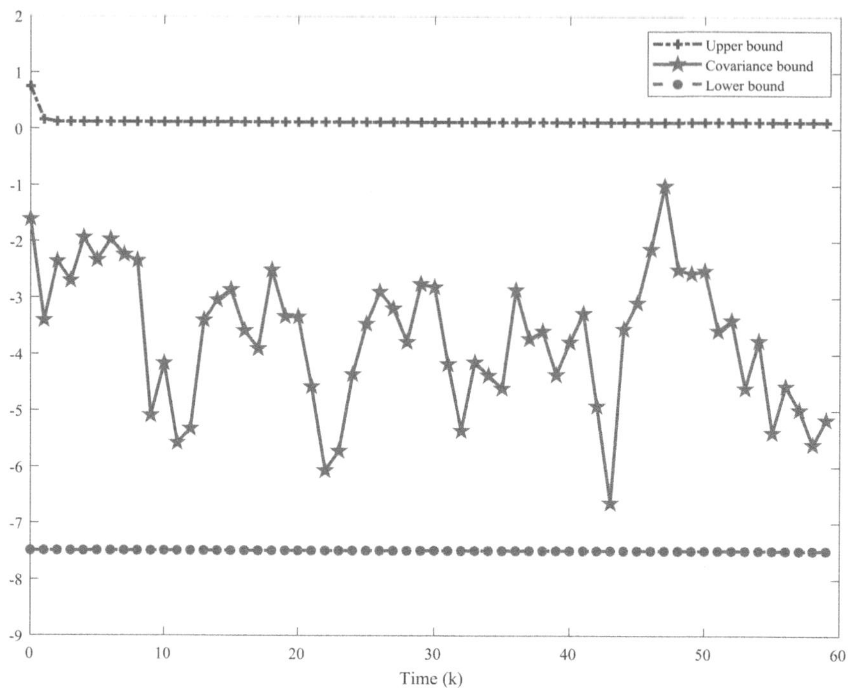

FIGURE 10.4 Comparison between $\log_{10}\left(\mathrm{tr}\left\{\mathcal{P}_{\tilde{x}_s}^{\min}\right\}\right)$ and its bounds.

TABLE 10.2
AMSE Comparison under Different Success Rates $\left[\bar{\beta}_1,\bar{\beta}_2\right]$

$\left[\bar{\beta}_1,\bar{\beta}_2\right]$	S1-AMSE1	S1-AMSE2	S2-AMSE1	S2-AMSE2	AMSE1	AMS2
[0.1,0.2]	0.1800	0.1220	0.3789	0.2380	0.1483	0.0743
[0.2,0.3]	0.2117	0.1543	0.6532	0.4564	0.1642	0.0869
[0.3,0.4]	0.3804	0.2037	0.8502	0.6448	0.3163	0.1630
[0.5,0.6]	16.4873	18.4755	148.6533	157.1435	14.8588	16.5562

TABLE 10.3
Average MSE Comparison under Different $\left[\bar{\alpha}_1,\bar{\alpha}_2\right]$

$\left[\bar{\alpha}_1,\bar{\alpha}_2\right]$	S1-MSE1	S1-MSE2	S2-MSE1	S2-MSE2	MSE1	MSE2
[0.6,0.4]	0.2117	0.1543	0.6532	0.4564	0.1642	0.0869
[0.8,0.2]	0.1943	0.1246	0.8433	0.6784	0.1794	0.0792
[0.4,0.6]	0.4336	0.3067	0.5286	0.4197	0.2044	0.0765

$$e_{3,m} \triangleq \tilde{\gamma}_m \left(1 + d_{3,m}^{-1} + d_{5,m}\right), e_{4,m} \triangleq \overline{\gamma}_m^2 \left(\tilde{\gamma}_m + \left(1 - d_{2,m}\right)\right),$$

$$e_{5,m} \triangleq c_m^2 \left(\tilde{\gamma}_m + \overline{\gamma}_m^2 + \tilde{\gamma}_m \left(d_{4,m}^{-1} + d_{5,m}^{-1}\right) - d_{1,m}\overline{\gamma}_m^2\right)$$

where, by recalling Assumption 10.1 and

$$\mathbb{E}\left\{\phi_{m,\alpha_k}\right\} = \sum_{j=1}^{p} \overline{\alpha}_j \delta\left(j - m\right) = \overline{\alpha}_m,$$

$\overline{\gamma}_m$ in (10.10) and $\tilde{\gamma}_m$ in (10.12) can be easily derived as follows

$$\overline{\gamma}_m = \overline{\alpha}_m \left(1 - \overline{\beta}_m\right), \tag{10.15}$$

$$\tilde{\gamma}_m = \overline{\alpha}_m \left(1 - \overline{\beta}_m\right)\left[1 - \overline{\alpha}_m + \overline{\alpha}_m\overline{\beta}_m\right]. \tag{10.16}$$

We first present the calculation formula of the local error covariance $P_{\tilde{x}_{m,k}}$. For brevity, in the subsequent derivation and analysis, all conditional expectations $\mathbb{E}\left\{\cdot \mid y_{m,1:k}\right\}$ and $\mathbb{E}\left\{\cdot \mid y_{m,1:k-1}\right\}$ are simplified as $\mathbb{E}\left\{\cdot\right\}$.

Theorem 10.1 The local error covariance $P_{\tilde{x}_{m,s}}$ of estimator (10.6) is given by

$$\begin{aligned}
P_{\tilde{x}_{m,s}} &= \left(I - K_{m,s}\overline{\gamma}_m C_m\right) P_{\tilde{x}_{m,s}^-} \left(I - K_{m,s}\overline{\gamma}_m C_m\right)^T \\
&\quad + \tilde{\gamma}_m K_{m,s} C_m P_{\tilde{x}_{m,s}^-} C_{m,s}^T K_{m,s}^T \\
&\quad + \tilde{\gamma}_m K_{m,s} h\left(\hat{x}_{m,s}^-\right) h^T\left(\hat{x}_{m,s}^-\right) K_{m,s}^T \\
&\quad + \left(\tilde{\gamma}_m + \overline{\gamma}_m^2\right) K_{m,s} \mathbb{E}\left\{\varsigma_{m,s}\varsigma_{m,s}^T\right\} K_{m,s}^T \\
&\quad + \left(\tilde{\gamma}_m + \overline{\gamma}_m^2\right) K_{m,s} R_{m,s} K_{m,s}^T \\
&\quad - \mathbb{E}\left\{\left(I - K_{m,s}\overline{\gamma}_m C_m\right) \tilde{x}_{m,s}^- \varsigma_{m,s}^T \overline{\gamma}_{m,s} K_{m,s}^T\right\} \\
&\quad - \mathbb{E}^T\left\{\left(I - K_{m,s}\overline{\gamma}_m C_m\right) \tilde{x}_{m,s}^- \varsigma_{m,s}^T \overline{\gamma}_{m,s} K_{m,s}^T\right\} \\
&\quad - \mathbb{E}\left\{\left(I - K_{m,s}\gamma_{m,s} C_m\right) \tilde{x}_{m,s}^- \upsilon_{m,s}^T \overline{\gamma}_m K_{m,s}^T\right\} \\
&\quad - \mathbb{E}^T\left\{\left(I - K_{m,s}\gamma_{m,s} C_m\right) \tilde{x}_{m,s}^- \upsilon_{m,s}^T \overline{\gamma}_m K_{m,s}^T\right\} \\
&\quad + \mathbb{E}\left\{K_{m,s}\left(\gamma_{m,s} - \overline{\gamma}_m\right)^2 C_m\tilde{x}_{m,s}^- h^T\left(\hat{x}_{m,s}^-\right) K_{m,s}^T\right\} \\
&\quad + \mathbb{E}^T\left\{K_{m,s}\left(\gamma_{m,s} - \overline{\gamma}_m\right)^2 C_m\tilde{x}_{m,s}^- h^T\left(\hat{x}_{m,s}^-\right) K_{m,s}^T\right\} \\
&\quad + \mathbb{E}\left\{K_{m,s}\left(\gamma_{m,s} - \overline{\gamma}_m\right)^2 C_m\tilde{x}_{m,s}^- \varsigma_{m,s}^T K_{m,s}^T\right\} \\
&\quad + \mathbb{E}^T\left\{K_{m,s}\left(\gamma_{m,s} - \overline{\gamma}_m\right)^2 C_m\tilde{x}_{m,s}^- \varsigma_{m,s}^T K_{m,s}^T\right\} \\
&\quad + \mathbb{E}\left\{K_{m,s}\left(\gamma_{m,s} - \overline{\gamma}_m\right)^2 h\left(\hat{x}_{m,s}^-\right) \varsigma_{m,s}^T K_{m,s}^T\right\} \\
&\quad + \mathbb{E}^T\left\{K_{m,s}\left(\gamma_{m,s} - \overline{\gamma}_m\right)^2 h\left(\hat{x}_{m,s}^-\right) \varsigma_{m,s}^T K_{m,s}^T\right\}
\end{aligned} \tag{10.17}$$

where

$$P_{\tilde{x}_{m,s+1}^-} = \mathbb{E}\left\{\xi_{m,s}\xi_{m,s}^T\right\} + Q_{m,s} + AP_{\tilde{x}_{m,s}}A^T$$
$$+ \mathbb{E}\left\{\xi_{m,s}\tilde{x}_{m,s}^T\right\}A^T + A\mathbb{E}^T\left\{\xi_{m,s}\tilde{x}_{m,s}^T\right\}. \tag{10.18}$$

Here, the cross-terms in (10.17)–(10.18), and redistributed noise covariance $Q_{m,s}$ as well as the gain matrix $K_{m,s}$ are to be determined later.

Proof According to (10.1), (10.3), (10.5), Assumption 10.2 and the definitions of $\gamma_{m,s}$, $\xi_{m,s}$ and $\zeta_{m,s}$, it is obtained that

$$f(x_s) = \xi_{m,s} + f(\hat{x}_{m,s}) + A\tilde{x}_{m,s},$$

$$y_{m,s} = \gamma_{m,s}h(x_s) + \gamma_{m,s}\upsilon_{m,s},$$

$$h_m(x_s) = \zeta_{m,s} + h(\hat{x}_{m,s}^-) + C_m\tilde{x}_{m,s}^-.$$

In line with (10.1) and (10.6), the local prediction and estimation errors are

$$\tilde{x}_{m,s}^- = x_s - \hat{x}_{m,s}^-$$
$$= \xi_{m,s-1} + \omega_{s-1} + A\tilde{x}_{m,s-1}, \tag{10.19}$$

$$\tilde{x}_{m,s} = x_s - \hat{x}_{m,s}$$
$$= \tilde{x}_{m,s}^- - K_{m,s}\left(y_{m,s} - \hat{y}_{m,s}^-\right)$$
$$= \left[I - K_{m,s}\bar{\gamma}_m C_m\right]\tilde{x}_{m,s}^- - K_{m,s}(\gamma_{m,s} - \bar{\gamma}_m)C_m\tilde{x}_{m,s}^-$$
$$- K_{m,s}\left(\gamma_{m,s} - \bar{\gamma}_m\right)h(\hat{x}_{m,s}^-) \tag{10.20}$$
$$- K_{m,s}\left(\gamma_{m,s} - \bar{\gamma}_m\right)\zeta_{m,s} - K_{m,s}\bar{\gamma}_m\zeta_{m,s} - K_{m,s}\left(\gamma_{m,s} - \bar{\gamma}_m\right)\upsilon_{m,s}$$
$$- K_{m,s}\bar{\gamma}_m\upsilon_{m,s}.$$

On one hand, based on (10.19), $P_{\tilde{x}_{m,s+1}^-}$ is derived as

$$P_{\tilde{x}_{m,s+1}^-} = \mathbb{E}\left\{\tilde{x}_{m,s+1}^-\left(\tilde{x}_{m,s+1}^-\right)^T\right\}$$
$$= \mathbb{E}\left\{\xi_{m,s}\xi_{m,s}^T\right\} + Q_{m,s} + AP_{\tilde{x}_{m,s}}A^T + \mathbb{E}\left\{\xi_{m,s}\tilde{x}_{m,s}^T\right\}A^T \tag{10.21}$$
$$+ A\mathbb{E}^T\left\{\xi_{m,s}\tilde{x}_{m,s}^T\right\},$$

which is the same as (10.18). On the other hand, based on error dynamics (10.21) and Assumption 10.1, $P_{\tilde{x}_{m,s}} \triangleq \mathbb{E}\left\{\tilde{x}_{m,s}\tilde{x}_{m,s}^T\right\}$ is derived easily as (10.17). The proof is now complete.

Theorem 10.2 Let the positive constants a_m, b_m, c_m and $d_{i,m}$, matrices A and C_m, and initial conditions $P_{\tilde{x}_{m,0}} > 0$ be given. Suppose there are matrix sequences $\mathcal{P}_{\tilde{x}_{m,s}}$ satisfying

$$
\begin{aligned}
\mathcal{P}_{\tilde{x}_{m,s}} &= e_{1,m}\left(I - K_{m,s}\bar{\gamma}_m C_m\right)\mathcal{P}_{\tilde{x}_{m,s}^-}\left(I - K_{m,s}\bar{\gamma}_m C_m\right)^T \\
&\quad + e_{2,m} K_{m,s} C_m \mathcal{P}_{\tilde{x}_{m,s}^-} C_{m,s}^T K_{m,s}^T \\
&\quad + e_{3,m} K_{m,s} h\left(\hat{x}_{m,s}^-\right)h^T\left(\hat{x}_{m,s}^-\right)K_{m,s}^T \\
&\quad + e_{4,m} K_{m,s} R_{m,s} K_{m,s}^T \\
&\quad + e_{5,m} K_{m,s} tr\left\{\mathcal{P}_{\tilde{x}_{m,s}^-}\right\} I K_{m,s}^T
\end{aligned}
\tag{10.22}
$$

where

$$
\begin{aligned}
\mathcal{P}_{\tilde{x}_{m,s}^-} &= \left(1 + b_m\right) a_m^2 tr\left\{\mathcal{P}_{\tilde{x}_{m,s}}\right\} I \\
&\quad + \left(1 + b_m^{-1}\right) A \mathcal{P}_{\tilde{x}_{m,s}} A^T + Q_{m,s}
\end{aligned}
\tag{10.23}
$$

with $Q_{m,s}$ to be determined later. Then, $\mathcal{P}_{\tilde{x}_{m,s}}$ are upper bounds on local error covariances $P_{\tilde{x}_{m,s}}$, i.e.

$$
P_{\tilde{x}_{m,s}} \leq \mathcal{P}_{\tilde{x}_{m,s}}
\tag{10.24}
$$

holds for all $s \geq 0$. Moreover, $\mathcal{P}_{\tilde{x}_{m,s}}$ are minimized by gains:

$$
K_{m,s} = e_{1,m} \bar{\gamma}_m \mathcal{P}_{\tilde{x}_{m,s}^-} C_m^T U_{m,s}^{-1}
\tag{10.25}
$$

where

$$
\begin{aligned}
U_{m,s} &= \left(e_{1,m}\bar{\gamma}_m^2 + e_{2,m}\right) C_m \mathcal{P}_{\tilde{x}_{m,s}^-} C_{m,s}^T \\
&\quad + e_{3,m} h\left(\hat{x}_{m,s}^-\right)h^T\left(\hat{x}_{m,s}^-\right) \\
&\quad + e_{4,m} R_{m,s} + e_{5,m} tr\left\{\mathcal{P}_{\tilde{x}_{m,s}^-}\right\} I.
\end{aligned}
\tag{10.26}
$$

Furthermore, the minimal upper bounds are calculated by

$$
\mathcal{P}_{\tilde{x}_{m,s}}^{\min} = e_{1,m} \mathcal{P}_{\tilde{x}_{m,s}^-}^{\min} - e_{1,m} \bar{\gamma}_m \mathcal{P}_{\tilde{x}_{m,s}^-}^{\min} C_m^T K_{m,s}^T,
\tag{10.27}
$$

where

$$
\begin{aligned}
\mathcal{P}_{\tilde{x}_{m,s}^-}^{\min} &= \left(1 + b_m\right) a_m^2 tr\left\{\mathcal{P}_{\tilde{x}_{m,s-1}}^{\min}\right\} I \\
&\quad + \left(1 + b_m^{-1}\right) A \mathcal{P}_{\tilde{x}_{m,s-1}}^{\min} A^T + Q_{m,s-1}.
\end{aligned}
\tag{10.28}
$$

Proof The cross-terms in (10.21) can be obtained as follows by recurring to the elementary inequalities.

$$
\begin{aligned}
&\mathbb{E}\left\{\xi_{m,s}\tilde{x}_{m,s}^{T}\right\}A^{T}+A\mathbb{E}^{T}\left\{\xi_{m,s}\tilde{x}_{m,s}^{T}\right\} \\
&\leq b_{m}\mathbb{E}\left\{\xi_{m,s}\xi_{m,s}^{T}\right\}+b_{m}^{-1}AP_{\tilde{x}_{m,s}}A^{T}
\end{aligned}
\tag{10.29}
$$

where $b_m > 0$ are any positive constants. Keeping in mind Assumption 10.2 and definitions of $\xi_{m,s}$ and $\zeta_{m,s}$, we have

$$
\begin{aligned}
\mathbb{E}\left\{\xi_{m,s}\xi_{m,s}^{T}\right\} &\leq \mathbb{E}\left\{\left\|\xi_{m,s}\right\|^{2}\right\}I \\
&\leq \mathbb{E}\left\{\left(a_{m}\left\|\tilde{x}_{m,s}^{-}\right\|^{2}\right)\right\}I \\
&\leq a_{m}^{2}\mathbb{E}\left\{\left\|\tilde{x}_{m,s}^{-}\right\|^{2}\right\}I \\
&= a_{m}^{2}tr\left\{P_{\tilde{x}_{m,s}^{-}}\right\}I,
\end{aligned}
\tag{10.30}
$$

$$
\begin{aligned}
\mathbb{E}\left\{\zeta_{m,s}\zeta_{m,s}^{T}\right\} &\leq \mathbb{E}\left\{\left\|\zeta_{m,s}\right\|^{2}\right\}I \\
&\leq \mathbb{E}\left\{\left(c_{m}\left\|\tilde{x}_{m,s}^{-}\right\|\right)^{2}\right\}I \\
&\leq c_{m}^{2}\mathbb{E}\left\{\left\|\tilde{x}_{m,s}^{-}\right\|^{2}\right\}I \\
&= c_{m}^{2}tr\left\{P_{\tilde{x}_{m,s}^{-}}\right\}I
\end{aligned}
\tag{10.31}
$$

where a_m and c_m are constants satisfying Assumption 10.2. Substituting (10.29)–(10.31) into (10.21) yields

$$
\begin{aligned}
P_{\tilde{x}_{m,s+1}^{-}} &\leq \left(1+b_{m}^{-1}\right)AP_{\tilde{x}_{m,s}^{-}}A^{T}+Q_{m,s} \\
&\quad +\left(1+b_{m}\right)a_{m}^{2}tr\left\{P_{\tilde{x}_{m,s}^{-}}\right\}I
\end{aligned}
\tag{10.32}
$$

where $tr\{\cdot\}$ is the trace of matrix "$\cdot$."

Recurring to the elementary inequalities, for any positive constants $d_{1,m}$, $d_{2,m}$, $d_{3,m}$, $d_{4,m}$, $d_{5,m}$, the cross-terms on the right-hand side of (10.17) are manipulated as follows:

$$
\begin{aligned}
&-\mathbb{E}\left\{\left(I-K_{m,s}\bar{\gamma}_{m}C_{m}\right)\tilde{x}_{m,s}^{-}\zeta_{m,s}^{T}\bar{\gamma}_{m}K_{m,s}^{T}\right\} \\
&-\mathbb{E}^{T}\left\{\left(I-K_{m,s}\bar{\gamma}_{m}C_{m}\right)\tilde{x}_{m,s}^{-}\zeta_{m,s}^{T}\bar{\gamma}_{m}K_{m,s}^{T}\right\} \\
&\leq d_{1,m}\bar{\gamma}_{m,s}^{2}K_{m,s}\mathbb{E}\left\{\zeta_{m,s}\zeta_{m,s}^{T}\right\}K_{m,s}^{T} \\
&\quad +d_{1,m}^{-1}\left(I-K_{m,s}\bar{\gamma}_{m}C_{m}\right)P_{\tilde{x}_{m,s}^{-}}\left(I-K_{m,s}\bar{\gamma}_{m}C_{m}\right)^{T},
\end{aligned}
\tag{10.33}
$$

$$
\begin{aligned}
&- \mathbb{E}\left\{ \left(I - K_{m,s}\gamma_{m,s}C_m \right) \tilde{x}_{m,s}^- \upsilon_{m,s}^T \bar{\gamma}_m K_{m,s}^T \right\} \\
&- \mathbb{E}^T \left\{ \left(I - K_{m,s}\gamma_{m,s}C_m \right) \tilde{x}_{m,s}^- \upsilon_{m,s}^T \bar{\gamma}_m K_{m,s}^T \right\} \\
&\le d_{2,m} \bar{\gamma}_m^2 K_{m,s} R_{m,s} K_{m,s}^T \\
&\quad + d_{2,m}^{-1} \left(I - K_{m,s}\bar{\gamma}_m C_m \right) P_{\tilde{x}_{m,s}^-} \left(I - K_{m,s}\bar{\gamma}_m C_m \right)^T ,
\end{aligned}
\tag{10.34}
$$

$$
\begin{aligned}
&\mathbb{E}\left\{ K_{m,s} \left(\gamma_{m,s} - \bar{\gamma}_m \right)^2 C_m \tilde{x}_{m,s}^- h^T(\hat{x}_{m,s}^-) K_{m,s}^T \right\} \\
&+ \mathbb{E}^T \left\{ K_{m,s} (\gamma_{m,s} - \bar{\gamma}_m)^2 C_m \tilde{x}_{m,s}^- h^T(\hat{x}_{m,s}^-) K_{m,s}^T \right\} \\
&\le d_{3,m} \tilde{\gamma}_m K_{m,s} C_m P_{\tilde{x}_{m,s}^-} C_{m,s}^T K_{m,s}^T \\
&\quad + d_{3,m}^{-1} \tilde{\gamma}_m K_{m,s} h(\hat{x}_{m,s}^-) h^T(\hat{x}_{m,s}^-) K_{m,s}^T ,
\end{aligned}
\tag{10.35}
$$

$$
\begin{aligned}
&\mathbb{E}\left\{ K_{m,s} \left(\gamma_{m,s} - \bar{\gamma}_m \right)^2 C_m \tilde{x}_{m,s}^- \zeta_{m,s}^T K_{m,s}^T \right\} \\
&+ \mathbb{E}^T \left\{ K_{m,s} \left(\gamma_{m,s} - \bar{\gamma}_m \right)^2 C_m \tilde{x}_{m,s}^- \zeta_{m,s}^T K_{m,s}^T \right\} \\
&\le d_{4,m} \tilde{\gamma}_m K_{m,s} C_m P_{\tilde{x}_{m,s}^-} C_{m,s}^T K_{m,s}^T \\
&\quad + d_{4,m}^{-1} \tilde{\gamma}_m K_{m,s} \mathbb{E}\left\{ \zeta_{m,s} \zeta_{m,s}^T \right\} K_{m,s}^T ,
\end{aligned}
\tag{10.36}
$$

$$
\begin{aligned}
&\mathbb{E}\left\{ K_{m,s} \left(\gamma_{m,s} - \bar{\gamma}_m \right)^2 h\left(\hat{x}_{m,s}^- \right) \zeta_{m,s}^T K_{m,s}^T \right\} \\
&+ \mathbb{E}^T \left\{ K_{m,s} \left(\gamma_{m,s} - \bar{\gamma}_m \right)^2 h\left(\hat{x}_{m,s}^- \right) \zeta_{m,s}^T K_{m,s}^T \right\} \\
&\le d_{5,m} \tilde{\gamma}_m K_{m,s} h\left(\hat{x}_{m,s}^- \right) h^T\left(\hat{x}_{m,s}^- \right) K_{m,s}^T \\
&\quad + d_{5,m}^{-1} \tilde{\gamma}_m K_{m,s} \mathbb{E}\left\{ \zeta_{m,s} \zeta_{m,s}^T \right\} K_{m,s}^T .
\end{aligned}
\tag{10.37}
$$

Putting (10.33)–(10.37) into (10.17) leads to

$$
\begin{aligned}
P_{\tilde{x}_{m,s}} \le{}& e_{1,m} \left(I - K_{m,s}\bar{\gamma}_m C_m \right) P_{\tilde{x}_{m,s}^-} \left(I - K_{m,s}\bar{\gamma}_m C_m \right)^T \\
&+ e_{2,m} K_{m,s} C_m P_{\tilde{x}_{m,s}^-} C_{m,s}^T K_{m,s}^T \\
&+ e_{3,m} K_{m,s} h(\hat{x}_{m,s}^-) h^T(\hat{x}_{m,s}^-) K_{m,s}^T \\
&+ e_{4,m} K_{m,s} R_{m,s} K_{m,s}^T \\
&+ e_{5,m} K_{m,s} \, tr\{ P_{\tilde{x}_{m,s}^-} \} I K_{m,s}^T .
\end{aligned}
\tag{10.38}
$$

Inspired by (10.38), define a matrix function $\phi_s\left(\mathcal{P}_{\tilde{x}_{m,s}}\right): \mathbb{R}^{n\times n} \to \mathbb{R}^{n\times n}$ by

$$
\begin{aligned}
\mathcal{P}_{\tilde{x}_{m,s}} &\triangleq \phi_s\left(\mathcal{P}_{\tilde{x}_{m,s-1}}\right) \\
&= e_{1,m}\left(I - K_{m,s}\bar{\gamma}_m C_m\right)\mathcal{P}_{\tilde{x}_{m,s}^-}\left(I - K_{m,s}\bar{\gamma}_m C_m\right)^T \\
&\quad + e_{2,m}K_{m,s}C_m \mathcal{P}_{\tilde{x}_{m,s}^-} C_{m,s}^T K_{m,s}^T \\
&\quad + e_{3,m}K_{m,s}h(\hat{x}_{m,s}^-)h^T(\hat{x}_{m,s}^-)K_{m,s}^T \\
&\quad + e_{4,m}K_{m,s}R_{m,s}K_{m,s}^T \\
&\quad + e_{5,m}K_{m,s}tr\left\{\mathcal{P}_{\tilde{x}_{m,s}^-}\right\}IK_{m,s}^T
\end{aligned}
\tag{10.39}
$$

where

$$
\begin{aligned}
\mathcal{P}_{\tilde{x}_{m,s}^-} &\triangleq \left(1+b_m^{-1}\right)A\mathcal{P}_{\tilde{x}_{m,s-1}}A^T + Q_{m,s-1} \\
&\quad + \left(1+b_m\right)a_m^2 tr\left\{\mathcal{P}_{\tilde{x}_{m,s-1}}\right\}I.
\end{aligned}
\tag{10.40}
$$

Bearing (10.39)–(10.40) in mind, it is verified that function $\phi_s\left(\mathcal{P}_{\tilde{x}_{m,s}}\right)$ satisfies conditions in the matrix inequality lemma in [184]. Thus, given initial conditions $\mathcal{P}_{\tilde{x}_{m,0}} = P_{\tilde{x}_{m,0}}$, there exist solutions $\mathcal{P}_{\tilde{x}_{m,s}}$ to equations

$$
\mathcal{P}_{\tilde{x}_{m,s+1}} = \phi_s\left(\mathcal{P}_{\tilde{x}_{m,s}}\right)
$$

such that

$$
P_{\tilde{x}_{m,s}} \leq \mathcal{P}_{\tilde{x}_{m,s}}
$$

is true for all $s \geq 0$, that is, $\mathcal{P}_{\tilde{x}_{m,s}}$ are the upper bounds on $P_{\tilde{x}_{m,s}}$.

Additionally, one has

$$
\begin{aligned}
\frac{\partial tr\left(\mathcal{P}_{\tilde{x}_{m,s}}\right)}{\partial K_{m,s}} &= -2e_{1,m}\left(I - K_{m,s}\bar{\gamma}_m C_m\right)\mathcal{P}_{\tilde{x}_{m,s}^-}C_m^T\bar{\gamma}_m \\
&\quad + 2e_{2,m}K_{m,s}C_m \mathcal{P}_{\tilde{x}_{m,s}^-} C_{m,s}^T \\
&\quad + 2e_{3,m}K_{m,s}h\left(\hat{x}_{m,s}^-\right)h^T\left(\hat{x}_{m,s}^-\right) \\
&\quad + 2e_{4,m}K_{m,s}R_{m,s} \\
&\quad + 2e_{5,m}K_{m,s}tr\left\{\mathcal{P}_{\tilde{x}_{m,s}^-}\right\}I \\
&= 0,
\end{aligned}
\tag{10.41}
$$

which leads to

$$K_{m,s} = e_{1,m}\bar{\gamma}_m P_{\tilde{x}_{m,s}^-} C_m^T U_{m,s}^{-1}, \tag{10.42}$$

where

$$U_{m,s} = \left(e_{1,m}\bar{\gamma}_m^2 + e_{2,m}\right) C_m P_{\tilde{x}_{m,s}^-} C_{m,s}^T$$
$$+ e_{3,m} h\left(\hat{x}_{m,s}^-\right) h^T\left(\hat{x}_{m,s}^-\right) \tag{10.43}$$
$$+ e_{4,m} R_{m,s} + e_{5,m} tr\left\{P_{\tilde{x}_{m,s}^-}\right\} I.$$

Here, (10.42) and (10.43) are the same as (10.25) and (10.26), respectively.

Expanding out (10.39) yields

$$P_{\tilde{x}_{m,s}}$$
$$= e_{1,m} P_{\tilde{x}_{m,s}^-} - e_{1,m}\bar{\gamma}_m K_{m,s} C_m P_{\tilde{x}_{m,s}^-} \tag{10.44}$$
$$- e_{1,m}\bar{\gamma}_m \left(K_{m,s} C_m P_{\tilde{x}_{m,s}^-}\right)^T + K_{m,s} U_{m,s} K_{m,s}^T$$

which, together with (10.25)–(10.26), results in

$$P_{\tilde{x}_{m,s}}^{\min} = e_{1,m} P_{\tilde{x}_{m,s}^-}^{\min} - e_{1,m}\bar{\gamma}_m P_{\tilde{x}_{m,s}^-}^{\min} C_m^T K_{m,s}^T,$$

i.e. (10.27), where $P_{\tilde{x}_{m,s}^-}^{\min}$ is calculated by

$$P_{\tilde{x}_{m,s}^-}^{\min} = (1 + b_m) a_m^2 tr\left\{P_{\tilde{x}_{m,s-1}}^{\min}\right\} I$$
$$+ \left(1 + b_m^{-1}\right) A P_{\tilde{x}_{m,s-1}}^{\min} A^T + Q_{m,s-1},$$

i.e. (10.28). This completes the proof.

Remark 10.3 Making a close observation on Theorem 10.2 and the expression $\bar{\gamma}_m = \bar{\alpha}_m\left(1 - \bar{\beta}_m\right)$, it is easily concluded that (1) a higher occurrence probability $\bar{\alpha}_m$ of the SCP and (2) a lower success rate $\bar{\beta}_m$ of DoS attack will both give rise to a smaller upper bound $P_{\tilde{x}_{m,s}^-}^{\min}$ on the filtering error covariance of the mth local filter, thereby producing better local filtering performance. This is reasonable because a higher occurrence probability $\bar{\alpha}_m$ means that more information from the mth sensor is transmitted to the filter when tracking the system state, while a lower success rate $\bar{\beta}_m$ implies that measurements from the mth sensor are less likely to be prohibited from sending information to its remote filter.

With the availability of the minimized upper bounds $P_{\tilde{x}_{m,s}^-}^{\min}$ in (10.27), the weight matrices $W_{m,s}$ can now be determined as follows according to the federated fusion rule.

$$W_{m,s} = P_{\tilde{x}_s}^{\min}\left(P_{\tilde{x}_{m,s}}^{\min}\right)^{-1}, \tag{10.45}$$

where $\mathcal{P}_{x_s}^{\min}$ is the minimized fused upper bound of the following form:

$$\mathcal{P}_{\tilde{x}_s}^{\min} = \left(\sum_{m=1}^{p} \left(\mathcal{P}_{\tilde{x}_{m,s}}^{\min} \right)^{-1} \right)^{-1}. \tag{10.46}$$

We now accomplish the design procedure of the desired variance-constrained non-linear fusion estimator (10.6). In this chapter, due to its distributed yet effective fusion manner, the federated fusion rule is adopted to integrate the filtering results from all local filters in order to make full use of all available information and produce better filtering results than the ones generated by the local filters. Following the federated fusion rule, the local estimates, the upper bounds and the noise covariances at time $s-1$ should be initialized as follows:

$$\begin{cases} \hat{x}_{m,s-1} \overset{\Delta}{=} \hat{x}_{s-1}, \\ \mathcal{P}_{\tilde{x}_{m,s-1}}^{\min} \overset{\Delta}{=} \epsilon_m^{-1} \mathcal{P}_{\tilde{x}_{s-1}}^{\min}, \\ Q_{m,s-1} \overset{\Delta}{=} \epsilon_m^{-1} Q_s \end{cases} \tag{10.47}$$

where ϵ_m are the information redistribution weights satisfying $\sum_{m=1}^{p} \epsilon_m = 1$.

Putting Theorems 10.1–10.2 together, we have the desired variance-constrained nonlinear filtering fusion algorithm, whose pseudocode is outlined in Table 10.1.

With the developed variance-constrained nonlinear filtering fusion paradigm in hand, we are now ready to assess its fusion performance by exploring the boundedness property of the minimal fused upper bound $\mathcal{P}_{\tilde{x}_s}^{\min}$.

Assumption 10.3 There exist non-negative constants $\bar{a}$, $\bar{c}_m$, $\bar{q}_m$, $\underline{c}_m$, $\underline{h}_m$, $\underline{q}_m$, $\underline{r}_m$ and $\underline{q}_m$ such that

$$\underline{q}_m I \leq Q_{m,s} \leq \bar{q}_m I, \underline{r}_m I \leq R_{m,s}, AA^T \leq \bar{a} I,$$

$$\underline{c}_m I \leq C_m C_m^T \leq \bar{c}_m I, \underline{h}_m I \leq h_m \left(\hat{x}_{m,s}^- \right) h_m^T \left(\hat{x}_{m,s}^- \right),$$

are true for all $s \geq 0$.

For the sake of notation brevity, we denote

$$\underline{u}_m \overset{\Delta}{=} e_{2,m} \underline{c}_m \underline{q}_m + e_{3,m} \underline{h}_m + e_{4,m} \underline{r}_m + e_{5,m} \underline{q}_m.$$

Theorem 10.3 Under Assumptions 10.1–10.3, there exist bounds $\underline{p}$ and $\bar{p}_s$ such that the minimal fused upper bound $\mathcal{P}_{\tilde{x}_s}^{\min}$ satisfies

$$\underline{p} I \leq \mathcal{P}_{\tilde{x}_s}^{\min} \leq \bar{p}_s I \tag{10.48}$$

with

$$\underline{p} = \left(\sum_{m=1}^{p} \underline{p}_m^{-1} \right)^{-1}, \tag{10.49}$$

$$\overline{P}_s = \left(\sum_{m=1}^{p} \overline{P}_{m,s}^{-1} \right)^{-1} \tag{10.50}$$

where

$$\underline{p}_m = \left[\frac{1}{\underline{q}_m e_{1,m}} + \frac{\overline{\gamma}_m^2 \overline{c}_m}{\underline{u}_m} \right]^{-1}, \tag{10.51}$$

$$\begin{aligned}
\overline{P}_{m,s} &= e_{1,m} \overline{q}_m \sum_{i=0}^{s} \left(e_{1,m} \left[(1+b_m) a_m^2 n + \left(1 + b_m^{-1} \right) \overline{a} \right] \right)^{i} \\
&\quad + \overline{P}_{m,0} \left(e_{1,m} \left[(1+b_m) a_m^2 n + \left(1 + b_m^{-1} \right) \overline{a} \right] \right)^{s+1}.
\end{aligned} \tag{10.52}$$

Here, $\overline{P}_{m,0} \triangleq \lambda_{\max} \left(\mathcal{P}_{\tilde{x}_{m,0}}^{\min} \right)$ and $\lambda_{\max}(\cdot)$ is the maximal eigenvalue of "$\cdot$".

Proof Parallel to (10.26), let us define

$$U_{m,s}^{\min} \triangleq e_{1,m} \overline{\gamma}_m C_m \mathcal{P}_{\tilde{x}_{m,s}}^{\min} C_{m,s}^T \overline{\gamma}_m + \tilde{U}_{m,s}, \tag{10.53}$$

$$\begin{aligned}
\tilde{U}_{m,s} &\triangleq e_{2,m} C_m \mathcal{P}_{\tilde{x}_{m,s}}^{\min} C_{m,s}^T + e_{3,m} h\left(\hat{x}_{m,s}^- \right) h^T \left(\hat{x}_{m,s}^- \right) + e_{4,m} R_{m,s} \\
&\quad + e_{5,m} tr\left\{ P_{\tilde{x}_{m,s}^-} \right\} I.
\end{aligned} \tag{10.54}$$

In the sequel, we first explore the lower bound $\underline{p}_m I$ on $\mathcal{P}_{\tilde{x}_{m,s}}^{\min}$. Putting (10.53)–(10.54) into (10.27) gives

$$\begin{aligned}
\mathcal{P}_{\tilde{x}_{m,s}}^{\min} &= e_{1,m} \mathcal{P}_{\tilde{x}_{m,s}^-}^{\min} - e_{1,m} \mathcal{P}_{\tilde{x}_{m,s}^-}^{\min} C_m^T \overline{\gamma}_m \left[e_{1,m} \overline{\gamma}_m C_m \mathcal{P}_{\tilde{x}_{m,s}^-}^{\min} C_{m,s}^T \overline{\gamma}_m + \tilde{U}_{m,s} \right]^{-1} \\
&\quad \times C_m \mathcal{P}_{\tilde{x}_{m,s}^-}^{\min} \overline{\gamma}_m e_{1,m} \\
&= \left[e_{1,m}^{-1} \left(\mathcal{P}_{\tilde{x}_{m,s}^-}^{\min} \right)^{-1} + \overline{\gamma}_m C_m^T \tilde{U}_{m,s}^{-1} C_m \overline{\gamma}_m \right]^{-1}.
\end{aligned} \tag{10.55}$$

Taking into account Assumption 10.3 and the positive definiteness of $\mathcal{P}_{\tilde{x}_{m,s-1}}^{\min}$, we have

$$\begin{aligned}
\mathcal{P}_{\tilde{x}_{m,s}^-}^{\min} &\geq (1+b_m) a_m^2 tr\left\{ \mathcal{P}_{\tilde{x}_{m,s-1}}^{\min} \right\} I + Q_{m,s-1} \\
&\geq \underline{q}_m I,
\end{aligned} \tag{10.56}$$

$$\begin{aligned}
\tilde{U}_{m,s} &\geq \left(e_{2,m} \underline{c}_m \underline{q}_m + e_{3,m} \underline{h}_m + e_{4,m} \underline{r}_m + e_{5,m} \underline{q}_m \right) I \\
&= \underline{u}_m I.
\end{aligned} \tag{10.57}$$

Putting (10.56)–(10.57) into (10.55) results in

$$
\begin{aligned}
\mathcal{P}_{\tilde{x}_{m,s}}^{\min} &= \left[e_{1,m}^{-1} \left(\mathcal{P}_{\tilde{x}_{m,s}^{-}}^{\min} \right)^{-1} + \bar{\gamma}_m C_m^T \tilde{U}_{m,s}^{-1} C_m \bar{\gamma}_m \right]^{-1} \\
&\geq \left[\frac{1}{\underline{q}_m e_{1,m}} + \frac{\bar{\gamma}_m^2 \bar{c}_m}{\underline{u}_m} \right]^{-1} I \\
&= \underline{p}_m I
\end{aligned}
\tag{10.58}
$$

where $\underline{p}_m$ is defined by

$$
\underline{p} I \leq \mathcal{P}_{\tilde{x}_s}^{\min} \leq \bar{p}_s I,
$$

i.e. (10.51).

Now, let us explore the upper bound $\bar{p}_{m,s} I$ on $\mathcal{P}_{\tilde{x}_{m,s}}^{\min}$ by using mathematical induction.
Initial Step. For s = 0, we have

$$
\mathcal{P}_{\tilde{x}_{m,0}}^{\min} \leq \lambda_{\max}\left(\mathcal{P}_{\tilde{x}_{m,0}}^{\min} \right) I = \bar{p}_{m,0} I,
\tag{10.59}
$$

where $\bar{p}_{m,0}$ is defined by $\bar{p}_{m,0} \triangleq \lambda_{\max}\left(\mathcal{P}_{\tilde{x}_{m,0}}^{\min} \right)$.
Inductive Step. Initial step assumes

$$
\mathcal{P}_{\tilde{x}_{m,t}}^{\min} \leq \bar{p}_{m,t} I
$$

is true for $t=0$. Letting $\mathcal{P}_{\tilde{x}_{m,t}}^{\min} \leq \bar{p}_{m,t} I$ be true for t=s. We then need to prove

$$
\mathcal{P}_{\tilde{x}_{m,t}}^{\min} \leq \bar{p}_{m,t} I
$$

is true for t = s + 1. Paying attention to (10.55), we have

$$
\left(\mathcal{P}_{\tilde{x}_{m,s+1}}^{\min} \right)^{-1} = e_{1,m}^{-1} \left(\mathcal{P}_{\tilde{x}_{m,s+1}^{-}}^{\min} \right)^{-1} + \bar{\gamma}_m C_m^T \tilde{U}_{m,s+1}^{-1} C_m \bar{\gamma}_m,
\tag{10.60}
$$

which implies that

$$
\begin{aligned}
\mathcal{P}_{\tilde{x}_{m,s+1}}^{\min} &\leq e_{1,m} \mathcal{P}_{\tilde{x}_{m,s+1}^{-}}^{\min} \\
&= e_{1,m} \left(1+b_m\right) a_m^2 tr\left\{ \mathcal{P}_{\tilde{x}_{m,s}}^{\min} \right\} I + e_{1,m}\left(1+b_m^{-1}\right) A \mathcal{P}_{\tilde{x}_{m,s}}^{\min} A^T + e_{1,m} Q_{m,s} \\
&\leq e_{1,m}\left[n\left(1+b_m\right) a_m^2 + \left(1+b_m^{-1}\right)\bar{a} \right] \bar{p}_{m,s} I + e_{1,m} \bar{q}_m I \\
&= e_{1,m} \bar{q}_m \sum_{i=0}^{s} \left(e_{1,m}\left[n\left(1+b_m\right)a_m^2 + \left(1+b_m^{-1}\right)\bar{a} \right] \right)^i \\
&\quad + \bar{p}_{m,0}\left(e_{1,m}\left[n\left(1+b_m\right)a_m^2 + \left(1+b_m^{-1}\right)\bar{a} \right] \right)^{s+1} \\
&= \bar{p}_{m,s+1},
\end{aligned}
\tag{10.61}
$$

where $\bar{P}_{m,s}$ is defined by

$$\underline{P}_m = \left[\frac{1}{\underline{q}_m e_{1,m}} + \frac{\bar{\gamma}_m^2 \bar{c}_m}{\underline{u}_m}\right]^{-1},$$

$$\bar{P}_{m,s} = e_{1,m}\bar{q}_m \sum_{i=0}^{s}\left(e_{1,m}\left[(1+b_m)a_m^2 n + (1+b_m^{-1})\bar{a}\right]\right)^i$$
$$+ \bar{P}_{m,0}\left(e_{1,m}\left[(1+b_m)a_m^2 n + (1+b_m^{-1})\bar{a}\right]\right)^{s+1},$$

i.e. (10.52).

Therefore, we conclude that

$$\mathcal{P}_{\tilde{x}_{m,t}}^{\min} \leq \bar{P}_{m,t} I$$

is true for $t = s + 1$.

Based on the initial and inductive steps, we now conclude that

$$\mathcal{P}_{\tilde{x}_{m,s}}^{\min} \leq \bar{P}_{m,s} I$$

is true for all $s \geq 0$, where $\bar{P}_{m,0} = \lambda_{\max}\left(\mathcal{P}_{\tilde{x}_{m,0}}^{\min}\right)$ and $\bar{P}_{m,s}$ is given by (10.52).

At last, based on (10.46), we have

$$\begin{aligned}
\mathcal{P}_{\tilde{x}_s}^{\min} &= \left(\sum_{m=1}^{p}\left(\mathcal{P}_{\tilde{x}_{m,s}}^{\min}\right)^{-1}\right)^{-1} \\
&\geq \left(\sum_{m=1}^{p}\left(\underline{P}_m I\right)^{-1}\right)^{-1} \\
&= \underline{p} I
\end{aligned} \tag{10.62}$$

where $\underline{p}$ is defined by

$$\underline{P}_m = \left[\frac{1}{\underline{q}_m e_{1,m}} + \frac{\bar{\gamma}_m^2 \bar{c}_m}{\underline{u}_m}\right]^{-1},$$

i.e. (10.49).

On the other hand, we have

$$\begin{aligned}
\mathcal{P}_{\tilde{x}_s}^{\min} &= \left(\sum_{m=1}^{p}\left(\mathcal{P}_{\tilde{x}_{m,s}}^{\min}\right)^{-1}\right)^{-1} \\
&\leq \left(\sum_{m=1}^{p}\left(\bar{P}_{m,s} I\right)^{-1}\right)^{-1} \\
&= \bar{P}_s I
\end{aligned} \tag{10.63}$$

where $\bar{p}_s$ is defined by

$$\bar{p}_{m,s} = e_{1,m}\bar{q}_m \sum_{i=0}^{s}\left(e_{1,m}\left[(1+b_m)a_m^2 n + (1+b_m^{-1})\bar{a}\right]\right)^i$$
$$+ \bar{p}_{m,0}\left(e_{1,m}\left[(1+b_m)a_m^2 n + (1+b_m^{-1})\bar{a}\right]\right)^{s+1},$$

i.e. (10.50), which completes the proof.

Remark 10.4 It is seen from the proof of Theorem 10.3 that Assumption 10.3 plays a pivotal role in ensuring the boundedness of the minimized fused upper bound $\mathcal{P}_{\tilde{x}_s}^{\min}$. Accounting for possible energy constraints in physical systems, such an assumption is reasonable, and appropriate positive numbers satisfying $\underline{q}_m I \le Q_{m,s} \le \bar{q}_m I$, $\underline{r}_m I \le R_m$, $AA^T \le \bar{a}I$, $\underline{c}_m I \le C_m C_m^T \le \bar{c}_m I$ can certainly be found.

Remark 10.5 In comparison to the available filtering literature on CPSs, our primary results own the following distinctive merits: (1) the addressed problem is new as multiple engineering-oriented phenomena (e.g. the system nonlinearity, cyber-attack and communication protocol) are comprehensively considered; (2) the devised fusion paradigm is new as the variance-constrained estimator is purposely built by means of intensive stochastic analysis; and (3) the performance evaluation is new where boundedness of minimal upper bounds with respect to both fused and local error covariances is rigorously guaranteed.

Remark 10.6 It can be seen from the design and analysis of the proposed variance-constrained filtering fusion algorithm that the implementation of such an algorithm involves the calculation of a few matrix equations. Consequently, the computational complexity of the proposed algorithm is highly dependent on both the dimensions of equations and number of decision variables. Generally speaking, there are two effective methods that can be used to reduce the computational complexity of filtering algorithm involving mass matrix operations, i.e. the model reduction method and the suboptimal method. To be specific, the mode reduction refers to the exploration of a possible lower-dimensional model that can be used to replace the original model without adding significant errors in practical applications, while the suboptimal method means that the development of certain suboptimal filtering algorithms might have worse performance but lower computation cost in contrast to the original algorithm. Obviously, in both methods, a proper trade-off between the filtering performance and the computation complexity should be found in order to accomplish the filtering task in an efficient and accurate way.

10.3 AN ILLUSTRATIVE EXAMPLE

Denote the mean squared error (MSE) and average MSE of x_k^j (the jth entry of x_k, $j = 1, 2, \ldots, n$) as

$$\text{MSE}j = \frac{1}{M}\sum_{i=1}^{M}\left(x_s^{j(i)} - \hat{x}_s^{j(i)}\right)^2,$$

$$\text{AMSE}j = \frac{1}{MK}\sum_{i=1}^{M}\sum_{s=1}^{K}\left(x_s^{j(i)} - \hat{x}_s^{j(i)}\right)^2$$

Where $K = 60$ and $M = 1000$ are the numbers of Monte Carlo trials and simulation time steps, respectively. In the following simulations, several performance comparisons are made between the proposed variance-constrained nonlinear fusion estimator (10.6) and its local counterpart to elaborate its performance superiority.

Throughout all simulations, we set the information redistribution weight as $\epsilon_m = 0.5$ ($m = 1,2$), the SCP scheduling probabilities as $\bar{\alpha}_1 = 0.6$ and $\bar{\alpha}_2 = 0.4$ and the success rates of launched DoS attacks as $\bar{\beta}_1 = 0.2$ and $\bar{\beta}_2 = 0.3$. The rest of the parameters are

$$h_1(x_s) = 0.95x_{1,s} + 0.2\sin(x_{2,s}), Q_s = 0.5I_2,$$

$$h_2(x_s) = 0.98x_{1,s} + 0.1\sin(x_{2,s}), R_{1,s} = R_{2,s} = 0.5,$$

$$f(x_s) = \begin{bmatrix} -0.56x_{1,s} + 0.02x_{2,s} + 0.04\sin(x_{1,s} + 0.2x_{2,s}) \\ 0.2x_{1,s} - 0.64x_{2,s} + 0.06\cos(x_{2,s}) \end{bmatrix}.$$

Consequently, we have

$$\bar{\gamma}_1 = 0.48, \bar{\gamma}_2 = 0.28, \tilde{\gamma}_1 = 0.25, \tilde{\gamma}_2 = 0.2,$$

according to (10.15)–(10.16). Meanwhile, the parameters in Assumptions 10.2–10.3 are obtained as

$$\underline{c}_1 = \underline{c}_2 = 0.9, \bar{c}_1 = \bar{c}_2 = 1, \underline{h}_m = \underline{h}_2 = 0.$$

$$A = \begin{bmatrix} -0.56 & 0.02 \\ 0.2 & -0.64 \end{bmatrix}, C_1 = 0.95, C_2 = 0.98, \underline{r}_1 = \underline{r}_2 = 0.1,$$

$$a_1 = a_2 = 0.074, c_1 = c_2 = 0.23, \underline{q}_1 = \underline{q}_2 = 0.001, \bar{q}_1 = \bar{q}_2 = 1, \bar{a} = 0.01.$$

In addition, we set the initial values as

$$\mathcal{P}_{\tilde{x}_0}^{\min} = 0.05I_2, \mathcal{P}_{\tilde{x}_{m,0}}^{\min} = 0.1I_2, x_0 = \hat{x}_0 = \hat{x}_{m,0} = \begin{bmatrix} 0.5 & 0.4 \end{bmatrix}^T,$$

and constants (with respect to matrix inequalities) as

$$d_{1,1} = d_{1,2} = 4, d_{2,1} = d_{2,2} = 5, b_1 = 1, b_2 = 1,$$

$$d_{3,1} = d_{3,2} = 1, d_{4,1} = d_{4,2} = 0.05, d_{5,1} = d_{5,2} = 0.10.$$

As a result, we have

$$e_{1,1} = e_{1,2} = 0.55, e_{2,1} = 0.31, e_{2,2} = 0.25, e_{3,1} = 0.52,$$

$$e_{3,2} = 0.25, e_{4,1} = -0.86, e_{4,2} = -0.3, e_{5,1} = 0.37, e_{5,2} = 0.3.$$

Figure 10.1 sketches the phenomena of DoS attacks on two sensors with the afore-mentioned success rates $\bar{\beta}_1 = 0.2$ and $\bar{\beta}_2 = 0.3$ of launched DoS attacks. After implementing the developed variance-constrained filtering fusion algorithm and its two local counterparts (based on sensor 1 and sensor 2 measurements, respectively) for 1000 Monte Carlo trials, the associate $\log_{10}$ (MSE) results are shown in Figures 10.2–10.3, where the blue diamond line, green circle line and red pentagram line represent the $\log_{10}$ (MSE) values based on sensor 1 measurements, sensor 2 measurements and all measurements, respectively. It is observed from Figures 10.2–10.3 that the tracking performance based on all information outperforms that based on either sensor information. This is because (1) all available data is properly fused to achieve the tracking target in the fusion algorithm; and (2) only partial sensor information is used to carry out the tracking task in both local filtering algorithms.

In order to reach the boundedness conclusion presented in Theorem 10.3, Figure 10.4 sketches the variation trends of the fused covariance bound $\log_{10}\left(\mathrm{tr}\left\{\mathcal{P}_{\tilde{x}_s}^{\min}\right\}\right)$, where its lower bound is $\log_{10}\left(\mathrm{tr}\left\{\underline{p}I\right\}\right)$ and its upper bound is $\log_{10}\left(\mathrm{tr}\left\{\underline{p}I\right\}\right)$. It is seen from Figure 10.4 that the $\log_{10}\left(\mathrm{tr}\left\{\mathcal{P}_{\tilde{x}_s}^{\min}\right\}\right)$ curve is located between the $\log_{10}\left(\mathrm{tr}\left\{\underline{p}I\right\}\right)$ and $\log_{10}\left(\mathrm{tr}\left\{\bar{p}_s I\right\}\right)$ curves, which indicates the correctness of the boundedness conclusion. In summary, all these simulation results verify that the proposed variance-constrained filtering fusion framework is effective in providing satisfactory performance in cases of the system nonlinearity, SCP, DoS attacks and variance constraints.

Next, we discuss the influences from the DoS attacks on the developed variance-constrained nonlinear fusion estimation algorithm. For the different success rates $\left[\bar{\beta}_1, \bar{\beta}_2\right]$ of launched DoS attacks, the associated AMSEs of both the local and fusion estimator are clearly outlined in Table 10.2 after 1000 Monte Carlo trials, where Sm-AMSE j $(m = 1, 2, j = 1, 2)$ is the AMSE with respect to the jth entry of x_s and is obtained from sensor m. Looking at Table 10.2, we draw conclusions: (1) the performance of the fusion estimator always outperforms that of local estimators under all considered success rates $\left[\bar{\beta}_1, \bar{\beta}_2\right]$; (2) the local estimator using S1 measurements has higher tracking accuracy than that using S2 measurements; (3) as the success rates $\left[\bar{\beta}_1, \bar{\beta}_2\right]$ increase, the accuracy of both local and fusion estimators deteriorates; and (4) when the success rate $\bar{\beta}_m$ of any sensor exceeds 0.5, the accuracy of the corresponding local estimator degrades significantly.

At last, we discuss the influences from the stochastic scheduling behavior on the developed variance-constrained nonlinear fusion estimation algorithm. For different occurrence probabilities $\left[\bar{\alpha}_1, \bar{\alpha}_2\right]$ of the scheduling behavior, the associated AMSEs of both local and fusion estimator are clearly outlined in Table 10.3 after 1000 Monte Carlo trials. Looking at Table 10.3, we draw conclusions: (1) the performance of the fusion estimator always outperforms that of local estimators under all occurrence probabilities $[\bar{\alpha}_1, \bar{\alpha}_2]$ and (2) the local estimator corresponding to the sensor with a larger (smaller) occurrence probability (of the scheduling behavior) has higher (lower) tracking accuracy.

10.4 SUMMARY

In this chapter, we have addressed the variance-constrained filtering fusion problem for a nonlinear CPS under the DoS attacks and the SCP. A set of local nonlinear filters has been adopted under variance constraints, and strict upper bounds on both local and fused error covariances that have been guaranteed, where filter gains and weight matrices have been parameterized by minimizing such upper bounds. By resorting to the federated fusion criterion, all local estimates have been incorporated to obtain a fusion estimate. Further evaluation has been given by analyzing boundedness of gained upper bounds. Finally, the applicability of the proposed nonlinear fusion estimator has been validated by a simulation experiment. Some future research directions include (1) solving the variance-constrained filtering fusion problem for nonlinear cyber-physical systems under other communication protocols, e.g. the Round-Robin protocol and the try-once-discard protocol, and (2) solving the variance-constrained filtering fusion problem for nonlinear cyber-physical systems with other communication protocols, e.g. the Round-Robin protocol and the try-once-discard protocol.

11 Conclusions and Future Topics

In this book, the state estimation and MSFF problems with censored data have been thoroughly investigated under a constrained network environment. Latest results about multi-sensor fusion estimator design issues have been, firstly, surveyed under different types of censored data subject to various network constraints originated from real-world engineering practice. Then, in each chapter, there is discussion regarding the addressed MSFF with different censored data (e.g. one-side censored data and two-side censored data) and network constraints (e.g., communication delays, redundant channels and protocol scheduling). Meanwhile, a set of sufficient conditions has been secured for designed MSFF schemes to achieve certain pre-specified fusion performances including the minimum variance specification, uniform boundedness index and federated fusion criterion. Subsequently, both numerical and practical illustrative demonstrations (e.g. distributed target tracking, ballistic roll rate tracing, longitudinal flight control, etc.) have been presented that are capable of verifying the feasibility and efficacy of all developed MSFF frameworks.

This book has established a unified theoretical framework for MSFF synthesis/analysis with censored data under a constrained network environment. A great variety of engineering-oriented censoring phenomena and network constraints, especially those induced by limited bandwidth, have been taken into account in a systematical yet effective procedure. It is worth mentioning that the acquired MSFF results are still quite limited, and some concluding remarks and prospective topics are highlighted as follows.

1 The accurate characterization of censored measurements is still a challenging problem since it is not only intricate to find random variables (apart from Bernoulli random variables) that fully reflect the censoring phenomenon but also hard to formulate an equivalent measurement model that benefits the subsequent filter design. A possible approach to describing the censoring phenomenon is to introduce a set of random events, where each event indicates whether the measurement is censored or not at a specific time instant. Nevertheless, this might add more difficulties in calculating the PDF with respect to the system state. As such, a promising research trend is to propose certain new technologies to characterize the censoring measurements.

2 It has been pointed out in the previous section that one of the underlying challenges is the selection of an appropriate fusion rule. Centralized fusion yields the optimal filtering fusion performance but the largest computation burden. Although distributed fusion has smaller computation burden and stronger reliability, it might sometimes output the suboptimal rather than

DOI: 10.1201/9781003461623-11

the optimal filtering fusion performance. Moreover, the computation of the cross-covariance is also a hard nut to crack, especially in case of multiple network-induced phenomena. Hence, a prospective research topic is to develop suitable fusion rules capable of reaching the optimal filtering fusion performance with moderate computation burden under multiple network-induced phenomena.

3 So far, most performance analysis results on the TKF have built themselves on the Bernoulli interpretation of the Tobit measurement model. As a matter of fact, it is theoretically possible to design recursive filtering algorithms using the original Tobit measurement model. Accordingly, an interesting research topic is to examine the performance of such recursive filtering algorithms where network-induced phenomena are also accommodated.

4 Another prospective research direction is to study MSFF with censored measurements over sensor and/or complex networks, as the relative results are very few, not to mention the case where constrained communication between neighboring sensor nodes is concerned. Basically, due to the topology structure (coupling manner) between different sensor nodes, the presence of sensor networks (complex networks) will inevitably render substantial challenges on the structure design and performance analysis of the designed filtering algorithm. As such, a particularly attractive direction is to explore MSFF with censored measurements and constrained communication over sensor networks and complex networks.

5 With the increasing research attention on higher safety and reliability standards, the network security is of utmost importance in real-world engineering. In contrast to the discussed communication constraints (e.g. nonlinear disturbances and probabilistic sensor failures), such cyber-attack behaviors could be more malicious to the desired MSFF structure and even lead to destructive behaviors of the target system. Furthermore, because of the attack proliferation through data fusion, the resultant destruction to both system and filter might be doubled, giving rise to immeasurable losses of industrial production. Therefore, it would be interesting to investigate how the malicious attacks affect the multi-sensor Tobit Kalman filtering framework and performance under a constrained network environment.

6 There is no doubt that data-based fusion plays a more and more important role in a great variety of engineering practice. Compared with traditional model-based fusion, data-based fusion could provide a feasible solution to fusion problems where the studied process/system is unfamiliar or hard to be modeled. Hence, it leads to a particularly attractive area for developing some data-based MSFF schemes or improving the established model-based fusion schemes to deal with the MSFF problems of unfamiliar or unmodeled engineering systems.

Bibliography

[1] J. Huang, D. W. C. Ho, F. Li, W. Yang, and Y. Tang, Secure remote state estimation against linear man-in-the-middle attacks using watermarking, *Automatica*, vol. 121, art. no. 109182, 2020.

[2] H. Lin, J. Lam, and Z. Wang, Secure state estimation for systems under mixed cyber-attacks: Security and performance analysis, *Information Sciences*, vol. 546, pp. 943–960, 2021.

[3] J. Qin, M. Li, J. Wang, L. Shi, Y. Kang, and W. X. Zheng, Optimal denial-of-service attack energy management against state estimation over an SINR-based network, *Automatica*, vol. 119, art. no. 109090, 2020.

[4] H. Darvishi, D. Ciuonzo, E. R. Eide, and P. S. Rossi, Sensor-fault detection, isolation and accommodation for digital twins via modular data-driven architecture, *IEEE Sensors Journal*, vol. 21, no. 4, pp. 4827–4838, 2021.

[5] H. Geng, Y. Liang, Y. Liu and F. E. Alsaadi, Bias estimation for asynchronous multi-rate multi-sensor fusion with unknown inputs, *Information Fusion*, vol. 39, pp. 139–153, 2018.

[6] R. Gravina, P. Alinia, H. Ghasemzadeh, and G. Fortino, Multi-sensor fusion in body sensor networks: State-of-the-art and research challenges, *Information Fusion*, Vol. 35, pp. 68–80, 2017.

[7] Y. Niu, L. Sheng, M. Gao, and D. Zhou, Dynamic event-triggered state estimation for continuous-time polynomial nonlinear systems with external disturbances, *IEEE Transactions on Industrial Informatics*, vol. 17, no. 6, pp. 3962–3970, 2021.

[8] B. Sun, S. Li, Q. Gu, and J. Ou, Coupling of peridynamics and numerical substructure method for modeling structures with local discontinuities, *CMES-Computer Modeling in Engineering & Sciences*, vol. 120, no. 3, pp. 739–757, 2019.

[9] B. Sun, J. Yan, S. Li, and Z. Kang, Peridynamic modeling and simulation of ice craters by impact, *CMES-Computer Modeling in Engineering & Sciences*, vol. 121, no. 2, pp. 465–492, 2019.

[10] B. Chen, G. Hu, D. W. C. Ho, W. Zhang, and L. Yu, Distributed robust fusion estimation with application to state monitoring systems, *IEEE Transactions on Systems, Man, and Cybernetics: Systems*, vol. 47, no. 11, pp. 2994–3005, 2017.

[11] W. Xu, D. W. C. Ho, L. Li, and J. Cao, Event-triggered schemes on leader-following consensus of general linear multiagent systems under different topologies, *IEEE Transactions on Cybernetics*, vol. 47, no. 1, pp. 212–223, 2017.

[12] L. Jiang, L. Yan, Y. Xia, Q. Guo, M. Fu, and K. Lu, Asynchronous multirate multisensor data fusion over unreliable measurements with correlated noise, *IEEE Transactions on Aerospace and Electronic Systems*, vol. 53, no. 5, pp. 2427–2437, 2017.

[13] S. Li, Z. Deng, X. Feng, R. He, and F. Pan, Joint parameter and state estimation for stochastic uncertain system with multivariate skew t noises, *Chinese Journal of Aeronautics*, vol. 35, no. 5, pp. 69–82, 2022.

[14] L. Xu, X. R. Li, and Z. Duan, Hybrid grid multiple-model estimation with application to maneuvering target tracking, *IEEE Transactions on Aerospace and Electronic Systems*, vol. 52, no. 1, pp. 122–136, 2016.

[15] L. Chen, K. Bo, F. Lee, and Q. Chen, Advanced feature fusion algorithm based on multiple convolutional neural network for scene recognition, *CMES-Computer Modeling in Engineering & Sciences*, vol. 122, no. 2, pp. 505–523, 2020.

[16] S. P. Yadav and S. Yadav, Fusion of medical images in wavelet domain: A hybrid implementation, *CMES-Computer Modeling in Engineering & Sciences*, vol. 122, no. 1, pp. 303–321, 2020.

[17] R. Caballero-Águila, A. Hermoso-Carazo, and J. Linares-Pérez, Networked fusion estimation with multiple uncertainties and time-correlated channel noise, *Information Fusion*, vol. 54, pp. 161–171, 2020.

[18] D. Ciuonzo, A. Aubry, and V. Carotenuto, Rician MIMO channel- and jamming-aware decision fusion, *IEEE Transactions on Signal Processing*, vol. 65, no. 15, pp. 3866–3880, 2017.

[19] L. Liu, L. Ma, J. Guo, J. Zhang, and Y. Bo, Distributed set-membership filtering for time-varying systems: A coding-decoding-based approach, *Automatica*, vol. 129, art. no. 109684, 2021.

[20] Y. S. Shmaliy, F. Lehmann, S. Zhao, and C. K. Ahn, Comparing robustness of the Kalman, H_∞, and UFIR filters, *IEEE Transactions on Signal Processing*, vol. 66, no. 13, pp. 3447–3458, 2018.

[21] Y. Shen, Z. Wang, B. Shen, and F. E. Alsaadi, H_∞ filtering for multi-rate multi-sensor systems with randomly occurring sensor saturations under the p-persistent CSMA protocol, *IET Control Theory & Applications*, vol. 14, no. 10, pp. 1255–1265, 2020.

[22] Y. Shen, Z. Wang, B. Shen, and H. Dong, Outlier-resistant recursive filtering for multi-sensor multirate networked systems under weighted Try-Once-Discard protocol, *IEEE Transactions on Cybernetics*, vol. 51, no. 10, pp. 4897–4908, 2021.

[23] K. Zhu, Y. Song, and D. Ding, Resilient RMPC for polytopic uncertain systems under TOD protocol: A switched system approach, *International Journal of Robust and Nonlinear Control*, vol. 28, no. 16, pp. 5103–5117, 2018.

[24] Y. Cui, Y. Liu, W. Zhang, and F. E. Alsaadi, Sampled-based consensus for nonlinear multiagent systems with deception attacks: The decoupled method, *IEEE Transactions on Systems, Man and Cybernetics: Systems*, vol. 51, no. 1, pp. 561–573, 2021.

[25] L. Wang, Z. Wang, G. Wei, and F. E. Alsaadi, Variance-constrained H_∞ state estimation for time-varying multi-rate systems with redundant channels: The finite-horizon case, *Information Sciences*, vol. 501, pp. 222–235, 2019.

[26] W. Xu, G. Hu, D. W. C. Ho, and Z. Feng, Distributed secure cooperative control under denial-of-service attacks from multiple adversaries, *IEEE Transactions on Cybernetics*, vol. 50, no. 8, pp. 3458–3467, 2020.

[27] Y. Gao, X. Li, and E. Song, Robust linear estimation fusion with allowable unknown cross-covariance, *IEEE Transactions on Systems, Man and Cybernetics: Systems*, vol. 46, no. 9, pp. 1314–1325, 2016.

[28] C. Wen, Z. Wang, Q. Liu, and F. E. Alsaadi, Recursive distributed filtering for a class of state-saturated systems with fading measurements and quantization effects, *IEEE Transactions on Systems, Man, and Cybernetics: Systems*, vol. 48, no. 6, pp. 930–941, 2018.

[29] H. Hashmipour, S. Roy, and A. Laub, Decentralized structures for parallel Kalman Filtering, *IEEE Transactions on Automatic Control*, vol. 33, no. 1, pp. 88–93, 1988.

[30] E. Song, Y. Zhu, J. Zhou, and Z. You, Optimal Kalman filtering fusion with cross-correlated sensor noises, *Automatica*, vol. 43, no. 8, pp. 1450–1456, 2007.

[31] Y. Zhu, Z. You, J. Zhao, K. Zhang, and X. Li, The optimality for the distributed Kalman filtering fusion with feedback, *Automatica*, vol. 37, no. 9, pp. 1489–1493, 2001.

[32] T. Tian, S. Sun, and N. Li, Multi-sensor information fusion estimators for stochastic uncertain systems with correlated noises, *Information Fusion*, vol. 27, pp. 126–137, 2016.

[33] N. Carlson, Federated square root filter for decentralized parallel processes, *IEEE Transactions on Aerospace and Electronic Systems*, vol. 25, no. 3, pp. 517–525, 1990.

[34] L. Yan, X. Li, Y. Xia, and M. Fu, Optimal sequential and distributed fusion for state estimation in cross-correlated noise, *Automatica*, vol. 49, no. 12, pp. 3607–3612, 2013.

[35] B. Allik, C. Miller, M. J. Piovoso, and R. Zurakowski, The Tobit Kalman filter: An estimator for censored measurements, *IEEE Transactions on Control Systems Technology*, vol. 24, no. 1, pp. 365–371, 2016.

[36] J. Tobin, Estimation of relationships for limited dependent variables, *Econometrica*, vol. 26, no. 1, pp. 24–36, 1958.

[37] T. Amemiya, Regression analysis when the dependent variable is truncated normal, *Econometrica*, vol. 41, no. 6, pp. 997–1016, 1973.

[38] C. Rago, P. Willett, and Y. Bar-Shalom, Censoring sensors: A low-communication-rate scheme for distributed detection, *IEEE Transactions on Aerospace and Electronic Systems*, vol. 32, no. 1, pp. 554–568, 1996.

[39] A. Assa and F. Janabi-Sharifi, A robust vision-based sensor fusion approach for real-time pose estimation, *IEEE Transactions on Cybernetics*, vol. 44, no. 2, pp. 217–227, 2014.

[40] F. Han, H. Dong, Z. Wang, G. Li, and F E. Alsaadi, Improved Tobit Kalman filtering for systems with random parameters via conditional expectation, *Signal Processing*, vol. 147, pp. 33–45, 2018.

[41] S. Li, X. Feng, Z. Deng, and F. Pan, Tobit Kalman filter with channel fading and dead-zone-like censoring, *International Journal of Systems Science*, vol. 52, no. 11, pp. 2183–2200, 2020.

[42] Z. Du and X. Li, Strong tracking Tobit Kalman filter with model uncertainties, *International Journal of Control, Automation and Systems*, vol. 17, no. 2, pp. 345–355, 2019.

[43] H. Geng, Z. Wang, Y. Liang, Y. Cheng, and F. E. Alsaadi, Tobit Kalman filter with time-correlated multiplicative sensor noises under redundant channel transmission, *IEEE Sensors Journal*, vol. 17, no, 24, pp. 8367–8377, 2017.

[44] H. Geng, Z. Wang, Y. Liang, Y. Cheng, and F. E. Alsaadi, Tobit Kalman filter with fading measurements, *Signal Processing*, vol. 140, pp. 60–68, 2017.

[45] H. Geng, Z. Wang, Y. Cheng, F. E. Alsaadi, and A. M. Dobaie, State estimation under non-Gaussian Levy and time-correlated additive sensor noises: A modified Tobit Kalman filtering approach, *Signal Processing*, vol. 154, pp. 120–128, 2019.

[46] H. Geng, Z. Wang, L. Zou, A. Mousavi, and Y. Cheng, Protocol-based Tobit Kalman filter under integral measurements and probabilistic sensor failures, *IEEE Transactions on Signal Processing*, vol. 69, pp. 546–559, 2021.

[47] H. Geng, Z. Wang, X. Yi, F. E. Alsaadi, and Y. Cheng, Tobit Kalman filtering for fractional-order systems with stochastic nonlinearities under Round-Robin protocol, *International Journal of Robust and Nonlinear Control*, vol. 31, pp. 2348–2370, 2021.

[48] J. Huang and X. He, Detection of intermittent fault for discrete-time systems with output dead-zone: A variant Tobit Kalman filtering approach, *Journal of Control Science and Engineering*, vol. 2017, art. no. 7849841, 9 pages, 2017.

[49] W. Li, Y. Jia, and J. Du, Tobit Kalman filter with time-correlated multiplicative measurement noise, *IET Control Theory & Applications*, vol. 11, no. 1, pp. 122–128, 2017.

[50] K. Loumponias, N. Vretos, G. Tsaklidis, and P. Daras, Using Kalman filter and Tobit Kalman filter in order to improve the motion recorded by Kinect sensor, in *Proceedings of the Panhellenic Statistics Conference*, Naousa, Greece, May 2016, pp. 322–334.

[51] K. Loumponias, N. Vretos, G. Tsaklidis, and P. Daras, An improved Tobit Kalman filter with adaptive censoring limits, *Circuits, Systems, and Signal Processing*, vol. 39, pp. 5588–5617, 2020.

[52] H. Geng, Z. Wang, and Y. Cheng, Distributed federated Tobit Kalman filter fusion over a packet-delaying network: A probabilistic perspective, *IEEE Transactions on Signal Processing*, vol. 66, no. 17, pp. 4477–4489, 2018.

[53] H. Geng, Z. Wang, F. E. Alsaadi, K. H. Alharbi, and Y. Cheng, Federated Tobit Kalman filtering fusion with dead-zone-like censoring and dynamical bias under the Round-Robin protocol, *IEEE Transactions on Signal and Information Processing Over Networks*, vol. 7, pp. 1–16, 2021.

[54] G. Wang, N. Li, and Y. Zhang, An event based multi-sensor fusion algorithm with dead-zone like measurements, *Information Fusion*, vol. 42, pp. 111–118, 2018.

[55] X. Geng, Y. Liang, and L. Jiao, EARC: Evidential association rule-based classification, *Information Sciences*, vol. 547, pp. 202–222, 2021.

[56] M. Wang, Z. Wang, H. Dong, and Q.-L. Han, A novel framework for backstepping-based control of discrete-time strict-feedback nonlinear systems with multiplicative noises, *IEEE Transactions on Automatic Control*, vol. 66, no. 4, pp. 1484–1496, 2021.

[57] Y. Yuan, H. Yuan, Z. Wang, L. Guo, and H. Yang, Optimal control for networked control systems with disturbances: A delta operator approach, *IET Control Theory and Applications*, vol. 11, no. 9, pp. 1325–1332, 2017.

[58] L. Zou, Z. Wang, Q.-L. Han, and D. H. Zhou, Moving horizon estimation of networked nonlinear systems with random access protocol, *IEEE Transactions on Systems, Man, and Cybernetics: Systems*, vol. 51, no. 5, pp. 2937–2948, 2021.

[59] D. Ding, Q.-L. Han, Z. Wang, and X. Ge, A survey on model-based distributed control and filtering for industrial cyber-physical systems, *IEEE Transactions on Industrial Informatics*, vol. 15, no. 5, pp. 2483–2499, 2019.

[60] D. Ding, Z. Wang, and Q.-L. Han, A set-membership approach to event-triggered filtering for general nonlinear systems over sensor networks, *IEEE Transactions on Automatic Control*, vol. 65, no. 4, pp. 1792–1799, 2020.

[61] X. Ge, Q.-L. Han, and Z. Wang, A threshold-parameter-dependent approach to designing distributed event-triggered H_∞ consensus filters over sensor networks, *IEEE Transactions on Cybernetics*, vol. 49, no. 4, pp. 1148–1159, 2019.

[62] J. Hu, Z. Wang, G.-P. Liu, C. Jia, and J. Williams, Event-triggered recursive state estimation for dynamical networks under randomly switching topologies and multiple missing measurements, *Automatica*, vol. 115, art. no. 108908, 2020.

[63] Q. Li, B. Shen, Z. Wang, T. Huang, and J. Luo, Synchronization control for a class of discrete time-delay complex dynamical networks: A dynamic event-triggered approach, *IEEE Transactions on Cybernetics*, vol. 49, no. 5, pp. 1979–1986, 2019.

[64] E. Tian, Z. Wang, L. Zou, and D. Yue, Chance-constrained H_∞ control for a class of time-varying systems with stochastic nonlinearities: The finite-horizon case, *Automatica*, vol. 107, pp. 296–305, 2019.

[65] X.-M. Zhang, Q.-L. Han, Z. Wang, and B.-L. Zhang, Neuronal state estimation for neural networks with two additive time-varying delay components, *IEEE Transactions on Cybernetics*, vol. 47, no. 10, pp. 3184–3194, 2017.

[66] L. Zou, Z. Wang, J. Hu, and D. Zhou, Moving horizon estimation with unknown inputs under dynamic quantization effects, *IEEE Transactions on Automatic Control*, vol. 65, no. 12, pp. 5368–5375, 2020.

[67] H. Geng, Y. Liang, and Y. Cheng, Target state and Markovian jump ionospheric height bias estimation for OTHR tracking systems, *IEEE Transactions on Systems, Man, and Cybernetics: Systems*, vol. 50, no. 7, pp. 2599–2611, 2020.

[68] Q. Guan, G. Wei, L. Wang, and Y. Song, A novel feature points tracking algorithm in terms of IMU-aided information fusion, *IEEE Transactions on Industrial Informatics*, vol. 17, no. 8, pp. 5304–5313, 2021.

[69] C. Huang, Y. Lan, G. Xu, X. Zhai, J. Wu, F. Lin, N. Zeng, Q. Hong, E. Y. K. Ng, Y. Peng, F. Chen, and G. Zhang, A deep segmentation network of multi-scale feature fusion based on attention mechanism for IVOCT lumen contour, *IEEE/ACM Transactions on Computational Biology and Bioinformatics*, vol. 18, no. 1, pp. 62–69, 2020.

[70] B. Chen, W. Zhang, and L. Yu, Distributed finite-horizon fusion Kalman filtering for bandwidth and energy constrained wireless sensor networks, *IEEE Transactions on Signal Processing*, vol. 62, no. 4, pp. 797–812, 2014.

[71] C. Wen, Z. Wang, T. Geng, and F. E. Alsaadi, Event-base distributed recursive filtering for state-saturated systems with redundant channels, *Information Fusion*, vol. 39, pp. 96–107, 2018.

[72] H. Geng, Z. Wang, Y. Chen, F. Alsaadi, and Y. Cheng, Multi-Sensor filtering fusion with parametric uncertainties and measurement censoring: Monotonicity and boundedness, *IEEE Transactions on Signal Processing*, vol. 69, pp. 5875–5890, 2021.

[73] B. Allik, C. Miller, M. J. Piovoso, and R. Zurakowski, Nonlinear estimators for censored data: A comparison of the EKF, the UKF and the Tobit Kalman filter, in *Proceedings of the American Control Conference*, Chicago, IL, USA, Jul. 2015, pp. 5146–5151.

[74] J. Hampshire and J. Strohbehn, Tobit maximum-likelihood estimation for stochastic time series affected by receiver saturation, *IEEE Transactions on Information Theory*, vol. 38, no. 2, pp. 457–469, 1992.

[75] B. Allik, C. Miller, M. J. Piovoso, and R. Zurakowski, Estimation of saturated data using the Tobit Kalman filter, in *Proceedings of American Control Conference*, Portland, OR, USA, Jun. 2014, pp. 4151–4156.

[76] K. Zhu, Y. Song, D. Ding, G. Wei, and H. Liu, Robust MPC under event-triggered mechanism and Round-Robin protocol: An average dwell-time approach, *Information Sciences*, vol. 457, pp. 126–140, 2018.

[77] K. Zhu, Z. Wang, Q.-L. Han, and G. Wei, Distributed set-membership fusion filtering for nonlinear 2-D systems over sensor networks: An encoding-decoding scheme, *IEEE Transactions on Cybernetics*, vol. 53, no. 1, pp. 416–427, 2023.

[78] K. Zhu, Z. Wang, Y. Chen, and G. Wei, Neural-network-based set-membership fault estimation for 2-D systems under encoding-decoding mechanism, *IEEE Transactions on Neural Networks and Learning Systems*, vol. 34, no. 2, pp. 786–798, 2023.

[79] L. Zou, Z. Wang, H. Geng, and X. Liu, Set-membership filtering subject to impulsive measurement outliers: A recursive algorithm, *IEEE/CAA Journal of Automatica Sinca*, vol. 8, no. 2, pp. 377–388, 2021.

[80] L. Zou, Z. Wang, Q.-L. Han, and D. Zhou, Full information estimation for time-varying systems subject to round-robin scheduling: A recursive filter approach, *IEEE Transactions on Systems, Man, and Cybernetics: Systems*, vo. 51, no. 3, pp. 1904–1916, 2021.

[81] X. Ge, Q.-L. Han, L. Ding, Y.-L. Wang, and X.-M. Zhang, Dynamic event-triggered distributed coordination control and its applications: A survey of trends and techniques, *IEEE Transactions on Systems, Man, and Cybernetics: Systems*, vol. 50, no. 9, pp. 3112–3125, 2020.

[82] X. Ge, Q.-L. Han, M. Zhong, and X.-M. Zhang, Distributed Krein space-based attack detection over sensor networks under deception attacks, *Automatica*, vol. 109, art. no. 108557, 2019.

[83] X.-M. Zhang, Q.-L. Han, and X. Ge, A novel approach to H_∞ performance analysis of discrete-time networked systems subject to network-induced delays and malicious packet dropouts, *Automatica*, vol. 136, art. no. 110010, 2022.

[84] Y. S. Moon, P. Park, W. H. Kwon, and Y. S. Lee, Delay-dependent robust stabilization of uncertain state-delayed systems, *International Journal of Control*, vol. 74, no. 14, pp. 1447–1455, 2001.

[85] D. Peaucelle, D. Arzelier, D. Henrion, and F. Gouaisbaut, Quadratic separation for feedback connection of an uncertain matrix and an implicit linear transformation, *Automatica*, vol. 43, no. 5, pp. 795–804, 2007.

[86] E. Fridman and U. Shaked, A descriptor system approach H_∞ control of linear time-delay systems, *IEEE Transactions on Automatic Control*, vol. 47, no. 2, pp. 253–270, 2002.

[87] M. Wu, Y. He, J. She, and G. Liu, Delay-dependent criteria for robust stability of time-varying delay systems, *Automatica*, vol. 40, no. 8, pp. 1435–1439, 2004.

[88] H. Liu, Z. Wang, W. Fei, and J. Li, H_∞ and $l_2 - l_\infty$ state estimation for discrete-time delayed memristive neural networks on finite horizon: The Round-Robin protocol, *Neural Networks*, vol. 132, pp. 121–130, 2020.

[89] H. Liu, Z. Wang, B. Shen, and H. Dong, Delay-distribution-dependent H_∞ state estimation for discrete-time memristive neural networks with mixed time-delays and fading measurements, *IEEE Transactions on Cybernetics*, vol. 50, no. 2, pp. 440–451, 2020.

[90] B. Shen, Z. Wang, D. Wang, J. Luo, H. Pu, and Y. Peng, Finite-horizon filtering for a class of nonlinear time-delayed systems with an energy harvesting sensor, *Automatica*, vol. 100, pp. 144–152, 2019.

[91] L. Zou, Z. Wang, H. Gao, and X. Liu, State estimation for discrete-time dynamical networks with time-varying delays and stochastic disturbances under the Round-Robin protocol, *IEEE Transactions on Neural Networks and Learning Systems*, vol. 28, no. 5, pp. 1139–1151, 2017.

[92] H. Geng, Y. Liang, F. Yang, L. Xu, and Q. Pan, Joint estimation of target state and ionospheric height bias in over-the-horizon radar target tracking, *IET Radar, Sonar & Navigation*, vol. 10, no. 7, pp. 1153–1167, 2016.

[93] R. Berry and R. Gallager, Communication over fading channels with delay constraints, *IEEE Transactions on Information Theory*, vol. 48, no. 5, pp. 1135–1149, 2002.

[94] E. Bigllieri, J. Proakis, and S. Shamai, Fading channels: Information-theoretic and communications aspects, *IEEE Transactions on Information Theory*, vol. 44, no. 6, pp. 2619–2692, 1998.

[95] M. Omidi, S. Pasupathy, and P. Gulak, Joint data and Kalman estimation for Rayleigh fading channels, *Wireless Personal Communications*, vol. 10, pp. 319–339, 1999.

[96] Q. Li, B. Shen, Z. Wang, and W. Sheng, Recursive distributed filtering over sensor networks on Gilbert-Elliott channels: A dynamic event-triggered approach, *Automatica*, vol. 113, art. no. 108681, 2020.

[97] W. Li, Z. Wang, Q. Liu, and L. Guo, An information aware event-triggered scheme for particle filter based remote state estimation, *Automatica*, vol. 103, pp. 151–158, 2019.

[98] W. Li, Z. Wang, Y. Yuan, and L. Guo, Two-stage particle filtering for non-Gaussian state estimation with fading measurements, *Automatica*, vol. 115, art. no. 108882, 12 pages, 2020.

[99] S. Zhang, Z. Wang, D. Ding, and H. Shu, Fuzzy filtering with randomly occurring parameter uncertainties, interval delays, and channel fadings, *IEEE Transactions on Cybernetics*, vol. 44, no. 3, pp. 406–417, 2014.

[100] S. Zhang, Z. Wang, D. Ding, and H. Shu, Fuzzy control with randomly occurring infinite distributed delays and channel fadings, *IEEE Transactions on Fuzzy Systems*, vol. 22, no. 1, pp. 189–200, 2014.

[101] E. Garone, B. Sinopoli, A. Goldsmith, and A. Casavola, LQG control for MIMO systems over multiple erasure channels with perfect acknowledgment, *IEEE Transactions on Automatic Control*, vol. 57, no. 2, pp. 450–456, 2012.

[102] N. Elia, Remote stabilization over fading channels, *Systems & Control Letters*, vol. 54, pp. 237–249, 2005.

[103] N. Xiao, L. Xie, and L. Qiu, Feedback stabilization of discrete-time networked systems over fading channels, *IEEE Transactions on Automatic Control*, vol. 57, no. 9, pp. 2176–2189, 2012.

[104] C. Xiao, J. Wu, S.-Y. Leong, Y. R. Zheng, and K. B. Letaief, A discrete-time model for triply selective MIMO Rayleigh fading channels, *IEEE Transactions on Wireless Communications*, vol. 3, no. 5, pp. 1678–1688, 2004.

[105] A. Alimohammad and B. F. Cockburn, Modeling and hardware implementation aspects of fading channel simulators, *IEEE Transactions on Vehicular Technology*, vol. 57, no. 4, pp. 2055–2069, 2008.

[106] F. Ren and Y. Zheng, A novel emulator for discrete-time MIMO triply selective fading channels, *IEEE Transactions on Circuits and Systems I: Regular Papers*, vol. 57, no. 9, pp. 2542–2551, 2010.

[107] J. Hu, H. Zhang, X. Yu, H. Liu, and D. Chen, Design of sliding-mode-based control for nonlinear systems with mixed-delays and packet losses under uncertain missing probability, *IEEE Transactions on Systems, Man, and Cybernetics: Systems*, vol. 51, no. 5, pp. 3217–3228, 2021.

[108] L. Ma, Z. Wang, Y. Liu, and F. E. Alsaadi, Distributed filtering for nonlinear time-delay systems over sensor networks subject to multiplicative link noises and switching topology, *International Journal of Robust and Nonlinear Control*, vol. 29, no. 10, pp. 2941–2959, 2019.

[109] H. Fang, M. A. Haile, and Y. Wang, Robust extended Kalman filtering for systems with measurement outliers, *IEEE Transactions on Control Systems Technology*, vol. 30, no. 2, pp. 795–802, 2022.

[110] H. Geng, M. A. Haile, and H. Fang, SSUE: Simultaneous state and uncertainty estimation for dynamical systems, *International Journal of Robust and Nonlinear Control*, vol. 31, pp. 1068–1083, 2020.

[111] X. Hao, Y. Liang, L. Xu, and X. Wang, Mode separability-based state estimation for uncertain constrained dynamic systems, *Automatica*, vol. 115, art. no. 108905, 2020.

[112] H. Movahedi, N. Tian, H. Fang, and R. Rajamani, Hysteresis compensation and nonlinear observer design for state-of-charge estimation using a nonlinear double-capacitor Li-Ion battery model, *IEEE/ASME Transactions on Mechatronics*, vol. 27, no. 1, pp. 596–604, 2022.

[113] C. Yan and H. Fang, Observer-based distributed leader-follower tracking control: A new perspective and results, *International Journal of Control*, vol. 94, no. 1, pp. 39–48.

[114] B.-L. Zhang, Q.-L. Han, and X.-M. Zhang, Event-triggered H_∞ reliable control for offshore structures in network environments, *Journal of Sound and Vibration*, vol. 368, pp. 1–21, 2016.

[115] D. Zhang, Q.-L. Han, and X.-M, Zhang, Network-based modeling and proportional-integral control for direct-drive-wheel systems in wireless network environments, *IEEE Transactions on Cybernetics*, vol. 50, no. 6, pp. 2462–2474, 2020.

[116] X.-M. Zhang and Q.-L. Han, Event-triggered H_∞ control for a class of nonlinear networked control systems using novel integral inequalities, *International Journal of Robust Nonlinear Control*, vol. 27, no. 4, pp. 679–700, 2017.

[117] X.-M. Zhang, Q.-L. Han, X. Ge, and L. Ding, Resilient control design based on a sampled-data model for a class of networked control systems under denial-of-service attacks, *IEEE Transactions on Cybernetics*, vol. 50, no. 8, pp. 3616–3626, 2020.

[118] X.-M. Zhang, Q.-L. Han, and J. Wang, Admissible delay upper bounds for global asymptotic stability of neural networks with time-varying delays, *IEEE Transactions on Neural Networks and Learning Systems*, vol. 29, no. 11, pp. 5319–5329, 2018.

[119] H. Gao, F. Han, B. Jiang, H. Dong and G. Li, Recursive filtering for time-varying systems under duty cycle scheduling based on collaborative prediction, *Journal of the Franklin Institute*, vol. 357, no. 17, pp. 13189–13204, Nov. 2020.

[120] C. Hu, Y. Liang, X. Wang, and L. Xu, A particle filter via constrained sampling for nonlinear dynamic systems, *International Journal of Robust and Nonlinear Control*, vol. 30, no. 13, pp. 4944–4959.

[121] J. Hu, G.-P. Liu, H. Zhang, and H. Liu, On state estimation for nonlinear dynamical networks with random sensor delays and coupling strength under event-based communication mechanism, *Information Sciences*, vol. 511, pp. 265–283, 2020.

[122] B. Jiang, H. Gao, F. Han, and H. Dong, Recursive filtering for nonlinear systems subject to measurement outliers, *Science China Information Sciences*, vol. 64, art. no. 172206, 2021.

[123] J. Li, Z. Wang, H. Dong, and G. Ghinea, Outlier-resistant remote state estimation for recurrent neural networks with mixed time-delays, *IEEE Transactions on Neural Networks and Learning Systems*, vol. 32, no. 5, pp. 2266–2273, 2021.

[124] J. Li, Z. Wang, H. Dong, and W. Fei, Delay-distribution-dependent state estimation for neural networks under stochastic communication protocol with uncertain transition probabilities, *Neural Networks*, vol. 130, pp. 143–151, 2020.

[125] H. Song, D. Ding, H. Dong, G. Wei, and Q.-L. Han, Distributed entropy filtering subject to DoS attacks in non-Gauss environments, *International Journal of Robust and Nonlinear Control*, vol. 30, no. 3, pp. 1240–1257, 2020.

[126] J. Suo, Z. Wang, B. Shen, and F. E. Alsaadi, Event-triggered stabilisation for switched delayed differential systems: The input-to-state stability, *IET Control Theory & Applications*, vol. 14, no. 13, pp. 1711–1721, 2020.

[127] E. E. Yaz and Y. I. Yaz, State estimation of uncertain nonlinear stochastic systems with general criteria, *Applied Mathematics Letters*, vol. 14, no. 5, pp. 605–610, 2001.

[128] Q. Liu, Z. Wang, X. He, G. Ghinea, and F. E. Alsaadi, A resilient approach to distributed filter design for time-varying systems under stochastic nonlinearities and sensor degradation, *IEEE Transactions on Signal Processing*, vol. 65, no. 5, pp. 1300–1309, 2017.

[129] L. Ma, Z. Wang, Q.-L. Han, and H. K. Lam, Envelope-constrained H_∞ filtering for nonlinear systems with quantization effects: The finite horizon case, *Automatica*, vol. 93, pp. 527–534, 2018.

[130] Y. Luo, Z. Wang, Y. Chen, and X. Yi, H_∞ state estimation for coupled stochastic complex networks with periodical communication protocol and intermittent nonlinearity switching, *IEEE Transactions on Network Science and Engineering*, vol. 8, no. 2, pp. 1414–1425, 2021.

[131] X. Jiang, F. Xia, and Z. Feng, Resilient H_∞ filtering for stochastic systems with randomly occurring gain variations, nonlinearities and channel fadings, *Circuits, Systems, and Signal Processing*, vol. 38, pp. 4548–4571, 2019.

[132] Y. Gao, J. Hu, and D. Chen, Variance-constrained resilient H_∞ state estimation for time-varying neural networks with randomly varying nonlinearities and missing measurements, *Advances in Difference Equations*, vol. 2019, art. no. 380, 2019.

[133] D. Ding, Z. Wang, and Q.-L. Han, A set-membership approach to event-triggered filtering for general nonlinear systems over sensor networks, *IEEE Transactions on Automatic Control*, vol. 65, no. 4, pp. 1792–1799, 2020.

[134] D. Ding, Z. Wang Z, Q.-L. Han, and G. Wei, Neural-network-based output-feedback control under Round-Robin scheduling protocols, *IEEE Transactions on Cybernetics*, vol. 49, nol. 6, pp. 2372–2384, 2019.

[135] R. W. Brockett and D. Liberzon, Quantized feedback stabilization of linear systems, *IEEE Transactions on Automatic Control*, vol. 45, no. 7, pp. 1279–1289, 2000.

[136] D. Liberzon, On stabilization of linear systems with limited information, *IEEE Transactions on Automatic Control*, vol. 48, no. 2, pp. 304–307, 2003.

[137] S. W. Yun, Y. J. Chol, and P. Park, Dynamic output-feedback guaranteed cost control for linear systems with uniform input quantization, *Nonlinear Dynamics*, vol. 62, no. 1, pp. 95–104, 2010.

[138] M. Fu and C. E. de Souza, State estimation for linear discrete-time systems using quantized measurements, *Automatica*, vol. 45, no. 12, pp. 2937–2945, 2009.

[139] M. Fu and L. Xie, The sector bound approach to quantized feedback control, *IEEE Transactions on Automatic Control*, vol. 50, no. 11, pp. 1698–1711, 2005.

[140] L. Chen, Y. Chen, and N. Zhang, Synchronization control for chaotic neural networks with mixed delays under input saturations, *Neural Processing Letters*, vol. 53, pp. 3735–3755, 2021.

[141] Q.-L. Han, Y. Liu, and F. Yang, Optimal communication network-based H_∞ quantized control with packet dropouts for a class of discrete-time neural networks with distributed time delay, *IEEE Transactions on Neural Networks and Learning Systems*, vol. 27, no. 2, pp. 426–434, 2016.

[142] L. Wang, Z. Wang, B. Shen, and G. Wei, Recursive filtering with measurement fading: A multiple description coding scheme, *IEEE Transactions on Automatic Control*, vol. 66, no. 11, pp. 5144–5159, 2021.

[143] L. Wang, Z. Wang, G. Wei, and F. E. Alsaadi, Observer-based consensus control for discrete-time multiagent systems with coding-decoding communication protocol, *IEEE Transactions on Cybernetics*, vol. 49, no. 12, pp. 4335–4345, 2019.

[144] Y. Zhao, X. He, J. Zhang, H. Ji, D. Zhou, and M. G. Pecht, Detection of intermittent faults based on an optimally weighted moving average T^2 control chart with stationary observations, *Automatica*, vol. 123, art. no. 109298, 2021.

[145] D. Liu, Z. Wang, Y. Liu, and F. E. Alsaadi, Extended Kalman filtering subject to random transmission delays: Dealing with packet disorders, *Information Fusion*, vol. 60, pp. 80–86, 2020.

[146] J. Li, M. J. Er, and H. Yu, Sampling and control strategy: Networked control systems subject to packet disordering, *IET Control Theory & Applications*, vol. 10, no. 6, pp. 674–683.

[147] A. Liu, W. Zhang, L. Yu, S. Liu, and M. Chen, New results on stabilization of net-worked control systems with packet disordering, *Automatica*, vol. 52, pp. 255–259.

[148] Y.-B. Zhao, J. Kim, G.-P. Liu, and D. Rees, Compensation and stochastic modeling of discrete-time networked control systems with data packet disorder, *International Journal of Control, Automation and Systems*, vol. 10, pp. 1055–1063.

[149] D. Liu, Z. Wang, Y. Liu, and F. E. Alsaadi, Recursive filtering for stochastic parameter systems with measurement quantizations and packet disorders, *Applied Mathematics and Computation*, vol. 398, art. no. 125960, 2020.

[150] W. Chen, D. Ding, X. Ge, Q.-L. Han, and G. Wei, H_∞ containment control of multi-agent systems under event-triggered communication scheduling: The finite-horizon case, *IEEE Transactions on Cybernetics*, vol. 50, no. 4, pp. 1372–1382, 2020.

[151] D. Ding, Q.-L. Han, X. Ge, and J. Wang, Secure state estimation and control of cyber-physical systems: A survey, *IEEE Transactions on Systems, Man, and Cybernetics: Systems*, vol. 51, no. 1, pp, 176–190, 2021.

[152] X. Ge, Q.-L. Han, and Z. Wang, A dynamic event-triggered transmission scheme for distributed set-membership estimation over wireless sensor networks, *IEEE Transactions on Cybernetics*, vol. 49, no. 1, pp. 171–183, 2019.

[153] X. Ge, S. Xiao, Q.-L. Han, X.-M. Zhang, and D. Ding, Dynamic event-triggered scheduling and platooning control co-design for automated vehicles over vehicular ad-hoc networks, *IEEE/CAA Journal of Automatica Sinica*, vol. 9, no. 1, pp. 31–46, 2022.

[154] X. Ge, Q.-L. Han, X.-M. Zhang, D. Ding, and F. Yang, Resilient and secure remote monitoring for a class of cyber-physical systems against attacks, *Information Science*, vol. 512, pp. 1592–1605, 2020.

[155] D. Ding, Z. Wang, J. Lam, and B. Shen, Finite-horizon H_∞ control for discrete time-varying systems with randomly occurring nonlinearities and fading measurements, *IEEE Transactions on Automatic Control*, vol. 60, no. 9, pp. 2488–2493, 2015.

[156] R. Kalman and R. Bucy, A new approach to linear filtering and prediction problems, *Journal of Basic Engineering*, vol. 82, no. 1, pp. 35–45, 1960.

[157] R. M. Alexandre, P. H. Jo ã o, and N. N. Girish, Redundant data transmission in control/estimation over lossy networks, *Automatica*, vol. 48, no. 8, pp. 1612–1620, 2012.

[158] H. Nam, Y. C. Ko, and M. S. Alouini, Spectral efficiency enhancement in multi-channel systems using redundant transmission and diversity reception, *IEEE Transactions on Wireless Communications*, vol. 7, no. 6, pp. 2143–2153, 2008.

[159] Y. Song, Z. Wang, D. Ding, and G. Wei, Robust model predictive control under redundant channel transmission with applications in networked DC motor systems, *International Journal of Robust and Nonlinear Control*, vol. 26, no. 18, pp. 3937–3957, 2016.

[160] D. Applebaum, *Lévy Processes and Stochastic Calculus*, Cambridge University Press, 2009.

[161] A. Breton and M. Musiela, A generation of the Kalman filter to models with infinite variance, *Stochastic Processes and Applications*, vol. 47, pp. 75–94, 1993.

[162] A. Ahn and R. Feldman, Optimal filtering of a Gaussian signal in the presence of Lévy noise, *SIAM Journal of Applied Mathematics*, vol. 60, pp. 359–369, 1999.

[163] D. Sornette and K. Ide, The Kalman-Lévy filter, *Physica D*, vol. 151, pp. 142–174, 2001.

[164] A. Sinha, T. Kirubarajan, and Y. Bar-Shalom, Application of the Kalman-Levy filter for tracking maneuvering targets, *IEEE Transactions on Aerospace and Electronic Systems*, vol. 43, no. 3, pp. 1099–1107, 2007.

[165] Y. Sun, X. Wu, J. Cao, Z. Wei, and G. Sun, Fractional extended Kalman filtering for nonlinear fractional system with Lévy noises, *IET Control Theory & Applications*, vol. 11, no. 3, pp. 349–358, 2017.

[166] A. Bryson and L. Henrikson, Estimation using sampled data containing sequentially correlated noise, *Journal of Spacecraft and Rockets*, vol. 5, no. 6, pp. 662–665, 1968.

[167] P. Jiang, J. Zhou, and Y. Zhu, Globally optimal Kalman filtering with finite-time correlated noises, in *Proceedings of the 49th IEEE Conference on Decision and Control*, Atlanta, GA, USA, Dec. 2010, pp. 5007–5012.

[168] W. Liu, State estimation of discrete-time systems with arbitrarily correlated noises, *International Journal of Adaptive Control and Signal Processing*, vol. 28, pp. 949–970, 2014.

[169] M. Petovello, K. O'Keefe, G. Lachapelle, and M. Cannon, Consideration of time-correlated errors in a Kalman filter applicable to GNSS, *Journal of Geodesy*, vol. 83, no. 1, pp. 51–56, 2009.

[170] A. Gopalakrishnan, N. S. Kaisare, and S. Narasimhan, Incorporating delayed and infrequent measurements in extended Kalman filter based nonlinear state estimation, *Journal of Process Control*, vol. 21, no. 1, pp. 119–129, 2011.

[171] Y. Guo and B. Huang, State estimation incorporating infrequent, delayed and integral measurements, *Automatica*, vol. 58, pp. 32–38, 2015.

[172] L. Schenato, Optimal estimation in networked control systems subject to random delay and packet drop, *IEEE Transactions on Automatic Control*, vol. 53, no. 5, pp. 1311–1317, 2008.

[173] H. Dong, Z. Wang, and H. Gao, Robust H_∞ filtering for a class of nonlinear networked systems with multiple stochastic communication delays and packet dropouts, *IEEE Transactions on Signal Processing*, vol. 55, no. 4, pp. 1957–1966, 2010.

[174] X. Wang, Q. Pan, Y. Liang, and H. Li, Gaussian sum approximation filter for nonlinear dynamic time-delay system, *Nonlinear Dynamics*, vol. 82, pp. 501–517, 2015.

[175] S. Nakamori, R. Caballero-Aguila, A. Hermoso-Carazo, and J. Linares-Perez, Recursive estimators of signals from measurements with stochastic delays using covariance information, *Applied Mathematics and Computation*, vol. 162, pp. 65–79, 2005.

[176] L. Shi, L. Xie, and R. M. Murray, Kalman filtering over a packet-delaying network: A probabilistic approach, *Automatica*, vol. 45, no. 3, pp. 2134–2140, 2009.

[177] Z. Xiong, J. Chen, R. Wang, and J. Liu, A new dynamic vector formed information sharing algorithm in federated filter, *Aerospace Science and Technology*, vol. 29, pp. 37–46, 2013.

[178] C. -S. Hsieh and F. -C. Chen, Optimal solution of the two-stage Kalman estimator, *IEEE Transactions on Automatic Control*, vol. 44, no. 1, pp. 194–199, 1999.

[179] F. Wang, Z. Wang, J. Liang, and X. Liu, Recursive state estimation for two-dimensional shift-varying systems with random parameter perturbation and dynamical bias, *Automatica*, vol. 112, art. no. 108658, 2020.

[180] W. Liu, X. Wang, and Z. Deng, Robust fusion time-varying Kalman estimators for multisensor networked systems with mixed uncertainties, *International Journal of Robust and Nonlinear Control*, vol. 28, pp. 4139–4174, 2018.

[181] L. Ma, Z. Wang, Q.-L. Han, and H. K. Lam, Variance-constrained distributed filtering for time-varying systems with multiplicative noises and deception attacks over sensor networks, *IEEE Sensors Journal*, vol. 17, no. 7, pp. 2279–2288, 2017.

[182] P. Tichavsky, C. H. Muravchik, and A. Nehorai, Posterior Cramer–Rao bounds for discrete-time nonlinear filtering, *IEEE Transactions on Signal Processing*, vol. 46, no. 5, pp. 1386–1396, 1998.

[183] L. Zuo, R. Niu, and P. K. Varshney, Conditional posterior Cramer–Rao lower bounds for nonlinear sequential Bayesian estimation, *IEEE Transactions on Signal Processing*, vol. 59, no. 1, pp. 1–14, 2011.

[184] Y. Theodor and U. Shaked, Robust discrete-time minimum-variance filtering, *IEEE Transactions on Signal Processing*, vol. 44, no. 2, pp. 181–89, 1996.

[185] Y. Cui, Y. Liu, W. Zhang, and F. E. Alsaadi, Sampled-based consensus for nonlinear multiagent systems with deception attacks: The decoupled method, *IEEE Transactions on Systems, Man and Cybernetics: Systems*, vol. 51, no. 1, pp. 561–573, 2021.

[186] W. He, F. Qian, Q.-L. Han, and G. Chen, Almost sure stability of nonlinear systems under random and impulsive sequential attacks, *IEEE Transactions on Automatic Control*, vol. 65, no. 9, pp. 3879–3886, 2020.

[187] Y. Zhao, W. Yao, J. Nan, J. Fang, X. Ai, J. Wen, and S. Chen, Resilient adaptive wide-area damping control to mitigate false data injection attacks, *IEEE Systems Journal*, vol. 15, no. 4, pp. 4831–4842, 2021.

[188] W. Xu, D. W. C. Ho, J. Zhong, and B. Chen, Event/self-triggered control for leader-following consensus over unreliable network with DoS attacks, *IEEE Transactions on Neural Networks and Learning Systems*, vol. 30, no. 10, pp. 3137–3149, 2019.

[189] F. Han, Z. Wang, G. Chen, and H. Dong, Scalable consensus filtering for uncertain systems over sensor networks with Round-Robin protocol, *International Journal of Robust and Nonlinear Control*, vol. 31, no. 3, pp. 1051–1066, 2021.

[190] J. He, and Y. Luo, A Bayesian updating method for non-probabilistic reliability assessment of structures with performance test data, *CMES-Computer Modeling in Engineering & Sciences*, vol. 125, no. 2, pp. 777–800, 2020.

[191] N. Hou, Z. Wang, D. W. C. Ho, and H. Dong, Robust partial-nodes-based state estimation for complex networks under deception attacks, *IEEE Transactions on Cybernetics*, vol. 50, no. 6, pp. 2793–2802, 2020.

[192] S. Liu, Y. Liang, L. Xu, T. Li, and X. Hao, EM-based extended object tracking without a priori extension evolution model, *Signal Processing*, vol. 188, art. no. 108181, 2021.

[193] S. Liu, Z. Wang, G. Wei, and M. Li, Distributed set-membership filtering for multi-rate systems under the Round-Robin scheduling over sensor networks, *IEEE Transactions on Cybernetics*, vol. 50, no. 5, pp. 1910–1920, 2020.

[194] Y. Liu, Z. Wang, and D. Zhou, Resilient actuator fault estimation for discrete-time complex networks: A distributed approach, *IEEE Transactions on Automatic Control*, vol. 66, no. 9, pp. 4214–4221, 2021.

[195] Y. Liu, Z. Wang, and D. Zhou, Scalable distributed filtering for a class of discrete-time complex networks over time-varying topology, *IEEE Transactions on Neural Networks and Learning Systems*, vol. 31, no. 8, pp. 2930–2941, 2020.

[196] K. Loumponias and G. Tsaklidis, Kalman filtering with censored measurements, *Journal of Applied Statistics*, vol. 19, no. 2, pp. 317–335, 2022.

[197] Y. Sun, D. Ding, H. Dong, and H. Liu, Event-based resilient filtering for stochastic nonlinear systems via innovation constraints, *Information Sciences*, vol. 546, pp. 512–525, 2021.

[198] G. Wei, W. Li, D. Ding, and Y. Liu, Stability analysis of covariance intersection-based Kalman consensus filtering for time-varying systems, *IEEE Transactions on Systems, Man, and Cybernetics: Systems*, vol. 50, no. 11, pp. 4611–4622, 2020.

[199] L. Zou, Z. Wang, and D. H. Zhou, Moving horizon estimation with non-uniform sampling under component-based dynamic event-triggered transmission, *Automatica*, vol. 120, art. no. 109154, 2020.

Index